THE MECHANICAL UNIVERSE

THE MECHANICAL UNIVERSE

INTRODUCTION TO MECHANICS AND HEAT

RICHARD P. OLENICK
ASSISTANT PROFESSOR OF PHYSICS, UNIVERSITY OF DALLAS
VISITING ASSOCIATE, CALIFORNIA INSTITUTE OF TECHNOLOGY

TOM M. APOSTOL
PROFESSOR OF MATHEMATICS, CALIFORNIA INSTITUTE OF TECHNOLOGY

DAVID L. GOODSTEIN
PROFESSOR OF PHYSICS AND APPLIED PHYSICS, CALIFORNIA INSTITUTE OF TECHNOLOGY
AND PROJECT DIRECTOR, *THE MECHANICAL UNIVERSE*

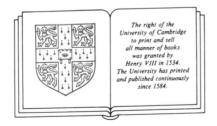

The right of the
University of Cambridge
to print and sell
all manner of books
was granted by
Henry VIII in 1534.
The University has printed
and published continuously
since 1584.

CAMBRIDGE UNIVERSITY PRESS
Cambridge
New York,   Port Chester
Melbourne,   Sydney

Published by the Press Syndicate of the University of Cambridge
The Pitt Building, Trumpington Street, Cambridge CB2 1RP
32 East 57th Street, New York, NY 10022, USA
10 Stamford Road, Oakleigh, Melbourne 3166, Australia

First published 1985
Reprinted 1986, 1989

Printed in the United States of America

Library of Congress Cataloging in Publication Data

Olenick, Richard P.

The mechanical universe.

To be used with the Mechanical universe television series.
   1. Mechanics.   2. Heat.   3. Physics—Study and
teaching (Higher)—United States—Audio-visual aids.
4. Mechanical universe (Television program)   I. Apostol,
Tom M.   II. Goodstein, David L., 1939–
III. Mechanical universe (Television program)   IV. Title.
QC125.2.044   1985   531   85-12835
ISBN 0-521-30429-6

CONTENTS

Preface   *xiii*

Chapter 1 INTRODUCTION TO THE MECHANICAL UNIVERSE (Program 1)   *1*

1.1 The Copernican Revolution   *1*
1.2 Units and Dimensions   *5*
1.3 A Final Word   *10*

Chapter 2 THE LAW OF FALLING BODIES (Program 2)   *13*

2.1 Aristotle's Description of Motion   *13*
2.2 Do Heavy Bodies Fall Faster Than Lighter Ones?   *15*
2.3 Medieval Laws of Falling Bodies   *15*
2.4 "The" Law of Falling Bodies   *20*
2.5 The Average Speed of a Falling Body   *25*

2.6 Instantaneous Speed   *26*
2.7 Acceleration   *32*
2.8 A Final Word   *36*

Chapter 3 DERIVATIVES (Program 3)   *39*

3.1 The Development of Differential Calculus   *39*
3.2 The Tangent Line–Derivative Connection   *41*
3.3 Rules of Differentiation   *47*
3.4 Derivatives of Special Functions   *58*
3.5 A Final Word   *67*

Chapter 4 INERTIA (Program 4)   *69*

4.1 If the Earth Moves: Aristotelian Objections   *69*
4.2 The Earth Moves: Galileo's Law of Inertia   *72*
4.3 Relative Motion   *75*
4.4 Projectile Motion: A Consequence of Inertia   *77*
4.5 A Final Word   *80*

Chapter 5 VECTORS (Program 5)   *81*

5.1 The Rise of Vector Analysis   *81*
5.2 Vectors – A Geometric View   *82*
5.3 Vectors: An Analytic View   *96*
5.4 The Cross Product   *106*
5.5 A Final Word   *110*

Chapter 6 NEWTON'S LAWS (Program 6)   *111*

6.1 The End of the Confusion   *111*
6.2 Newton's Laws of Motion   *112*
6.3 Units of Mass, Momentum, and Force   *117*
6.4 Projectile Motion: An Application of Newton's Second Law   *119*
6.5 A Final Word   *127*

Chapter 7 INTEGRATION (Program 7)   *129*

7.1 Antidifferentiation, the Reverse of Differentiation   *129*
7.2 Antidifferentiation and Quadrature   *133*
7.3 The Leibniz Integral Notation   *137*
7.4 Applications of the Second Fundamental Theorem to Physics   *145*
7.5 A Final Word   *150*
Appendix 1 The Fundamental Theorems of Calculus   *151*
Appendix 2 Quadrature of a Hyperbolic Segment. The Logarithm Function   *152*

Chapter 8 THE APPLE AND THE MOON (Program 8)   *159*

8.1 The Genesis of an Idea   *159*
8.2 The Law of Universal Gravitation   *160*
8.3 Acceleration of Gravity on the Earth   *165*
8.4 Why the Moon Doesn't Fall to the Earth   *168*
8.5 A Final Word   *172*

Chapter 9 MOVING IN CIRCLES (Program 9)   *175*

9.1 The Perfection of Circular Motion   *175*
9.2 Derivatives of Vector Functions   *177*
9.3 Uniform Circular Motion   *183*
9.4 Circular Orbits   *187*
9.5 Motion along Space Curves   *189*
9.6 A Final Word   *191*

Chapter 10 FORCES (Program 10)   *193*

10.1 The Fundamental Forces   *193*
10.2 Gravitational and Electric Forces   *195*
10.3 Contact Forces   *198*
10.4 Application of Newton's Laws   *202*
10.5 Friction   *210*
10.6 A Final Word   *215*

Chapter 11 GRAVITY, ELECTRICITY, AND MAGNETISM (Program 11)   *217*

11.1 Finding the Connection between Electricity and Magnetism   *217*
11.2 Faraday's Fields   *219*
11.3 A Prediction from Electromagnetism   *224*
11.4 A Final Word   *226*

Chapter 12 THE MILLIKAN OIL-DROP EXPERIMENT (Program 12)   *229*

12.1 The Discovery of the Electron   *229*
12.2 Motion in a Resistive Medium   *233*
12.3 The Oil-Drop Experiment   *239*
12.4 A Final Word   *243*

Chapter 13 THE LAW OF CONSERVATION OF ENERGY (Program 13)   *245*

13.1 Toward an Idea of Energy   *245*
13.2 Work and Potential Energy   *247*
13.3 The Law of Conservation of Energy   *252*

13.4 Heat and Energy   *260*
13.5 A Final Word   *263*

Chapter 14 ENERGY AND STABILITY (Program 14)   *267*

14.1 Forms of Energy   *267*
14.2 Gravitational Potential Energy   *274*
14.3 Potential Energy and Stability   *278*
14.4 A Final Word   *284*

Chapter 15 TEMPERATURE AND THE GAS LAWS (Program 45)   *287*

15.1 Temperature and Pressure   *287*
15.2 The Gas Laws of Boyle, Charles, and Gay-Lussac   *293*
15.3 The Ideal-Gas Law   *295*
15.4 Temperature and Energy   *297*
15.5 A Final Word   *299*

Chapter 16 THE ENGINE OF NATURE (Program 46)   *301*

16.1 The Age of Steam   *301*
16.2 Work and the Pressure–Volume Diagram   *303*
16.3 The First Law of Thermodynamics   *309*
16.4 Adiabatic and Isothermal Processes   *312*
16.5 The Second Law of Thermodynamics   *316*
16.6 The Carnot Engine   *319*
16.7 A Final Word   *324*

Chapter 17 ENTROPY (Program 47)   *327*

17.1 Toward an Understanding of Entropy   *327*
17.2 Engines and Entropy   *330*
17.3 Entropy and the Second Law of Thermodynamics   *336*
17.4 An Implication of the Entropy Principle   *338*
17.5 A Final Word   *341*

Chapter 18 THE QUEST FOR LOW TEMPERATURES (Program 48)   *343*

18.1 Cooling Off   *343*
18.2 The States of Matter   *344*
18.3 Behavior of Water   *349*
18.4 Liquefaction of Gases   *351*
18.5 The Joule–Thomson Effect   *355*
18.6 A Final Word   *357*

Chapter 19 THE CONSERVATION OF MOMENTUM (Program 15)   359

19.1 The Universe as a Machine   359
19.2 Newton's Laws in Retrospect   360
19.3 The Law of Conservation of Momentum   364
19.4 Applications of Conservation of Momentum   368
19.5 Collision Forces and Times   376
19.6 A Final Word   378

Chapter 20 HARMONIC MOTION (Program 16)   379

20.1 Finding a Clock That Wouldn't Get Seasick   379
20.2 Simple Harmonic Motion   381
20.3 Energy Conservation and Simple Harmonic Motion   387
20.4 Initial Conditions   390
20.5 The Simple Pendulum   393
20.6 A Final Word   396

Chapter 21 RESONANCE (Program 17)   399

21.1 Forced Oscillations   399
21.2 Describing Resonance   401
21.3 Swinging and Singing Wires in the Wind   408
21.4 A Final Word   409

Chapter 22 COUPLED OSCILLATORS AND WAVES (Program 18)   413

22.1 Newton and the Speed of Sound   413
22.2 Coupled Oscillators   415
22.3 Water Waves and Wave Characteristics   417
22.4 Wave Speed   420
22.5 Sound   426
22.6 Sound Intensities   428
22.7 Sound and Heat   429
22.8 A Final Word   430

Chapter 23 ANGULAR MOMENTUM (Program 19)   433

23.1 The Search for Order   433
23.2 The Law of Equal Areas   437
23.3 Angular Momentum   440
23.4 Torque and Angular Momentum   444
23.5 Vortices and Firestorms   447
23.6 Angular Momentum and the Architecture of the Heavens   450
23.7 A Final Word   454

Chapter 24 GYROSCOPES (Program 20)   *457*

24.1 An Ancient Question   *457*
24.2 The Gyroscope   *459*
24.3 The Gyrocompass   *466*
24.4 Angular Velocity of Precession   *469*
24.5 The Earth as a Gyroscope   *473*
24.6 A Final Word   *474*

Chapter 25 KEPLER'S LAWS AND THE CONIC SECTIONS (Program 21)   *477*

25.1 The Quest for Precision   *477*
25.2 Kepler's Laws   *479*
25.3 Conic Sections   *482*
25.4 The Ellipse   *485*
25.5 The Conics and Eccentricity   *489*
25.6 Cartesian Equations for Conic Sections   *492*
25.7 A Final Word   *494*

Chapter 26 SOLVING THE KEPLER PROBLEM (Program 22)   *497*

26.1 Setting the Stage   *497*
26.2 Polar Coordinates and the Unit Vectors $\hat{\mathbf{r}}$ and $\hat{\boldsymbol{\theta}}$ *499*
26.3 Solution of the Kepler Problem   *502*
26.4 A Final Word   *504*

Chapter 27 ENERGY AND ECCENTRICITY (Program 23)   *507*

27.1 Celestial Omens: Comets   *507*
27.2 Energy in Space   *508*
27.3 Energy and Eccentricity   *511*
27.4 Orbits and Eccentricity   *513*
27.5 Planetary Motion and Effective Potential   *516*
27.6 Calculating the Orbit from Initial Conditions   *519*
27.7 A Final Word   *522*

Chapter 28 NAVIGATING IN SPACE (Program 24)   *525*

28.1 Freeways in the Sky   *525*
28.2 Navigating in Space   *526*
28.3 Transfer Orbits   *529*
28.4 Gravity Assist   *534*
28.5 A Final Word   *537*

Chapter 29 LOOSE ENDS AND BLACK HOLES (Program 25) 539

29.1 Kepler's Third Law 539
29.2 The Earth–Sun and Earth–Moon Systems 541
29.3 The Tides 543
29.4 The Principle of Equivalence 546
29.5 Einstein's Theory of Gravity 548
29.6 Black Holes 549
29.7 A Final Word 550

Chapter 30 THE HARMONY OF THE SPHERES: AN OVERVIEW OF THE MECHANICAL UNIVERSE (Program 26) 553

30.1 Winding Up *The Mechanical Universe* 553
30.2 The Physics of the Heavens and Earth 554
30.3 The Language of Physics 556
30.4 The Conservation Laws of the Mechanical Universe 559
30.5 A Final Word 561

Appendix A THE INTERNATIONAL SYSTEM OF UNITS 563

Appendix B CONVERSION FACTORS 565

Appendix C FORMULAS FROM ALGEBRA, GEOMETRY, AND TRIGONOMETRY 569

Appendix D ASTRONOMICAL DATA 571

Appendix E PHYSICAL CONSTANTS 573

SELECTED BIBLIOGRAPHY 575

Index 579

PREFACE

I GENERAL INTRODUCTION

The Mechanical Universe is a project that encompasses fifty-two half-hour television programs, two textbooks in four volumes (including this one), teachers' manuals, specially edited videotapes for high school use, and much more. It seems safe to say that nothing quite like it has been attempted in physics (or any other subject) before. A few words about how all this came to be seem to be in order.

Caltech's dedication to the teaching of physics began fifty years ago with a popular introductory textbook written by Robert Millikan, Earnest Watson, and Duane Roller. Millikan, whose exploits are celebrated in Chapter 12 of this book, was Caltech's founder, president, first Nobel prizewinner, and all-around patron saint. Earnest Watson was dean of the faculty, and both he and Duane Roller were distinguished teachers.

Twenty years ago, the introductory physics courses at Caltech were taught by Richard Feynman, who is not only a scientist of historic proportions, but also a dramatic and highly entertaining lecturer. Feynman's words were lovingly recorded, transcribed, and published in a series of three volumes that have become genuine and indispensable classics of the science literature.

The teaching of physics at Caltech, like the teaching of science courses everywhere, is constantly undergoing transition. Caltech's latest effort to infuse new life in freshman physics was instituted by Professor David Goodstein and eventually led to the creation of *The Mechanical Universe*. Word reached the cloistered Pasadena campus that a fundamental tool of scientific research, the cathode-ray tube, had been adapted to new purposes, and in fact could be found in many private homes. Could it be that a large public might be introduced to the joys of physics by the flickering tube that sells us spray deodorants and light beer?

As the idea of using television to teach physics started to reach serious proportions, a gift was announced by Walter Annenberg, publisher and former U.S. Ambassador to Great Britain, to support the use of broadcast means for teaching at the college level. Ultimately, nearly $6 million of funds from the Annenberg School of Communications to the Corporation for Public Broadcasting through the Annenberg/CPB Project would be spent in support of *The Mechanical Universe*. That, in brief, is the story of how *The Mechanical Universe* came to be.

II PREFACE FOR STUDENTS

Each chapter of this book corresponds to one program of *The Mechanical Universe* television series. The book can also be used in the more traditional way as a physics textbook, without the television series. We anticipate that you will read each chapter, view each program one or more times, and take advantage of further guidance, instruction, practice, and other help provided by institutions that offer this course for academic credit.

It seems only fair to warn you that *The Mechanical Universe* does not follow the traditional guidelines of the college physics curriculum in the United States. Quite aside from the use of television, we have taken a somewhat unconventional view of the relationship between physics, the historical, philosophical, and social context in which it arose, and the mathematics on which it is based. A special word is in order about the use and level of mathematics that will be required of you.

It is assumed that every student of *The Mechanical Universe* begins with a solid secondary-school background in algebra and trigonometry. Certain other aspects of mathematics that are essential to the understanding and appreciation of physics are presented as they arose historically, as an intrinsic part of the development of physics. These include the derivative (Chapters 2 and 3), vectors (Chapter 5), the reverse of differentiation (Chapter 6), integrals (Chapter 7), and, throughout the book as needed, the use of analytic geometry and differential equations. This list may seem intimidating, but we firmly believe that the material is presented in a way that puts it within your grasp and makes it worth the effort to assimilate.

The traditional American college curriculum includes a physics course based on algebra and trigonometry that attempts to teach physics without the help of calculus. With rare exceptions, the popularity of such a course rivals that of compulsory military service. Our view is that the fault lies neither with the teachers nor with the students, but rather with the misguided attempt to bowdlerize physics of its essential mathematical underpinnings. We sincerely hope that students of *The Mechanical Universe* will be inspired to take additional courses in calculus as part of their education in mathematics (just as we hope they will want to pursue further the issues raised here in history and philosophy), but the mathematics presented in this course is sufficient for its intended purpose of use in the study of physics.

Although most important ideas in this course are presented in the television series, many of them cannot be learned by simply watching television any more than they can be learned by simply listening to a classroom lecture. Mastering physics requires the active mental and physical effort of asking and answering questions, and especially of solving problems. The examples and questions interspersed through every chapter are intended to play an essential role in the process of learning.

The backbone of *The Mechanical Universe* is a series of events that began in 1543 with a book published by Copernicus and culminated a century and a half later with a view of the universe that, consciously or not, has formed the basis of virtually all human intellectual activity since then. It is our firm view that a thorough understanding of that revolution in human thought is essential to any serious education. Offering that knowledge in an innovative way is the purpose of *The Mechanical Universe*.

III PREFACE FOR INSTRUCTORS AND ADMINISTRATORS

We expect that the ways in which *The Mechanical Universe* television series and textbooks are used will vary widely according to the circumstances and preferences of the institutions that offer it as a college course. The television programs can be viewed at home via broadcast or cable, presented in class, offered for viewing at the student's convenience at campus facilities, or even dispensed with altogether. However, we hope that no institution will imagine that the course can be presented without the services of live, flesh-and-blood college physics teachers. For most students, physics cannot be learned from a book alone, and it cannot be learned from a television screen either.

No laboratory component is offered as a part of *The Mechanical Universe* project. The reason is not that we judge a physics laboratory course to be unimportant or uninteresting, but rather that we judge its presentation by us to be impractical. We expect each institution offering the course to decide how it wishes to handle the laboratory component of learning physics.

This book is intended for use by students who would otherwise have taken the standard algebra- and trigonometry-based college physics course. Differential and integral calculus are treated just as vectors always have been in courses at this level: as subjects to be taught like any other portion of mechanics. For example, in Chapter 2 the mathematical concept of derivative is introduced in a natural way to explain the physical concepts of instantaneous velocity and acceleration. Chapter 3 describes further properties of derivatives, but is not to be considered a substitute for a course on differential calculus. Since this is a physics text rather than a mathematics text, our treatment of calculus is closely tied to intuitive and physical concepts. Thus, for example, in Chapter 7, the integral is defined in terms of area, and intuitive properties of area are used freely to derive corresponding properties of integrals. In a more rigorous treatment, as found in mathematics books, the integral is defined in terms of numbers (as a limit of a sum), and then the integral is used to define area. Care has been taken to present the mathematics in a way consistent with the views of mathematicians as well as physicists.

Some comments on the overall strategy of the presentation may be helpful. The necessary mathematics is mostly found in Chapters 3 (derivatives), 5 (vectors), 7 (integrals), and 9 (circular motion). These are interleaved with chapters that present physical ideas that, although profound and important (e.g., Chapter 4, inertia), require less homework practice than the mathematical ones. In addition, Chapters 2 and 6 purposely anticipate the ideas (respectively, derivatives and integrals) to be covered in the following chapters.

Chapter 6 seems to carry a particularly heavy burden. Not content with introducing Newton's laws, it also introduces some simple differential equations and their solutions. However, Chapter 4 has already laid the groundwork for understanding the trajectories worked out in Chapter 6, and the application of Newton's laws in other contexts is a persistent theme of following chapters, through Chapter 12 (the Millikan experiment).

There follow chapters on energy (13 and 14), heat, matter, and entropy (15–18), the conservation of momentum (19), harmonic motion (20–22), and angular momentum (23 and 24). Repetition and return to earlier ideas are not to be feared: Chapter 9 (circular motion) is designed partly as preparation for 23 and 24 (angular momentum), and Chapter 22 (waves) anticipates a series of chapters on aspects of waves to be found in the sequel to this volume, *Beyond the Mechanical Universe*.

As we stated above, each television program corresponds to one chapter of the book, and this is indicated in the table of contents. However, the programs for Chapters 15–18 (thermodynamics) will normally not be shown in a television broadcast of the first semester of *The Mechanical Universe*, although the videos will be made available to schools. Instead, in a broadcast "Energy and Stability" (Chapter 14/Program 14) will be followed immediately by "The Conservation of Momentum" (Chapter 19/Program 15). The textual material for these topics exluded from the programming sequence is included in the book because some instructors may wish to present a course covering both mechanics and heat.

The intellectual climax of this first volume is to be found in Chapters 25, 26, and 27, in which Newton's laws are used to solve the Kepler problem. In reality, Kepler's first law is derived in Chapter 23, and his third law must await Chapter 29. The deduction and analysis of elliptical orbits are the real payoff. This is where the student becomes master of *The Mechanical Universe*, so much so that Chapter 28 can be devoted to a discussion of how to navigate in space.

Throughout *The Mechanical Universe*, history is used as a means to humanize physics. It should go without saying that we don't expect students to memorize names and dates any more than we expect them to memorize detailed formulas and constants. *The Mechanical Universe* may or may not contribute to the vocational training of any given student. We hope it will contribute to their education.

IV ACKNOWLEDGMENTS

The Mechanical Universe textbooks, like the television series itself, would not have been possible without the cheerful and dedicated work of a long list of people who aided in its realization.

Number one of the list is Professor Steven Frautschi, of Caltech, lead author of the companion volume for science and engineering majors, who made countless valuable contributions to this book as well.

Special mention goes to *The Mechanical Universe* Local Advisory Committee, each member of which read and criticized in detail every chapter of the manuscripts, thus lending the benefit of their very considerable teaching experience: Keith Miller, Professor of Physics, Pasadena City College; Ronald F. Brown, Professor of Physics, California Polytechnic State University, San Luis Obispo; Eldred F. Tubbs, Member of the Technical Staff, Jet Propulsion Laboratory, Caltech; Elizabeth Hodes, Professor of Mathematics, Santa Barbara City College; and Eric J. Woodbury, Chief Scientist (retired), Hughes Aircraft Company.

In addition, parts of the manuscripts have been read and criticized by Margaret Osler (University of Calgary), Judith Goodstein (Caltech), and Robert Westman (UCLA), distinguished historians all; Dave Campbell (Saddleback Community College) and Jim Blinn (Jet Propulsion Laboratory), members of *The Mechanical Universe* team; Theodore Sarachman (Whittier College); and the entire 1983 and 1984 Caltech freshman classes.

In addition to all of these, homework problems were provided by Mark Muldoon and Brian Warr, and problems were checked for accuracy by Mark and Brian as well as by George Siopsis and Milan Mijic, all of Caltech.

Under the splendid direction of Project Secretary Renate Bigalke, the words and equations, mistakes and corrections of the authors were patiently and accurately rendered into the computer memory by Laurie Cornachio, Marcia Goodstein, and Sarb Nam Khalsa. All of the work was watched over anxiously by Hyman Field of the Annenberg/CPB Project (sponsors of *The Mechanical Universe*) and gently prodded along by David Tranah and Peter-John Leone of Cambridge University Press. We are especially pleased that Cambridge, which published Newton's *Principia,* has decided to follow it up with *The Mechanical Universe.* Sally Beaty, Executive Producer of *The Mechanical Universe* television series, was present and instrumental at every important juncture in the creation of these books. Geraldine Grant and Richard Harsh supervised an extensive formal evaluation of various components of *The Mechanical Universe* project, including drafts of the chapters from this volume; the results of that effort have had their due effect on the final work. Carol Harrison sniffed out many photos and their sources for us.

Finally, special thanks are due to Don Delson, Project Manager of *The Mechanical Universe,* who, through some miracle of organizational skill, cunning, and compulsive worrying, managed to keep the whole show going.

CHAPTER

INTRODUCTION TO THE MECHANICAL UNIVERSE

In the center of all the celestial bodies rests the sun. For who could in this most beautiful temple place this lamp in another or better place than that from which it can illuminate everything at the same time? Indeed, it is not unsuitable that some have called it the light of the world; others, its mind, and still others, its ruler. Trismegistus calls it the visible God; Sophocles' Electra, the all-seeing. So indeed, as if sitting on a royal throne, the Sun rules the family of the stars which surround it.

Nicolaus Copernicus, *De Revolutionibus Orbium Coelestium* (1543)

1.1 THE COPERNICAN REVOLUTION

We find it difficult to imagine the frame of mind of people who once firmly believed the earth to be the immovable center of the universe, with all the heavenly bodies revolving harmoniously around it. It is ironic that this view, inherited from the Middle Ages and handed down by the Greeks, particularly Greek thought frozen in the writings of Plato and Aristotle, was one designed to illustrate our insignificance amid the grand scheme of the universe – even while we resided at its center.

1

Aristotle's world consisted of four fundamental elements – fire, air, water, and earth – and each element was inclined to seek its own natural place. Flame leapt through air, bubbles rose in water, rain fell from the heavens, and rocks fell to earth: the world was ordered. Each element strove to return to its sphere surrounding the center of the universe. But even as Aristotle ordered the world, he did not deem it perfect. It was subject to death and decay just as were its inhabitants. Perfection was reserved for the heavens alone, which were serene and immutable.

Above the sphere of fire were the crystalline spheres of the moon, planets, sun, and the stars beyond. Each heavenly body was fixed in its orbiting sphere, traveling across the sky in a circle – the perfect shape Plato deemed as the ideal path all cosmic bodies should follow. Thus conceived, the universe was so simple that it was fully and adequately represented in great clocks constructed and painted by craftsmen of the Middle Ages. And the motions of the heavenly bodies were like the inner workings of clocks – regular, predictable, and in the mind of man, free of earthly decay.

This scheme, this grand plan, was an effort to describe the environment as it presented itself to the human senses. It was an attempt to find one simple, all-encompassing explanation for natural phenomena. The modern era began when people began to ask questions that Aristotle's world view could not answer.

Our purpose is to understand what has come to be known as classical mechanics – the science that arose to answer those new questions. No discovery in human thought is more important. Through it the temple of Aristotelian thought collapsed and no less than a new view of our place in the universe gradually rose from its ashes to replace it. So before we start to study physics, let's introduce some of the principal heroes of the story that we're going to see unfold.

Figure 1.1 Nicolaus Copernicus (1473–1543). (Courtesy of the Polish Cultural Institute in London.)

Nicolaus Copernicus, a timid monk who lived from 1473 to 1543, began the revolution with his book *De Revolutionibus Orbium Coelestium*, or *On the Revolutions of the*

Celestial Spheres, published in the year of his death, 1543. In this work, Copernicus froze the sun in the sky and set the earth in motion around it.

In an effort to simplify the Aristotelian model of the universe, with its circles upon circles to describe the complex motion of planets, Copernicus placed the sun at the center of the universe. In doing so, the earth was reduced to a common planet orbiting the sun, just like the five other known planets. This theory so upset the academic world that the word *revolution* has come to be associated with radical change. Thus we have not only the revolution of the earth around the sun, but also that of the colonies against Great Britain.

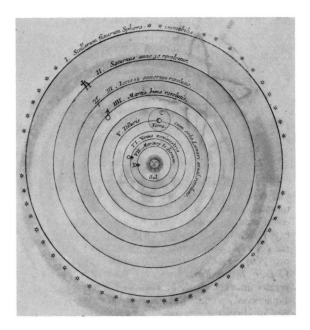

Figure 1.2 Diagram of the sun-centered system from Copernicus's *On Revolutions*. (Courtesy of the Archives, California Institute of Technology.)

Although Copernicus's ideas ultimately changed the Western view of the world, his book was at first largely ignored, and later considered heresy. One of the first scientists to seize upon the monk's revolutionary ideas was Johannes Kepler. Born a generation after the death of Copernicus and living from 1571 to 1630, Kepler ardently believed in the sun-centered system. In a battle to fit the best data on the seemingly wandering motions of the planets to the Copernican model, Kepler arrived at a set of mathematical equations that precisely described the motions. His results were as startling as they were elegant: planets tirelessly move around the sun in elliptical orbits.

At the beginning of the seventeenth century the discovery of the telescope drove the last nail in the coffin of the Aristotelian world view. Galileo Galilei, who lived from 1564 until 1642, explored the sky with the newly invented telescope. There, among other unexpected wonders, he found moons revolving around Jupiter. This sighting gave direct proof that the earth did not have to be the center of all heavenly motions. Through his

Figure 1.3 Johannes Kepler (1571–1630). (Courtesy of the Archives, California Institute of Technology.)

fertile experimentation and keen insight into the character of natural phenomena, Galileo's genius solidified the work that started with Copernicus's theory and hastened the destruction and replacement of Aristotelian phenomenology by the science of mechanics.

Figure 1.4 Galileo Galilei (1564–1642). (SCALA/Art Resource, N.Y.)

Like the early Greeks, the scientists of the new mechanics were intent on inventing an all-encompassing theory that would explain and describe every aspect of observable phenomena. Towering above all other scientists in the history of mechanics is Sir Isaac Newton. Born in the year Galileo died, 1642, Newton composed the grand synthesis that brought together the laws of the heavens and the laws of the earth. The physics that he established stood unchallenged until the beginning of the twentieth century.

Figure 1.5 Sir Isaac Newton (1642–1727). (Courtesy of the Mansell Collections.)

1.2 UNITS AND DIMENSIONS

One of the themes in the history of science is the great discovery that there is a connection between what goes on in the world – what we can observe – and mathematics. The discovery of the connection between mathematics and how the world works was first made by the Pythagoreans – followers of the Greek philosopher Pythagoras, in the fifth century B.C., a time when the *Iliad* and the *Odyssey* were cast into their final form, when Confucius walked the earth, and when the Greeks began to question nature rather than oracles for answers. The connection was preserved through time principally by astronomers who knew that the planets and stars followed courses that could be predicted by mathematical formulas and tables. However, it was believed during all of this time that the laws of the heavens, whatever they were, were in no way connected with the laws that govern the earth. And so even though the Pythagoreans knew that events on Earth obeyed mathematical laws, this idea was later forgotten as the views of Aristotle and Plato came to dominate Greek thought – and for that matter, all of Western thought – for nearly 2000 years.

Quantification in the natural sciences, the description of natural phenomena in mathematical terms, began to arise again at the end of the Middle Ages, about the same time as the invention of double-entry bookkeeping, an essential device for keeping track of the blossoming commerce of the period. Scholars still debate which of these two great discoveries inspired the other. No matter which came first, it is certainly true that commerce and science share a common need for standardized units of measure. Throughout history, establishing common units of time, distance, and weight for the sake of orderly agriculture and commerce has been one of the principal responsibilities of government, and the degradation of these standards, usually for the purpose of increasing effective taxes, has been one of the best-known symptoms of corrupt government. The nursery rhyme ''Jack and Jill'' started as a popular jibe at the inflation of standards of measure (a gill is still a unit of volume; up and down the hill has obvious meaning; keeping the units constant were the responsibility of ''the crown''; and so on).

Honest or not, units of measure were seldom the same in different political jurisdictions, and were generally based on some convenient or traditional magnitude. For example, still preserved by use in the United States today, the mile (from *milia*, thousand) was once 1000 standard paces of a Roman legion; the yard the distance from one's nose to one's outstretched fingers (see Fig. 1.6); the foot is obvious enough; and the inch was one thumb – from joint to tip.

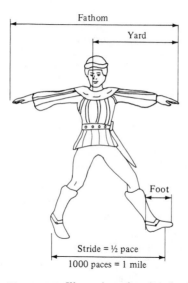

Figure 1.6 Illustration of ancient basic units of the British system.

Example 1
Determine the number of inches in 24 yards.

In order to convert from yards to inches we use a trick; we multiply 24 yards by 1, written in a creative way:

$$24 \text{ yd} = 24 \text{ yd} \times \left[\frac{3 \text{ ft}}{1 \text{ yd}} \right] \times \left[\frac{12 \text{ in.}}{1 \text{ ft}} \right],$$

or

$$24 \text{ yd} = 864 \text{ in.}$$

Each of the terms in brackets is equal to 1 by conversion. By canceling the units, we are left with inches as the final unit. Working scientists use this cancelation method to get their conversions straight.

One of the legacies of Napoleon's conquests in Europe was a new system of units, not based on tradition and whimsy, but on cool, precise French logic and on the decimal system. It is, nevertheless, firmly based on human magnitudes and on the properties of water, an essential ingredient of life. For example, the central unit of length is the *meter*, which is roughly a yard. But instead of dividing it into feet and thumbs, it is divided into tenths (decimeters), hundredths (centimeters), thousandths (millimeters), and so on. The unit of mass, the *gram*, is the mass of one cubic centimeter of water; the unit of volume, the *liter*, is 1000 cubic centimeters, so that a liter of water has a mass of one kilogram; and so on. The definitions of these quantities no longer vary from one country to another; they are fixed by international treaty, and are used almost everywhere except in the United States. This system is formally known as the *Système International d'Unités*, or SI for short. Americans who wish to learn science or engineering, or just to shop in Europe (or Canada) must learn to convert from units based on the size of a dead king's foot to units based on the distance between two scratch marks on a platinum–iridium bar kept in a refrigerated vault in Sèvres, France.

Example 2

Determine the number of centimeters in a football field.

We perform a conversion of 100 yd into centimeters using the cancelation method:

$$100 \text{ yd} = 100 \text{ yd} \times \left[\frac{3 \text{ ft}}{1 \text{ yd}}\right] \times \left[\frac{12 \text{ in.}}{1 \text{ ft}}\right] \times \left[\frac{2.54 \text{ cm}}{1 \text{ in.}}\right],$$

or, multiplying the numbers and canceling the units,

$$100 \text{ yd} = 9144 \text{ cm}.$$

Remembering that there are 2.54 cm in 1 in. is handy.

Mastering the flow of time and dividing it into units seems to have been a part of the growth of every civilization on earth. Astronomer-priests of agricultural societies were responsible for deciding when to begin the annual cycle of tilling, planting, and harvesting. Smaller divisions of time corresponded, at least roughly, to the death and rebirth of the moon (months) and, of course, the daily cycle of light and dark. Intermediate clusters of days, 5 or 10 or, by Roman times, 7 days per week, also came up. Dividing time into units smaller than a day proved more difficult, since it involved inventing time-keeping devices rather than mere counting. Hours, minutes, and seconds are relatively recent inventions, as is the idea that these units should have the same duration all year, regardless of the proportion of daylight and darkness in each day. Fortunately, unlike the units of length and mass, the same units of time are used everywhere, even in the United States. Units of one second and longer have the traditional names (minute, hour, day, week, month, year, century, millennium), whereas shorter times get metric-style prefixes (millisecond, microsecond, nanosecond, etc.).

Example 3

How many seconds are there in a fortnight?

To answer this, we again use the cancelation method:

$$1 \text{ fortnight} = 1 \text{ fortnight} \times \left[\frac{2 \text{ weeks}}{\text{fortnight}}\right] \times \left[\frac{7 \text{ days}}{\text{week}}\right] \times \left[\frac{24 \text{ h}}{\text{day}}\right] \times \left[\frac{3600 \text{ s}}{\text{h}}\right],$$

or

$$1 \text{ fortnight} = 1{,}209{,}600 \text{ s}.$$

Quantities like seconds, grams, and meters not only have units, which may vary from one jurisdiction to another, but also dimensions, meaning, respectively, time, mass, and distance. Quantities with the same dimensions, but different *units*, can easily be compared (see Examples 1–3): an inch is bigger than two centimeters, but less than a

Table 1.1 Metric Prefixes and Abbreviations

Multiple	Prefix	Abbreviation
10^{12}	tera	T
10^{9}	giga	G
10^{6}	mega	M
10^{3}	kilo	k
10^{-2}	centi	c
10^{-3}	milli	m
10^{-6}	micro	μ
10^{-9}	nano	n
10^{-12}	pico	p
10^{-15}	femto	f

light-year. But quantities with different dimensions cannot be compared at all, regardless of their units: a kilogram is neither bigger nor smaller than an hour or a yard.

For many years after the beginning of the quantification of nature, all quantification consisted of comparing quantities of the same dimension. The idea of compounding quantities of different dimensions – say, dividing a distance by a time to form a speed – is a rather recent but richly useful invention. A compound quantity like speed has a unique dimension (distance divided by time), but various units (cm/s, furlongs/fortnight, etc.). We can write this as an equation of words,

$$\bar{v} = \frac{\text{distance traveled}}{\text{time elapsed}} . \tag{1.1}$$

The bar over the v reminds us that we are talking about an average speed during that time interval. If distance is measured in either feet or meters and time is measured in seconds, the units of speed are feet per second, written ft/s, or meters per second, written m/s. The important lesson is that units can be compounded when new physical quantities are defined.

Another physical quantity that we will encounter is acceleration. We often hear about cars that can speed up, say from 0 to 60 mi/h, in so many seconds. That's an example of acceleration. The average acceleration during a time interval is the change in speed during that interval divided by the time:

$$\bar{a} = \frac{\text{change in speed}}{\text{time elapsed}} . \tag{1.2}$$

Just as speed can be measured in ft/s or m/s, the units of acceleration can be (ft/s)/s (read feet per second per second), usually abbreviated ft/s^2 (read feet per second squared), or (m/s)/s, abbreviated m/s^2.

Example 4
What is the speed limit of 55 mi/h in m/s?

Using the conversion that there are 1.6 km in 1 mi, we can write the following equation:

$$55\,\frac{\text{mi}}{\text{h}} = 55\,\frac{\text{mi}}{\text{h}} \times \left[\frac{1.6\ \text{km}}{1\ \text{mi}}\right] \times \left[\frac{1000\ \text{m}}{1\ \text{km}}\right] \times \left[\frac{1\ \text{h}}{60\ \text{min}}\right] \times \left[\frac{1\ \text{min}}{60\ \text{s}}\right],$$

or

$$55\ \text{mi/h} = 24\ \text{m/s}.$$

Questions

1. Determine which of the following distances is greatest:

(a) 560 yd

(b) 0.3 mi

(c) 0.5 km

(d) 723,000 cm.

2. Convert the following quantities into ft/s:

    (a) 23 km/h   **(b)** 88 nm/ms
    (c) 40 cm/yr   **(d)** 3 furlongs/fortnight

(a furlong is one-eighth of a mile).

3. Determine which of the following bizarre quantities have units of distance, speed, or acceleration:

    (a) 2.3 m yr/s   **(b)** 144 m yd s/mi yr^2
    (c) 4 cm/s century   **(d)** 54.2 m^2/ft s.

1.3 A FINAL WORD

The rise of modern science grows out of quantification. But quantification means much more than expressing observations in mathematical form. It is also a turning away from natural philosophy – grand schemes based on aesthetic preference – to detailed and precise observations and measurement. In other words, it is a turning to the accumulation of knowledge by small, detailed increments. "I would rather learn a single fact, no matter how ordinary," wrote Galileo, "than discourse endlessly on Great Issues." Before Copernicus, many speculated on a sun-centered universe, but Copernicus took the trouble to do the detailed calculations and produce the astronomical tables that made his system a serious competitor for the very successful Ptolemaic system that came before it. Others before Galileo speculated on the properties of matter in motion, but Galileo based his arguments on detailed observations. His experiments with balls rolling down smooth inclined planes led to the law of falling bodies (Chapter 2), the law of inertia (Chapter 4), and ultimately to the law of conservation of energy (Chapter 13). Before clocks as we know them were invented, he devised means of timing his experiments by weighing how much water flowed through a specially constructed pipe. These measurements were accurate to a tenth of a second. The ultimate result of this kind of careful attention to detail, together with ingenuity, was no less than a new view of the universe.

Quantification of physics tends to condense its ideas into mathematical formulas. As Galileo so delightfully expressed it, the great book of Nature lies ever open before our eyes, but it is written in mathematical characters. History teaches us that mathematics helps to advance physics, but it also shows that, like the tides, ideas flow in both directions. New discoveries in one field often lead to improvements in the other. For example, early in the seventeenth century the French mathematician Pierre de Fermat devised a crude method for drawing a tangent line to a curve. This gave Newton a hint for determining the velocity of a moving point, and this, in turn, led to Newton's version of differential calculus. The mathematics in this book is developed in the same spirit. When new mathematical concepts are introduced, such as derivatives, integrals, and vectors, they arise naturally from physical problems; and then these new concepts help us to read the great book of Nature and to write new chapters.

After Copernicus and his revolution and the events that led Europe through the years of Kepler and Galileo, and finally Newton, our view of the universe was, and still is today, that we live on a speck of dust in a lost corner, somewhere in the universe. Aristotle and Plato tried to teach us humility by placing us in a lowly sphere, isolated from the serene perfection of the heavens, but never in their wildest dreams could they have imagined the impact of the psychological change that occurred when we first realized that we are not at the center of the universe.

When we study history, we learn about kings and queens, about social problems, wars, economics, and so on. All of these things come and go. And when they're gone the world is pretty much the same as it was before. If you walk through an ancient Roman town today, say the town of Herculaneum in Italy, you can perceive exactly where you are and what the people were doing and why the town was built. The human condition has not changed very much in the past 2000 years. But there is one profound change that has altered the human race forever. That is the discovery of our real place in the universe. Studying history, we learn about the Renaissance and Reformation, the Counter Reformation and the Thirty Years' War – the events that dominated the history of Europe during the time of Copernicus, Kepler, Galileo, and Newton. But those events were minor readjustments in the social fabric compared with this one monumental change that was occurring – new ideas in the history of ideas – that changed the human race absolutely forever. Our job in this volume is to study that story, to see exactly how and why it happened that we found our real place in the universe.

CHAPTER

THE LAW OF FALLING BODIES

(Natural) philosophy is written in this enormous book, which is continuously open before our eyes (I mean the universe) but it cannot be understood without first learning to understand the language, and to recognize the characters in which it is written. It is written in a mathematical language . . .

Galileo Galilei, *Il Saggiatore* (1623)

2.1 ARISTOTLE'S DESCRIPTION OF MOTION

Before we began to learn to read Galileo's mathematical book of the universe, our descriptions of nature were qualitative and verbal. For centuries, only words were used to describe the motion of objects, words based on the writings of the Greek philosopher Aristotle from the fourth century B.C. Aristotle's descriptions of motion centered on the idea of a ''natural place.'' Each of the four elements – earth, air, water, fire – that composed all matter on Earth had a natural place. The lowest natural place was assigned to earth, at the center of the cosmos. The natural place for water was just above that of earth. Since air was lighter than earth or water, it was assigned a natural place above water. Fire was given the highest place. When any element or object made of that element was in its natural place, it remained at rest. According to Aristotle, when an object was removed from its natural place, it possessed a *potentia*, or tendency, to return to its place. If uninhibited, any object would be drawn back to its natural place. Natural motion resulted from an object seeking its natural place. Hence fire (or smoke), being naturally light, would rise, whereas rocks, being naturally heavy, would fall.

Furthermore, Aristotle asserted that any object, after it is released, quickly reaches some final speed, which it maintains to the end of its path. When we pick up a stone and release it, the stone strives to return to its natural place, the earth, and quickly gains a speed that it maintains during its entire fall.

In his work *De Caelo*, Aristotle further described the behavior of falling bodies: "the greater the mass of earth or fire the quicker always its movement toward its own place." From the common observation that a heavy stone falls faster than a feather, Aristotle reasoned that weight is a factor that governs the speed of the fall. Consequently, the heavier an object, the greater would be its *potentia* to return to the earth. In turn, this stronger tendency would cause a greater speed of fall.

Aristotle's description of motion agreed well with common observations of falling leaves, raindrops, and stones. In all cases, the body encounters resistance to its fall from the air. But what if there were no air to offer resistance and impede the fall? How would objects fall through a vacuum? Aristotle argued that in a vacuum, all bodies, being unresisted, would fall with the same infinite speed. But Aristotle, like most other ancient Greeks, regarded infinity as an incoherent concept. Motion through a void was dismissed by Aristotle because it was inconceivable; he concluded that the vacuum could not exist.

The ideas we have examined present a qualitative description of motion. Aristotle believed that mathematics was of little value in describing the world around him because its use demanded a degree of abstraction which lost the qualitative texture of reality. Only qualitative observations were necessary to form a theory. These ideas were seized by scholars in the fourteenth through sixteenth centuries, sometimes with the fervor of a dogma. Consequently physics in general consisted mainly of qualitative explanations until Galileo grasped the usefulness of mathematics in describing the world.

Questions

1. How would an Aristotelian explain the motion of bubbles in water?

2. Which of the following graphs of speed versus time correctly illustrates Aristotle's description of falling body motion?

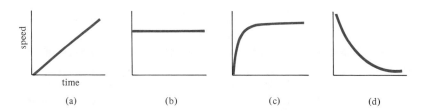

(a)   (b)   (c)   (d)

3. Which of the following graphs of distance fallen versus time for a heavy body H and light body L would an Aristotelian construct?

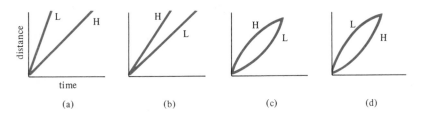

(a)   (b)   (c)   (d)

2.2 DO HEAVY BODIES FALL FASTER THAN LIGHTER ONES?

In 1638 Galileo published his work *Dialogues Concerning Two New Sciences*, in which he took a major step toward the understanding of motion. His insight was to imagine a body falling without any resistance. He realized that in a vacuum all bodies, heavy or light, would fall at the same rate.

This was a brilliant insight because Galileo could not produce a vacuum; he could only imagine one. We have already seen that the Aristotelians, who dominated physical thought in Galileo's era, believed that a vacuum, or void, was impossible. Galileo realized that the question of the existence or nonexistence of the void was unimportant. What was important was that to understand falling bodies, the effects of air resistance should be ignored.

The argument he used was this: Consider a heavy rock connected to a light one by a string. According to Aristotle, when released, the heavy one pulls the lighter one down and tries to make it fall faster than it would if unattached. The light one, on the other hand, tends to slow down the heavier one. Thus the combined body must fall faster than the lighter one alone, and slower than the heavy one. But the combined body is heavier than the heavy one. It should therefore fall faster than the heavy one. The Aristotelian view thus leads to a contradiction if we ignore air resistance. This kind of contradiction had led Aristotle and his followers to believe that a vacuum was impossible. Galileo concluded instead that all bodies would fall at the same rate in a vacuum.

Demonstrating that all bodies fall at the same rate has become one of the classic experiments of physics. Innumerable students have witnessed a feather and a penny fall at the same rate in an evacuated tube as a classroom demonstration. When astronaut David R. Scott of the Apollo 15 mission found himself on the airless surface of the moon, he could not resist repeating this classic experiment for the whole world to see. As he said then, he could not have got to where he was standing without Galileo's discovery.

2.3 MEDIEVAL LAWS OF FALLING BODIES

Even before Galileo, a number of medieval scholars tried to describe the falling motion of a heavy body. One of the earliest attempts is credited to a fourteenth-century scholar, Albert of Saxony. In trying to answer the question, "In what way does a body get faster as it falls?" Albert argued that the speed of a body is proportional to the distance it has fallen. This proportionality between speed and distance fallen means that compared to its speed after having fallen 1 ft, a body is traveling twice as fast after falling 2 ft, three

Figure 2.1 Apollo astronaut David R. Scott dropping a falcon feather and hammer on the moon.

times as fast after 3 ft, and so on. For centuries, anybody who thought about it at all assumed Albert was right.

Another fourteenth-century scholar, Nicole Oresme, had a different conjecture. From a mathematical study of various possible types of motion, he found a law that he at one time suspected could describe falling bodies, although he was not much concerned with anything so messy as the real world. What he suggested was a relation between speed and time rather than speed and distance: the speed is directly proportional to the time spent falling. This means that compared to its speed after falling for 1 s, an object is moving twice as fast after 2 s, three times as fast after 3 s, and so on.

Example 1

What would the laws of Albert of Saxony and Oresme imply if graphs were used to illustrate them?

Both laws involve direct proportionalities, which means that if you increase one of the quantities in the relation, the other should increase by a proportional amount. For example, in Oresme's law, if the time is tripled, the speed is also tripled. This implies a straight line when graphed. Consequently the two laws are represented by the following graphs:

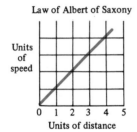

Law of Albert of Saxony

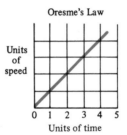

Oresme's Law

The two laws we have examined thus far focused on how the speed of a falling body changes. Leonardo da Vinci, in the fifteenth century, formulated a different kind of law. He expressed his law in terms of quantities that are easy to measure: intervals of distance and time. He proposed that the distances fallen in successive equal intervals of time are proportional to the consecutive integers. This statement needs to be explained. Suppose that, starting from rest, a body falls one unit of distance in the first time interval. After the second time interval, the body will have fallen an additional two units of distance; after the third time interval, it will have fallen three more units of distance, and so on according to the consecutive integers (1, 2, 3, 4, . . .).

Leonardo's law also makes a statement about the average speed of the object as it falls. We can discover this relation if we recall the definition of average speed and apply his law. The average speed for any time interval is the distance traveled divided by the time. According to his law, for equal successive time intervals, the distance traveled is proportional to the consecutive integers. The average speed for each successive time interval, therefore, will be the distance interval, which as he said goes as the consecutive integers, divided by the time interval. Since all the time intervals are equal, the average speed over consecutive time intervals is also proportional to the consecutive integers.

Example 2

If in a certain time interval a bowling ball falls 2.5 m, what total distance would it have fallen after ten such time intervals according to Leonardo's law?

From Leonardo's law, we know that after the second time interval, the ball would have fallen an additional 2 times 2.5 m. That makes the total distance fallen after two time intervals equal to 2.5 m plus 2 times 2.5 m. We can generalize this for ten successive time intervals as follows:

$$\text{total distance fallen} = 2.5 \text{ m} + 2 \times 2.5 \text{ m} + 3 \times 2.5 \text{ m}$$

$$+ 4 \times 2.5 \text{ m} + 5 \times 2.5 \text{ m}$$

$$+ 6 \times 2.5 \text{ m} + 7 \times 2.5 \text{ m} + 8 \times 2.5 \text{ m}$$

$$+ 9 \times 2.5 \text{ m} + 10 \times 2.5 \text{ m}$$

$$= 2.5 \text{ m} \times (1 + 2 + 3 + \cdots + 8 + 9 + 10)$$

$$= 2.5 \text{ m} \times 55$$

$$= 137.5 \text{ m}.$$

In other words, the total distance fallen after any number of equal time intervals is equal to the distance fallen in the first time interval times the sum of the consecutive integers up to the number of time intervals.

Galileo developed his own law of falling bodies, which he stated in a number of different ways. One of these resembled Leonardo's formulation. He said that the distance fallen in successive equal time intervals is proportional to the odd numbers. In other words, suppose that you measure the distance fallen by a body in the first time interval and find it to be one unit of distance. According to Galileo's law of odd numbers, in the next equal time interval, the body will fall an additional three units of distance; in the next interval it will fall five more units of distance; and so on according to the odd numbers (1, 3, 5, 7, . . .).

Does Galileo's law of odd numbers make any statement about how the average speed changes? It does, and we can figure it out just as we did for Leonardo's. According to Galileo, the average speed in successive equal time intervals is proportional to the consecutive odd numbers.

Example 3

What would graphs of *total* distance fallen as a function of time look like for Leonardo's and Galileo's laws?

Using the techniques of Example 2, we can first construct tables that give the total distance fallen for each successive time interval.

Leonardo's Law		Galileo's Law	
Number of time intervals	Total units of distance fallen	Number of time intervals	Total units of distance fallen
0	0	0	0
1	$0+1=1$	1	$0+1=1$
2	$0+1+2=3$	2	$0+1+3=4$
3	$0+1+2+3=6$	3	$0+1+3+5=9$
4	$0+1+2+3+4=10$	4	$0+1+3+5+7=16$
5	$0+1+2+3+4+5=15$	5	$0+1+3+5+7+9=25$
6	$0+1+2+3+4+5+6=21$	6	$0+1+3+5+7+9+11=36$
7	$0+1+2+3+4+5+6+7=28$	7	$0+1+3+5+7+9+11+13=49$

Once we have such tables, constructing the graphs is a simple matter. After labeling and numbering the axes, we plot the points on the appropriate axes then connect the points:

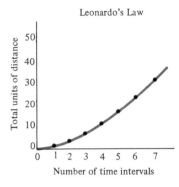

Leonardo's Law

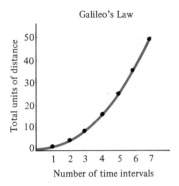

Galileo's Law

There is another law, formulated by Oresme, which also deserves to be considered. We have waited until this point to discuss it because it is different from the other laws. From geometrical constructions concerning his speed law, Oresme arrived at another description of falling bodies, which related distance to time. But the relation is not like the laws of Leonardo and Galileo. This law relates the total distance fallen to the time spent falling. Specifically, it states that the distance fallen is proportional to the square of the time spent falling.

We can understand this law as follows: Suppose that after one unit of time, you find a body to have fallen one unit of distance. According to the time-squared law, after two units of time, the object will have fallen four units of distance. And after three time units, it will have fallen nine distance units from its starting point. If you worked through Example 3, what we have just stated may give you a sense of déjà vu. In that example we found that the total distance fallen according to Galileo's law goes as 1, 4, 9, 16, 25, etc., units of distance as the time increases through 1, 2, 3, 4, 5, etc., time units. But $1^2 = 1$, $2^2 = 4$, $3^2 = 9$, $4^2 = 16$, $5^2 = 25$. In other words, Galileo's odd-number law predicts the same total distances fallen for each time unit as does Oresme's time-squared law. The two laws are equivalent descriptions of falling bodies.

A final word concerning the types of laws we have encountered. We have seen five possible descriptions for falling bodies that are different answers to different questions concerning this motion. Two of them proposed a relation between speed and either distance, as in the law of Albert of Saxony, or time, as in Oresme's law. Another pair expressed the distance fallen in successive time intervals; Leonardo thought that this distance should be proportional to the integers, whereas Galileo said that it went as the odd integers. And then there was the time-squared law, unique because it related the total distance fallen to the time spent falling. Before we can decide which law is best, we have to decide which question we really want to answer.

Asking the right questions is not easy. Neither is performing experiments to select the best description from a number of possibilities. Besides, the Aristotelian cast of mind did not make it seem important to experiment. Furthermore, there did not exist a good timepiece to measure fractions of a second. For example, even if Galileo had dropped a cannonball from the leaning tower of Pisa (he probably did not), it would have taken approximately 2 s to reach the ground. And given the technology available to him it would have been impossible to make any detailed measurements of the falling ball in such a short interval of time. Consequently, the dependence of distance and speed on time was difficult to measure and there was no definitive way to choose among the various laws proposed.

In the next section we shall determine which laws are valid descriptions of falling bodies.

Questions

4. Oresme's law relates the speed of a falling object at any instant to the time. In his day no one had thought of a way to measure the speed of a falling body. Is there any way to do it today?

5. In his work *Two New Sciences*, Galileo asserts that the law of Albert of Saxony cannot be correct because, when an object is released, it has not fallen any distance, and therefore, by Saxony's law, it has no speed. Having no speed, of course, it would not begin to fall. Consequently, he argues, if this law were correct, the body would never fall. Are you convinced by this argument? Why or why not?

6. A stone is dropped, and one unit of time later, another stone is dropped from the same point. What is the number of units of distance between the two stones after the first one has fallen for three units according to Leonardo's law? According to Galileo's law?

7. Construct graphs of average speed as a function of time for Leonardo's and Galileo's laws.

8. Followers of the sixth-century B.C. Greek philosopher Pythagoras believed that numbers had shapes, and one shape was a square. The first square number was 1 since you could draw a little square around one dot like this: $\boxdot$. The next square number was 4, which can be constructed from your original square by adding more dots to form a larger square like this: $\boxplus$. The next square number was 9, and you can continue to enlarge your square by adding more dots. (You can also do this with squares on graph paper instead of dots.) Examine this procedure for 1, 4, 9, 16, and 25, by considering how many dots you are adding to form each successive square. What is the general rule? Galileo knew of this Pythagorean discovery. How is it related to his law of falling bodies?

9. Assume an object falls according to the time-squared law. If it falls 16 ft in 1 s, how long will it take to fall 144 ft?

2.4 "THE" LAW OF FALLING BODIES

We have now encountered several possible laws of falling bodies. The time has come to let Nature choose among them.

Figure 2.2 shows a falling ball in successive instants of time. The picture was taken by a camera with an open shutter, while the ball was illuminated by an instrument called a stroboscope. The stroboscope emits flashes of light at regular intervals of time. It is our clock in this experiment. The same interval of time has elapsed between each successive image of the ball in our picture. A scale behind the ball allows us to measure the distance fallen.

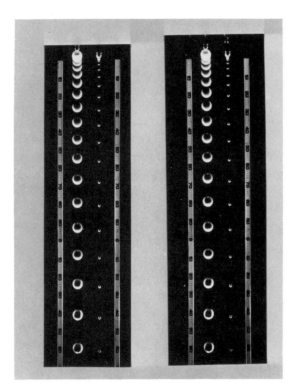

Figure 2.2 Stroboscopic picture of a falling ball. (*PSSC Physics*, 2nd ed., 1965; D.C. Heath and Company with Education Development Center, Inc., Newton, Mass.)

Data of this kind do not directly test Albert's law or Oresme's law about the speed of a falling body, because we have not measured the speed, but the position of the body. On the other hand, these data are ideal for choosing between Leonardo's law and Galileo's law for the distance fallen in successive intervals of time.

Before we try to choose, we should understand that these are real experimental data. Like all real experimental data, they are imperfect. For example, we can't tell exactly where the ball is, because the images are a bit fuzzy and some of them overlap. Besides, there really is air resistance, which Galileo (but not Leonardo) wanted to ignore. These are not the only possible sources of error. There is experimental error in all experiments, no matter how ingenious or carefully done.

We examine our data in the following way: we measure the distance fallen between the first and second flashes (the first flash illuminates the ball at the instant of release). This distance we call c. According to Leonardo's law, the next distance should be $2c$, the one after that should be $3c$, and so on. Figure 2.3 shows our picture with lines drawn on it to indicate where each image should be, constructed in this way.

Now we reconstruct the predictions of Galileo's law of odd numbers. If c is the distance between the first and second flashes, then the next distance should be $3c$, the one after that should be $5c$, and so on. In addition, the total distance fallen at each flash should be proportional to the square of the time because Galileo's law is the same as Oresme's time-squared law. The result is also shown in Fig. 2.3.

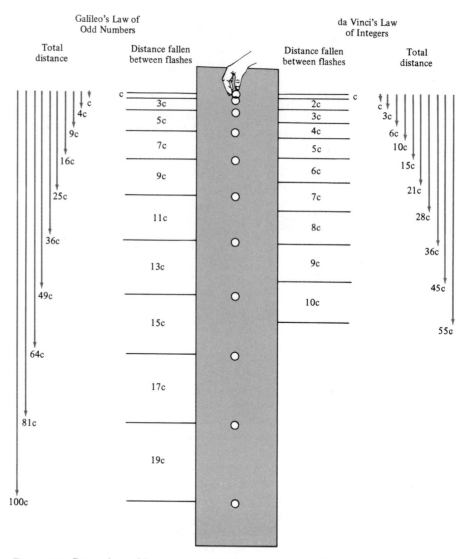

Figure 2.3 Comparison of Leonardo's and Galileo's predictions for falling bodies.

In spite of all experimental uncertainties, the comparison between the two figures is decisive. It is Galileo's law that we choose.

Galileo arrived at this law, or, at the very least, confirmed it experimentally, as we have done, but without the benefit of our stroboscope. Instead he used (among other techniques) a water clock to time balls rolling down inclined planes. As he made the plane steeper, the ball would roll faster, but the motion always had the same property we have observed here: the total distance was proportional to the square of the elapsed time. He concluded that the law would also hold if the plane were vertical, in other words, in free fall.

Questions

10. In Fig. 2.2 the first two images of the ball overlap. Can you think of any other way to construct our predictions without using the first two images?

11. What other sources of error are there in our experiment?

12. Suppose you were a seventeenth-century defender of Leonardo or Albert of Saxony trying to find a flaw in Galileo's demonstration of his law. How could you criticize what he had done?

We are now in a position to write an equation for freely falling bodies. We shall use the symbol s to represent distance fallen. Since this distance depends on time, or is a function of time, we shall use functional notation and write $s(t)$ to indicate the distance. Remember that this is read "s of t" and means that the distance s depends on t, or as we sometimes say, is a function of time t. Warning: mathematicians and physicists use parentheses to mean different things; sometimes parentheses are used to indicate multiplication, but in this notation, $s(t)$ does not mean s times t. With this notation, our free fall equation is

$$s(t) = ct^2. \tag{2.1}$$

The right-hand side of this equation states in symbols what we have said with words. It describes explicitly how the distance changes with the time. The distance is a constant times the square of the time. This equation describes the distance any body in a vacuum has fallen at every instant of time from the time of its release until the end of its fall.

Because $s(t)$ has units of length and t has units of time, the constant c must also have units. These units must be length divided by time squared; only in this way will the time-squared units cancel to yield units of length for $s(t)$.

When $t = 1$, $s(1) = c$, so c is numerically equal to the distance any object will fall in the first unit of time. We measure t in seconds, and then determine c by an experiment like the one we've already done with the ball. In U.S. conventional units such a measurement yields a value of 16 ft/s^2 for c. In SI units it has the value of 4.9 m/s^2. This value is sort of a local ordinance: it has the same value for every object falling in a vacuum near the surface of the earth. On another planet, c may have a different value, but the form of the law of falling bodies will not change. If you know c on the planet, you can still use $s(t) = ct^2$ to find the distance any object will fall on the surface of the planet. Figure 2.4 shows a computer simulation of stroboscope pictures on the moon and various planets.

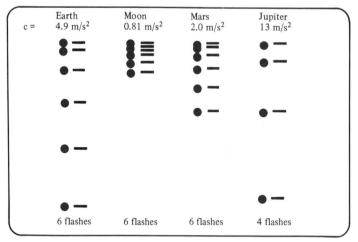

Figure 2.4 Computer simulation of stroboscopic pictures on the moon and various planets.

Example 4
A diver from a 15-m platform wishes to know how long he is in the air. What can we tell him?

In this case we know the distance fallen, $s(t) = 15$ m, and we want to determine the time required to fall this distance. From the free-fall equation, Eq. (2.1), we know that

$$s(t) = ct^2.$$

Solving for t algebraically, we obtain

$$t = \sqrt{\frac{s}{c}}.$$

We can now substitute for s and c in SI units to obtain the answer:

$$t = \sqrt{\frac{15 \text{ m}}{4.9 \text{ m/s}^2}},$$

$$t = 1.75 \text{ s}.$$

Of course, air resistance will cause the real time to be a bit longer. We neglect air resistance in the remaining examples and questions in this chapter.

2.5 THE AVERAGE SPEED OF A FALLING BODY

Now that we know how the distance changes with time, can we find how fast the body is falling? We must be very careful to pose our question correctly. How fast it's moving means how far it travels in a specified amount of time. We need to specify the time interval during which the object drops.

If we do specify the time interval, the free-fall equation will allow us to determine the average speed during that time. Here's how we can calculate the average speed over any time interval. First, we determine $s(t)$ at some time t using our free-fall equation, Eq. (2.1). Next we calculate the distance fallen after a later time, say t_1. That is, we calculate $s(t_1)$ from our equation. The length of the time interval is $t_1 - t$, and the distance fallen during this time interval is $s(t_1) - s(t)$. The quotient is the average speed for that interval:

$$\overline{v} = \frac{\text{distance traveled}}{\text{travel time}} = \frac{s(t_1) - s(t)}{t_1 - t}. \qquad (2.2)$$

Example 5

What is the average speed of an object during the third second of the fall?

In order to calculate the average speed, we follow the steps leading to Eq. (2.2). Let's call $t = 2$ s and $t_1 = 3$ s. From $s(t) = (4.9 \text{ m/s}^2)t^2$, we obtain

$$s(2) = (4.9 \text{ m/s}^2)\,(2 \text{ s})^2 = 19.6 \text{ m},$$

$$s(3) = (4.9 \text{ m/s}^2)\,(3 \text{ s})^2 = 44.1 \text{ m}.$$

Substituting into Eq. (2.2), we obtain

$$\overline{v} = \frac{s(3) - s(2)}{3 \text{ s} - 2 \text{ s}} = \frac{44.1 \text{ m} - 19.6 \text{ m}}{3 \text{ s} - 2 \text{ s}},$$

$$\overline{v} = 24.5 \text{ m/s}.$$

We have now seen how the average speed during a particular time interval can be calculated. But this is only the average speed. In Example 5, the object is moving slower than 24.5 m/s at the start of the interval and faster at the end of it. What we would really like to know is the speed, not on the average, but at some instant of time. But the speed at some instant means that there is no time interval. If we try to take the distance fallen in no time interval and divide that by zero time, we end up trying to divide zero by zero, which is a mathematical disaster. Yet we have an intuitive feeling that at every instant a falling body does have a definite speed.

This problem plagued mathematicians for thousands of years, and it was not until calculus was invented, a generation after the death of Galileo, that the solution was found. In the next section we shall ''reinvent'' one concept of the calculus, called the *derivative*. With this tool in hand we shall be able to calculate the speed of a falling body at any instant.

Questions [use Eqs. (2.1) and (2.2) in your work]

13. A ball is dropped from a very high tower. One second later another ball is dropped. Does the distance between the balls increase, decrease, or remain the same as they fall? Explain your reasoning.

14. A cliff diver is in the air for 4.5 s before hitting the water. Calculate the following for his dive:

  (a) From what height did he fall?
  (b) What was his average speed over the entire dive?
  (c) What was his average speed over the last half-second?

15. Calculate the average speed of a falling ball between the fourth and fifth seconds of its fall. Letting $t = 4$ s, $t_1 = 5$ s, compare your answer to $c(t_1 + t)$ and comment on your comparison.

16. Calculate the average speed of a falling object for the following time intervals in seconds:

  (a) 3 to 4   **(b)** 3 to 3.5
  (c) 3 to 3.25   **(d)** 3 to 3.1.

  Do you notice any particular trend of the average speed?

17. An experiment performed on the moon finds that a feather falls 20.75 m in 5 s. Can you find the value of c on the moon?

2.6 INSTANTANEOUS SPEED

Our goal in this section is to find the speed of a falling body at any instant. Our strategy will be to calculate the average speed for a certain time interval, and then at the very end of the calculation, allow the time interval to shrink to zero. This simple idea leads to one of the most important inventions in the history of mathematics. It is called the *derivative*.

The distance the body has fallen after time t is

$$s(t) = ct^2. \tag{2.1}$$

Suppose we ask what distance the body has fallen a short time later. We can answer this precisely if we know what that little time later is. Let's imagine a small amount of time and call it h. This small time interval can be one second or a millionth of a second; right now its value is unimportant. Having given it a name, h, we can now restate our question more precisely: what is the distance the object has fallen a time $t + h$ after it started falling? We use Eq. (2.1) to answer this question. The functional notation of Eq. (2.1) means that if we change t to some other value, we must insert that value into the right-hand side of the equation as well. In other words, the distance fallen after $t + h$ seconds, s at time $t + h$, is

$$s(t + h) = c(t + h)^2. \tag{2.3}$$

The $s(t + h)$ is functional notation, whereas $c(t + h)^2$ means the multiplication of c times $(t + h)$ times $(t + h)$.

At this point we rewrite our last result by explicitly multiplying out the square. Squaring, we obtain

$$s(t + h) = c(t + h)^2 = c(t + h)(t + h),$$
$$s(t + h) = c(t^2 + 2th + h^2). \tag{2.4}$$

Example 6

Calculate the distance a rock falls after 2.25 s using the results of both Eqs. (2.3) and (2.4) to show that they are equivalent.

According to Eq. (2.1), we can easily find the distance fallen to be

$$s(2.25) = (4.9 \text{ m/s}^2) (2.25 \text{ s})^2,$$

$$s(2.25) = 24.8 \text{ m}.$$

To use Eq. (2.4), let's call $t = 2.00$ s, and $h = 0.25$ s. Substituting, we obtain

$$s(2 + 0.25) = (4.9 \text{ m/s}^2) [(2 \text{ s})^2 + 2(2 \text{ s})(0.25 \text{ s}) + (0.25 \text{ s})^2]$$

$$s(2 + 0.25) = 24.8 \text{ m}.$$

We can now calculate the average speed of the falling body over the small time interval h. In order to accomplish this, we first recall the definition of average speed: distance fallen divided by time spent falling. We shall do precisely this division using the values we now have for $s(t)$ and $s(t + h)$. The distance fallen in the time interval h is $s(t + h) - s(t)$, so the average speed over this time interval is

$$\overline{v} = \frac{s(t + h) - s(t)}{h},$$

or, in other words,

$$\overline{v} = \frac{c(t^2 + 2th + h^2) - ct^2}{h}. \tag{2.5}$$

In this last expression the two ct^2 terms will cancel, so we perform that cancelation first:

$$\overline{v} = \frac{2cth + ch^2}{h}. \tag{2.6}$$

Now we can cancel the h in the denominator with an h in each of the terms in the numerator. Our result is

$$\overline{v} = 2ct + ch. \tag{2.7}$$

This is the average speed of the falling body over any time interval h after time t.

Example 7

Calculate the average speed of a falling rock in the 0.25-s interval between 2.00 s and 2.25 s.

We can calculate this average speed by two formulas. First, we go back to the definition of average speed and find

$$\overline{v} = \frac{s(2.25) - s(2)}{0.25 \text{ s}}.$$

From Example 6 we know that $s(2.25) = 24.8$ m, and it is easy to see that

$$s(2) = (4.9 \text{ m/s}^2)(2 \text{ s})^2 = 19.6 \text{ m}.$$

Therefore we have

$$\overline{v} = \frac{24.8 \text{ m} - 19.6 \text{ m}}{0.25 \text{ s}},$$

$$\overline{v} = 20.8 \text{ m/s}.$$

We can also check this result using Eq. (2.7) with $t = 2.00$ s, and $h = 0.25$ s:

$$\overline{v} = 2(4.9 \text{ m/s}^2)(2.00 \text{ s}) + (4.9 \text{ m/s}^2)(0.25 \text{ s}),$$

$$\overline{v} = 20.8 \text{ m/s}.$$

So far we haven't done anything new except to replace a time interval with the symbol h. But at this point in the calculation, we can ask another question: what happens to the average speed if we take smaller and smaller values of h? From our last equation,

$$\overline{v} = 2ct + ch. \tag{2.7}$$

If we allow h to shrink to zero, the term ch goes away, and we find the average speed approaches the value $2ct$. It seems natural to call this the instantaneous speed $v(t)$ and to write

$$v(t) = 2ct. \tag{2.8}$$

We have succeeded in finding what we wanted: the speed of a falling body at any instant of time; $v(t)$ is called the speed function, and the functional notation reminds us that the speed depends upon what time we're talking about. The bar is no longer written over the $v(t)$ because $v(t)$ is not an average speed but an instantaneous speed, the limiting value of the average speed.

The process of obtaining the speed from the distance is called *differentiation*. The function $v(t)$ is called the *derivative* of $s(t)$ or the *instantaneous rate of change* of $s(t)$. These are simply alternative names for the same thing. To emphasize that $v(t)$ comes from $s(t)$, we write

$$v(t) = s'(t), \tag{2.9}$$

where the prime ($'$) means the derivative of s with respect to t. The symbol $s'(t)$ is read "s prime of t."

Example 8

At first encounter the derivative seems almost miraculous. It is difficult to believe that as the time interval shrinks to zero, a simple number remains for the instantaneous speed. We provide a numerical example of this process in order to remove the mystery enshrouding it.

From the free-fall equation $s(t) = ct^2$, where $c = 4.9$ m/s^2, we can calculate the average speed for various time intervals. The table that follows represents these calculations.

Time (s)	Distance fallen (m)
2.5000	30.62500
2.2500	24.80625
2.1000	21.60900
2.0500	20.59225
2.0100	19.79649
2.0010	19.61960
2.0005	19.60980
2.0000	19.60000

From this table we may calculate the average speed for different time intervals h from $t = 2.0000$ s.

Time interval h (s)	Average speed (m/s)
0.5000	$\bar{v} = \dfrac{s(2.5000) - s(2.0000)}{0.5000} = 22.05000$
0.2500	$\bar{v} = \dfrac{s(2.2500) - s(2.0000)}{0.2500} = 20.82500$
0.1000	$\bar{v} = \dfrac{s(2.1000) - s(2.0000)}{0.1000} = 20.09000$
0.0500	$\bar{v} = \dfrac{s(2.0500) - s(2.0000)}{0.0500} = 19.84500$
0.0100	$\bar{v} = \dfrac{s(2.0100) - s(2.0000)}{0.0100} = 19.64900$
0.0010	$\bar{v} = \dfrac{s(2.0010) - s(2.0000)}{0.0010} = 19.60490$
0.0005	$\bar{v} = \dfrac{s(2.0005) - s(2.0000)}{0.0005} = 19.60246$

The instantaneous speed calculated from Eq. (2.8) is

$$v(2.0000) = 19.60000 \text{ m/s}.$$

We see from this example how, by shrinking the time interval, the average speed gets closer and closer to a definite value, the instantaneous speed.

We could have obtained the same result, $s'(t) = v(t) = 2ct$, by a slightly different method. Think of the distance fallen at t,

$$s(t) = ct^2,\tag{2.10}$$

and the distance fallen at a slightly later time t_1,

$$s(t_1) = ct_1^2.\tag{2.11}$$

We now write the average speed between t and t_1 just as before,

$$\bar{v} = \frac{s(t_1) - s(t)}{t_1 - t},$$

$$\bar{v} = \frac{ct_1^2 - ct^2}{t_1 - t}.\tag{2.12}$$

Note that the length of the interval now is $t_1 - t$. This is what we called h before. We remove the common factor c in the numerator so that it becomes $c(t_1^2 - t^2)$. Now we recall that the differences of two squares can be factored:

$$t_1^2 - t^2 = (t_1 + t)(t_1 - t).$$

Altogether we can rewrite Eq. (2.12) as

$$\bar{v} = c\frac{(t_1 + t)(t_1 - t)}{t_1 - t}.\tag{2.13}$$

The difference $t_1 - t$ cancels out top and bottom, leaving us with a new expression for the average speed,

$$\bar{v} = c(t_1 + t).\tag{2.14}$$

This is the average speed in the interval between t and t_1.

Observe what happens to the average speed in Eq. (2.14) if we let t_1 approach t. Letting t_1 approach t is the same as letting h shrink to zero. The result is that $\bar{v}$ approaches the value $c(t + t) = 2ct$, so

$$v(t) = 2ct\tag{2.8}$$

just as before.

Now that we know the instantaneous speed, we see that it is proportional to the time spent falling. This was Oresme's law. The law of Albert of Saxony, that speed is proportional to distance, turns out to be incorrect. We leave for you (Question 21) the fun of finding out how speed actually does depend on distance.

Example 9

We use Eq. (2.14) to calculate once again the average speed between $t = 2.00$ s and $t_1 = 2.25$ s:

$$\bar{v} = (4.9 \text{ m/s}^2)(2.00 \text{ s} + 2.25 \text{ s})$$

$$= 20.8 \text{ m/s}.$$

Example 10

A flower pot falls from a window ledge 144 ft above the pavement. What is the speed of the flower pot right before it strikes the ground?

With Eq. (2.8) we can calculate the speed of the flower pot right before impact if we know how long it takes it to fall. The time of the fall is not specified, but we can calculate it from the free-fall equation. Our plan, therefore, will be first to calculate the time of the fall, then to use that time to find the speed. Solving the free-fall equation for time t, we get

$$s(t) = ct^2,$$

$$t = \sqrt{\frac{s(t)}{c}},$$

$$t = \sqrt{\frac{144 \text{ ft}}{16 \text{ ft/s}^2}} = \sqrt{9 \text{ s}^2} = 3 \text{ s}.$$

Therefore the speed after the 3 s fall is

$$v(t) = 2ct,$$

$$v(2) = 2(16 \text{ ft/s}^2)(3 \text{ s}),$$

$$v(2) = 96 \text{ ft/s}.$$

Often in physics problems, two steps must be taken to reach the final answer.

Questions

18. A ball is dropped from a high tower. One second later another ball is dropped.

Does the ratio of their speeds (the speed of the first one compared to the speed of the second one) increase, decrease, or remain the same as they fall? Explain.

19. **(a)** Calculate the time required for a falling body to reach a speed of 80 m/s.
(b) If the body in (a) reaches that speed right before hitting the ground, from how high up was it dropped?

20. **(a)** Determine how long it would take an object to fall 65 m.
(b) Find the speed of the object when it has fallen 65 m.

21. From a knowledge of the distance and speed equations for free fall, obtain a relation between the instantaneous speed and the distance fallen. Compare your result to the law of Albert of Saxony.

22. Determine the ratio of the instantaneous speed of fall of an object in free fall for each second from 1 to 10 s to the speed at 1 s. Can you formulate any "law" based on your result?

2.7 ACCELERATION

We now have two descriptions for the motion of falling bodies, each representing an answer to a different question concerning this motion. One, $s(t)$, describes the distance fallen at any instant; the other, $v(t)$, expresses the speed at any instant. Furthermore, these two quantities are related by differentiation, a process that allows us to find the instantaneous rate of change of a function. We now ask, How fast does the speed of a falling body change?

Our starting point toward an answer is the definition of the average acceleration $\bar{a}$:

$$\bar{a} = \frac{\text{change in speed}}{\text{time elapsed}}.$$
(2.15)

Since we know the speed of a falling body at any instant, we can easily calculate the change in speed. At time t the freely falling body has a precise speed of

$$v(t) = 2ct.$$
(2.8)

What is the speed a short time later? Let us call that short time h; then the speed of the object at time $t + h$ will be

$$v(t + h) = 2c(t + h).$$
(2.16)

All we've done is substitute a different time, $t + h$, into the equation that predicts the speed at any instant.

We can now calculate the average acceleration over this small time interval by dividing the change in v by the time interval length h:

$$\bar{a} = \frac{v(t + h) - v(t)}{h}.$$
(2.17)

Putting in the values of v at $t + h$ and t, we have

$$\bar{a} = \frac{2c(t + h) - 2ct}{h}.$$
(2.18)

Now you may be itching to let h shrink to zero, but you can't do it yet. If you did, you would be dividing by zero. Before we can let h go to zero, we need to perform a couple of steps:

$$\bar{a} = \frac{2ct + 2ch - 2ct}{h},$$

$$\bar{a} = \frac{2ch}{h};$$

the h's in the numerator and denominator cancel to yield

$$\bar{a} = 2c. \tag{2.19}$$

Our last result tells us, on the average, how rapidly the speed changes. But here we note something new. The average acceleration doesn't depend on the value of the time t, nor on the interval length h. In other words, it is always the same. The average acceleration is constant.

Example 11

What is the average acceleration of a diver between 2.0 s and 2.5 s of the dive?

Using our speed function, we first calculate the speed of the diver at 2.0 s and at 2.5 s, then we substitute these values into the definition of the average acceleration.

From $v(t) = 2ct$, we obtain

$$v(2) = 2(4.9 \text{ m/s}^2)(2 \text{ s}) = 19.6 \text{ m/s}^2,$$

$$v(2.5) = 2(4.9 \text{ m/s}^2)(2.5 \text{ s}) = 24.5 \text{ m/s}^2.$$

Consequently, the average acceleration is

$$\bar{a} = \frac{v(2.5) - v(2)}{2.5 \text{ s} - 2 \text{ s}} = \frac{24.5 \text{ m/s} - 19.6 \text{ m/s}}{0.5 \text{ s}},$$

$$\bar{a} = 9.8 \text{ m/s}^2.$$

Having obtained an expression for the average acceleration of a falling body over a time interval of length h, we can find the instantaneous acceleration. We accomplish this by allowing h to shrink to zero. But h doesn't appear in that expression, Eq. (2.19). That means the average acceleration is not changed by letting h go to zero; the instantaneous acceleration is the same as the average acceleration. Writing $a(t)$ for the instantaneous acceleration, we have

$$a(t) = 2c. \tag{2.20}$$

This is another expression of the law of falling bodies: all bodies fall with the same constant acceleration.

In the fourteenth century, Nicole Oresme studied the properties of a kind of motion, which he called uniformly difform motion. By that he meant that the speed of the body changed at a constant rate. It was exactly the uniformly accelerated motion we have just discovered. He knew that, in such motion, the speed would be proportional to the time, the distance to the time squared, and even that the distance in successive intervals of time would go as the odd numbers. He may even have toyed with the idea that falling bodies actually performed his theoretical motion, but fundamentally, the fourteenth-century scholar was not very interested in the mere behavior of objects in the everyday world.

Nearly three centuries later, Galileo discovered that falling bodies are uniformly accelerated. It is not known whether Galileo ever heard of Oresme, but he did use strikingly similar arguments, and obtained all of the same results, to describe this special kind of motion. We have now duplicated the feat of Oresme and Galileo. We know everything there is to know about uniformly accelerated motion.

For Oresme and Galileo, discovering the properties of uniformly accelerated motion was an extraordinary intellectual feat. We have done it much more easily, but only by first discovering a new branch of mathematics, the differential calculus, which is ideally suited to this kind of analysis. The time has come to summarize both the physics and the mathematics we have learned in this chapter.

First, the physics: Any falling body (near the surface of the earth, and neglecting air resistance) falls with the same constant acceleration throughout its fall. The constant acceleration is given its own name g, the same "g" of g forces; we write

$$a(t) = g. \tag{2.21}$$

The other forms of the law of falling bodies may now be rewritten by using g in place of $2c$. The speed is proportional to the time, obeying

$$v(t) = gt \tag{2.22}$$

and the distance is proportional to the time squared,

$$s(t) = \frac{1}{2} gt^2. \tag{2.23}$$

These last three equations tell us what we need to know about falling bodies. Galileo's law of odd numbers is also correct, but it is far less useful than these expressions.

Finally, the mathematics: We have seen that the distance a body falls, $s(t)$, has an instantaneous rate of change, or derivative, equal to the speed $v(t)$:

$$v(t) = s'(t).$$

In turn, the speed has an instantaneous rate of change, the acceleration, $a(t)$:

$$a(t) = v'(t).$$

In other words, the acceleration is the derivative of the speed. Note that the acceleration is obtained from the distance by differentiating twice. This can also be written

$$a(t) = s''(t).$$

Then $s''(t)$, s double prime of t, is called the *second derivative* of s with respect to t.

The fact that acceleration is the rate of change of the rate of change of something makes it very hard to think about without using derivatives. For that reason, we must admire Oresme and Galileo, but we don't have to follow their difficult methods, because we do know how to differentiate. Note the mathematical effect a derivative has:

$s(t)$ goes as t^2:

$$s(t) = \frac{1}{2} gt^2.$$

Its derivative, $v(t)$, has one lower power of t (i.e., v is proportional to t to the power one),

$$v(t) = gt.$$

Taking another derivative knocks off another t, giving us

$$a(t) = g.$$

In the next chapter we'll learn the general rules of differentiation.

Questions

23. Suppose that the distance traveled as a function of time for falling bodies had turned out to be $s(t) = kt$, where k is a constant.

(a) What would be the speed of a falling body at any instant?
(b) What would be the instantaneous acceleration of the object?

24. Referring to the stroboscope pictures of the falling ball (Fig. 2.2), consider that for some picture at time t, the ball is at $s(t)$. Then for the next photo, it is at $s(t + h)$, and for the one after that, it is at $s(t + 2h)$, where h is the time between flashes.

(a) Write an expression for the average speed of the ball between the first two pictures. Obtain a similar expression for the average speed between the second and third pictures.
(b) Show that the average rate of change of the two average speeds in part (a) is given by

$$\overline{a} = \frac{s(t + 2h) - 2s(t + h) + s(t)}{h^2}.$$

25. For problem 24, use the explicit form $s(t) = \frac{1}{2} gt^2$ to show that when the time interval h is allowed to shrink to zero, the instantaneous acceleration is g.

2.8 A FINAL WORD

One of the jobs of physics is to find simple, economical underlying principles that explain the complicated world we live in. We have done that here. If we drop an object, it falls under the influence of the earth's gravity. As it falls, its motion is opposed, with varying degrees of success, by the air through which it must move. If we can imagine disposing of the air and describing the effect of gravity alone, we discover a dramatic and surprising fact: all bodies fall at the same rate.

We could be satisfied with that fact. After all, discovering it was quite an impressive accomplishment. But of course, we are not satisfied. We want to know why it is true. What is the nature of gravity that it leads to such strange behavior? That question has turned out to be one of the deepest in all the history of physics. It has persisted even into our own century. It was the starting point from which Albert Einstein built his celebrated general theory of relativity.

But we are getting ahead of our story. Once we knew there was one law for all falling bodies, the job was then to express that law with precision. We have done that, too. We learned that all bodies fall with the same constant acceleration.

Acceleration is the rate of change of speed. Besides constant acceleration, we know that the speed of the falling body increases in proportion to the time spent falling. Speed is the rate of change of distance. The speed is proportional to the time; the distance is proportional to the square of the time.

So we have in fact three precise, mathematical statements of the law of falling bodies. They are all true, and they are connected to each other by one of the great and crucial discoveries in the history of mathematics: the differential calculus.

Galileo Galilei discovered and expressed the law of falling bodies. He was a brilliant and arrogant man who managed to offend the ecclesiastical authorities of his time so much that he spent the last eight years of his life a prisoner, under house arrest at his estate near Florence.

The discovery of the calculus is usually credited to two men, Isaac Newton and Gottfried Wilhelm Leibniz. Actually its discovery was the result of a long evolutionary process in which Newton and Leibniz played decisive roles. It was a mighty triumph, the most important event in mathematics in thousands of years. But Newton and Leibniz sacrificed the joy of their discovery in a bitter, acrimonious dispute over who deserved credit for discovering it first.

On the other hand, Albert Einstein became a folk hero to a whole world that never pretended to understand what he had done or why he deserved honor.

All of these are threads in the story we are going to see unfold.

Figure 2.5 Galileo recanting before a tribunal of the Inquisition.
(Courtesy of the Mansell Collections.)

CHAPTER

DERIVATIVES

It is most useful that the true origins of memorable inventions be known, especially of those which were conceived not by accident but by an effort of meditation . . . One of the noblest inventions of our time has been a new kind of mathematical analysis, known as the differential calculus.

Gottfried Wilhelm von Leibniz, *Historia et Origo Calculi Differentialis* (1714)

3.1 THE DEVELOPMENT OF DIFFERENTIAL CALCULUS

After the advent of algebra in the sixteenth century, mathematical discoveries inundated Europe. The most important were *differential calculus* and *integral calculus*, bold new methods for attacking a host of problems that had challenged the world's best minds for more than 2000 years. Differential calculus deals with ideas such as *speed, rate of growth, tangent lines,* and *curvature,* whereas integral calculus treats topics such as *area, volume, arc length,* and *centroids.*

Work begun by Archimedes in the third century B.C. led ultimately to the birth of integral calculus in the seventeenth century. This development has a long and fascinating history to which we shall return in Chapter 7.

Differential calculus has a relatively short history. Its principles were first formulated early in the seventeenth century when a French mathematician, Pierre de Fermat, tried to devise a way of finding the smallest and largest values of a given function. He imagined the graph of a function having, at each of its points, a direction given by a tangent line, as suggested by the points labeled in Fig. 3.1. Fermat noted that the tangent line is horizontal at points like C and E where the function has a maximum or a minimum. This observation, in turn, made it seem worthwhile to have a general method for finding the slope of the tangent line at an arbitrary point. That method turned out to be differentiation.

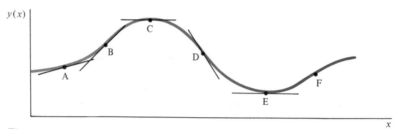

Figure 3.1 A simple function and tangent lines at various points.

Sir Isaac Newton and Baron Gottfried Wilhelm von Leibniz (Fig. 3.2) are generally regarded as the discoverers of calculus. Actually, many individual problems involving tangents and areas had been studied with some success since ancient times and with considerable success in the half century preceding Newton and Leibniz. One of the major contributions of Newton and Leibniz was to develop systematic techniques for solving whole classes of related problems by routine methods. They were also the first to appreciate the significance of a discovery made by Newton's teacher, Isaac Barrow, who had observed a geometric relationship between tangents and areas. This is now known as the *fundamental theorem*, which relates differential and integral calculus. Barrow did not realize the importance of his discovery. Newton and Leibniz not only recognized this theorem as a mathematical fact, but they exploited it to extract a powerful instrument that unified differential and integral calculus into one discipline, the *infinitesimal calculus*. With this new calculus, schoolchildren could routinely solve problems that previously required the genius of an Archimedes or a Newton.

Newton, influenced by his teacher Barrow, developed his version of differenital calculus in the years 1665–6 and wrote his *Tractatus de methodis serierum et fluxionum* in 1670–1, although it was not published until 1736, nine years after his death. Newton called the rate of change of a quantity a *fluxion* and wrote $\dot{x}$ for what we call the derivative $x'(t)$. But in his astonishing book *Principia Mathematica Philosophiae Naturalis* (1687) many propositions concerning speed, acceleration, and tangents, discovered with the help of fluxions, were proved by classical geometric methods. Leibniz was obviously following the path staked out by Fermat when he wrote "A New Method for Maxima and Minima, as Well as Tangents, Which Is Not Obstructed by Fractional or Irrational Quantities,"

Figure 3.2 Baron Gottfried Wilhelm von Leibniz. (Courtesy of Yeshiva College.)

published in 1684. It used a different notation for the derivative. For $x'(t)$, Leibniz wrote dx/dt. We'll see later how this notation arises. Notation plays a subtle but very influential role in the development of mathematics. All three notations, $\dot{x}$, x', and dx/dt, are still used today.

3.2 THE TANGENT LINE–DERIVATIVE CONNECTION

Suppose that we have a function like that shown as the graph of Fig. 3.3. If the curve represented a hillside, you might be interested in knowing the slope of the hill, especially if you were going to hike up it (or slide down it). We already know how to calculate the slope of a straight line (see Appendix C if you need a quick review of how), but here we don't have a straight line. How then can we even talk about the slope, let alone try to calculate it? The answer lies in approximating the curve between two specified points by a straight line.

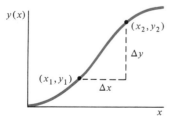

Figure 3.3 Graph of a function with points indicated for determining the slope.

Let's call the vertical distance between our two points Δy, and the horizontal distance we'll call Δx. The slope of the straight line between the two points (formally known as a chord) is given by

$$\text{average slope} = \frac{\Delta y}{\Delta x} = \frac{y_2 - y_1}{x_2 - x_1}. \tag{3.1}$$

Now you may object that what we have found is not really the slope. And you're right. We've found the *average* slope, which tells how steep the hill (or function) is on the average between the two points. What we really want to know is the slope, not on the average, but at one point.

Example 1

Find the average slope between $x = 0$ and $x = 2$ for the function $y(x) = x^2 + 3$.

We can calculate the average slope by using the definition of the slope of a straight line between the two points:

$$\text{average slope} = \frac{\Delta y}{\Delta x} = \frac{y(2) - y(0)}{2 - 0}, \tag{3.1}$$

$$\text{average slope} = \frac{(2^2 + 3) - (0 + 3)}{2},$$

$$\text{average slope} = 2.$$

This average slope represents the slope of the straight line drawn between our two points.

We encountered a similar problem in Chapter 2 when we wanted to know not only the average speed of a falling object in a time interval but also the speed at each instant of time. By calculating the average speed and allowing the time interval to shrink to zero, we found the instantaneous speed, which was a derivative. We follow a similar plan of attack to find the slope of the hill at a particular point.

To have a concrete example in mind, suppose the hill has the shape of a parabola, as in Fig. 3.4.

The equation of the parabola, $y = x^2$, describes the shape of the curve. At each point (x, y) of the curve, the number x is called the *abscissa* and y is called the *ordinate*. The ordinate y is a function of x because its value depends on the value we choose for x. We also write $y = y(x)$ to emphasize that y is a function of x. Our goal is to obtain the slope at an arbitrary point (x, y) of the curve.

Start with a point (x, y) and choose a nearby point $(x + h, y(x + h))$, where h is some small number (positive or negative, but not zero). The corresponding ordinates have the values $y(x) = x^2$ and $y(x + h) = (x + h)^2$, as shown in Fig. 3.3. The slope of the chord joining these points is

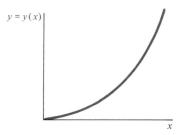

Figure 3.4   Graph of the parabola $y = x^2$.

$$\text{average slope} = \frac{\Delta y}{\Delta x} = \frac{y(x + h) - y(x)}{(x + h) - x}. \tag{3.1}$$

Since $y(x) = x^2$ and $y(x + h) = (x + h)^2$ this becomes

$$\text{average slope} = \frac{(x + h)^2 - x^2}{h}.$$

Now $(x + h)^2 = x^2 + 2hx + h^2$, so $(x + h)^2 - x^2 = 2hx + h^2$, and we get

$$\text{average slope} = \frac{2hx + h^2}{h} = 2x + h, \tag{3.2}$$

where we have canceled the common factor h. Now we can see what happens to the average slope when h gets smaller and smaller. As h approaches zero, the average slope $2x + h$ approaches $2x$. Therefore it seems reasonable to say that the slope of the curve *at* the point (x, y) is $2x$:

$$\text{slope at } (x, y) = 2x.$$

Since we get to choose x, we have found the slope at any point we like.

This last result may look familiar, and indeed it is: we have just calculated the derivative $y'(x)$ of the function $y(x) = x^2$ with respect to x [just as we calculated the derivative of the distance function $s(t) = ct^2$ with respect to t in Chapter 2]. In Chapter 2 we started with $s(t) = ct^2$ and obtained $s'(t) = 2ct$. Here we started with $y(x) = x^2$ and obtained $y'(x) = 2x$.

But our graph also give us a geometrical interpretation of the process of differentiation. Figure 3.5 shows how the position of the chord changes as h takes smaller and smaller values. The point $(x + h, y(x + h))$ moves along the curve toward the point (x, y) and

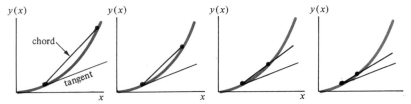

Figure 3.5 Illustrating that the tangent line is the limiting position of the chord as one end point approaches the other.

the chord moves closer and closer to a line through (x, y) with slope $2x$. This line is called the *tangent line* at (x, y).

In other words, the line through (x, y) with slope $y'(x)$ is, by definition, the tangent line to the curve at (x, y). Geometrically, then, *differentiation gives us the slope of the tangent line at each point of the curve*.

Example 2

For the function shown below, can you roughly sketch a graph of its derivative?

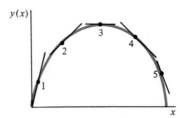

Using the idea that the derivative of a function at a point is the slope of the tangent line at that point, we can estimate the slopes of the tangents at points 1 through 5. At point 1, the slope is pretty steep, which means it's large and positive, at 2 it is still positive but smaller, at three it is zero, at 4 it is negative, and at 5 it is steep and negative. Plotting these points and smoothly connecting them, we obtain this graph:

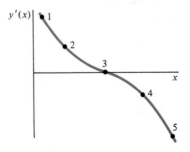

Another way of writing the derivative, instead of $y'(x)$, is dy/dx. This is the notation that Leibniz invented, and here we see where it came from. The derivative dy/dx is what $\Delta y/\Delta x$ becomes when Δx shrinks to zero:

$$\frac{dy}{dx} = \lim_{\Delta x \to 0} \frac{\Delta y}{\Delta x} = \lim_{h \to 0} \frac{y(x + h) - y(x)}{h}. \tag{3.3}$$

The abbreviation "$\lim_{\Delta x \to 0}$" here is read "the limit as Δx shrinks to zero." In other words, dy/dx is what happens to $\Delta y/\Delta x$ in the limit as we allow Δx (therefore Δy) to shrink to zero. Equation (3.3) is the definition of the derivative and can be used to calculate analytically the derivative of any function you will encounter in physics.

Leibniz thought of the derivative dy/dx as a quotient of "infinitesimal" quantities dy and dx, which he regarded as unimaginably small, but still not quite exactly equal to zero. This concept provoked a great deal of philosophical controversy during the early decades of the development of calculus because neither Leibniz nor his followers could give a satisfactory definition of infinitesimals. Eventually the controversy was resolved with the introduction of the theory of limits, which treated dy/dx as a single symbol rather than as a ratio of infinitesimals. Nevertheless, Leibniz's idea of treating dy/dx as if it were a fraction, dy divided by dx, is still popular today because it often leads quickly to correct results that require greater effort to obtain without the use of infinitesimals.

Sometimes the functional notation $y'(x)$ is preferable to the Leibniz notation dy/dx because it specifies precisely where the derivative is wanted. For example, if $y'(x) = 2x$, then $y'(4) = 8$. In the Leibniz notation the last equation $y'(4) = 8$ would be written in the less convenient form

$$\frac{dy}{dx}\bigg|_{x=4} = 8.$$

The vertical stroke next to the symbol dy/dx with $x = 4$ near the bottom specifies the point at which the derivative is evaluated.

Example 3

Calculate the derivative of the constant function $y(x) = c$ according to the definition of the derivative in Eq. (3.3).

In order to use the definition

$$y'(x) = \lim_{h \to 0} \frac{y(x + h) - y(x)}{h}, \tag{3.3}$$

we need to know $y(x + h)$. That's simple: $y(x + h) = c$, since $y(\text{anything}) = c$. Substituting into the definition we obtain

$$y'(x) = \lim_{h \to 0} \frac{c - c}{h} = \lim_{h \to 0} \frac{0}{h},$$

$$y'(x) = \lim_{h \to 0} 0 = 0.$$

This last step is the key to understanding calculus. We have a fraction whose numerator (top) is zero, while the denominator (bottom) gets smaller and smaller and smaller. We don't ask philosophical questions about what happens at the exact instant the bottom becomes zero – instead we ask, How does the fraction behave during this process? Where is it going? The answer is that during the whole process of shrinking h, the fraction $0/h$ is calmly, implacably, equal to zero. Consequently, the derivative of a constant function is zero. This result agrees (as it must!) with the slope of the graph of a constant function, which is a horizontal line and has zero slope.

Example 4

What is the derivative of the function $y(x) = mx + b$?

We know the answer to this question because the graph of the given function is a straight line; therefore the slope should be the constant m. Let's prove this by using the definition of the derivative, Eq. (3.3). Since $y(x + h) = m(x + h) + b = mx + mh + b$, we have

$$y'(x) = \lim_{h \to 0} \frac{y(x + h) - y(x)}{h}$$

$$= \lim_{h \to 0} \frac{mx + mh + b - mx - b}{h},$$

$$y'(x) = \lim_{h \to 0} \frac{mh}{h} = \lim_{h \to 0} m = m.$$

Example 5

Determine the slope of the function $y(x) = x^2 + 3$ at $x = 2$.

Retracing the steps leading to Eq. (3.2), we calculate the slope of a line between the points $2 + h$ and 2:

$$\text{average slope} = \frac{y(2 + h) - y(2)}{2 + h - 2},$$

$$\text{average slope} = \frac{[(2 + h)^2 + 3] - (2^2 + 3)}{h},$$

$$\text{average slope} = \frac{(4 + 4h + h^2 + 3) - (4 + 3)}{h},$$

$$\text{average slope} = \frac{4h + h^2}{h} = 4 + h.$$

Now we can let h go to zero and obtain in the limit the slope at $x = 2$:

$$y'(2) = \left. \frac{dy}{dx} \right|_{x=2} = 4.$$

Note that the slope of this function is the same as that for $y(x) = x^2$ at $x = 2$. Adding a constant does not change the value of the derivative of a function.

Questions

1. Determine the derivative of the function $y(x) = x^2$ at the points $x = 0, 1, 2, 3, 4,$ 5, and plot your results. What type of a function do you obtain? Can you find any connection between your graph and a graph of speed versus time for free fall?

2. Roughly sketch the graph of the derivative for each of the functions shown below using knowledge of the connection between the derivative and the tangent line.

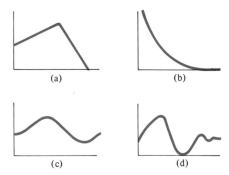

    (a)   (b)

    (c)   (d)

3. Find the average slope between the points $x = 2$ and $x = 5$ for each of the following functions:

  (a) $y(x) = x^2$ **(b)** $y(x) = x^2 - 3$
  (c) $y(x) = 3x^2 + 5$ **(d)** $y(x) = (x - 1)^2$.

4. For each of the functions in problem 3, calculate the derivative $y'(3)$.

5. Which of the following functions has the greatest slope at $x = -2$?

  (a) $y(x) = x$ **(b)** $y(x) = x^2 + 1$
  (c) $y(x) = 5$ **(d)** $y(x) = 3x^2 - 2$.

3.3 RULES OF DIFFERENTIATION

The power of the newly developed calculus of Newton and Leibniz lies in its wide range of applicability to complicated problems. The keys to this power are a few simple rules of differentiation. Newton and Leibniz realized that complicated functions are usually composed of simpler functions that are added, subtracted, multiplied, or divided. Consequently, the rules of differentiation concern these ways in which functions are built.

In Example 5, we encountered a function that was actually the sum of two functions: $y(x) = x^2 + 3$. Let's formulate a general rule to handle the derivative of the sum of two functions, which we'll call $y(x)$ and $z(x)$. What we want to know is

$$\frac{d}{dx} [y(x) + z(x)].$$

Suppose $y(x)$ represents the amount of water in a barrel as a function of the rainfall x, and $z(x)$ is the amount for a different-shaped barrel. Our intuition tells us that the change in the total amount of water in the barrels with a change in rainfall, $[y(x) + z(x)]'$, is simply the sum of the change in each, that is, $y'(x) + z'(x)$. But can we show this?

The most straightforward way to prove our hunch is to apply the definition of the derivative [Eq. (3.3)] to the sum:

$$\frac{d}{dx}[y(x) + z(x)] = \lim_{h \to 0} \frac{[y(x + h) + z(x + h)] - [y(x) + z(x)]}{h}.$$

We can rearrange terms in the numerator to read

$$\frac{d}{dx}[y(x) + z(x)] = \lim_{h \to 0} \frac{[y(x + h) - y(x)] + [z(x + h) - z(x)]}{h}.$$

Because the limit of a sum is the sum of the limits, we can break this down into two parts and get

$$\frac{d}{dx}[y(x) + z(x)] = \lim_{h \to 0} \frac{y(x + h) - y(x)}{h} + \lim_{h \to 0} \frac{z(x + h) - z(x)}{h}.$$

But our last result is the sum of two derivatives. In other words, *the derivative of the sum of functions is the sum of their derivatives*:

$$\frac{d}{dx}[y(x) + z(x)] = \frac{dy}{dx} + \frac{dz}{dx}. \tag{3.4}$$

Of course, this rule holds for the sum of any number of functions.

A second rule concerns the derivative of a product of two functions. Suppose $y(x)$ and $z(x)$ each depend on x. Then what is

$$\frac{d}{dx}[y(x) \cdot z(x)]?$$

Finding the answer to this question was one of the keys to making differential calculus a working tool. Isaac Newton worked it out using his geometric intuition, plus something else harder to describe.

He said, suppose we have a rectangle whose sides are y and z. Then the area is yz, as shown in Fig. 3.6a:

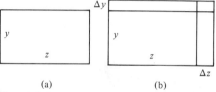

(a)   (b)

Figure 3.6 (a) Rectangle of sides y and z and of area yz. (b) Increase in the sides and area of the rectangle.

Now suppose further that y grows by a small amount Δy, and z by a small amount Δz, as illustrated in Fig. 3.6b. What is the new area? There are three extra pieces:

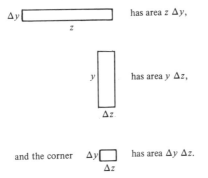

Δy ▭ z has area $z\ \Delta y$,

y ▯ Δz has area $y\ \Delta z$,

and the corner Δy ▯ Δz has area $\Delta y\ \Delta z$.

In other words, the change in the product yz is y times the change in z plus z times the change in y (each of these is small compared to the rectangle itself) and another little piece, much smaller than either of the changes, $\Delta y\ \Delta z$.

This led Newton to his solution: *The rate of change of yz is y times the rate of change of z, plus z times the rate of change of y.* In symbols,

$$\frac{d}{dx}[y(x)z(x)] = y\frac{dz}{dx} + z\frac{dy}{dx}. \tag{3.5}$$

What happened to the little piece in the corner? Newton chose to ignore it. He knew that was right, but he didn't offer any proof.

The theory of limits shows why. Suppose y and z have nice ordinary derivatives, dy/dx and dz/dx. These come about because when x increases by Δx, y increases by Δy and z increases by Δz. Then

$$\frac{\Delta(yz)}{\Delta x} = \frac{(y + \Delta y)(z + \Delta z) - yz}{\Delta x}.$$

Note that the first term in the numerator of this fraction is just the increase in the area of Newton's rectangle:

$$(y + \Delta y)(z + \Delta z) = yz + y\ \Delta z + z\ \Delta y + \Delta y\ \Delta z.$$

When we subtract off the term yz we have

$$\frac{\Delta(yz)}{\Delta x} = \frac{y\ \Delta z}{\Delta x} + \frac{z\ \Delta y}{\Delta x} + \frac{\Delta y\ \Delta z}{\Delta x}.$$

Now, when we take the limit as $\Delta x \to 0$, Δy and Δz will, of course, also go to zero. We then get

$$\lim_{\Delta x \to 0} y \frac{\Delta z}{\Delta x} = y \frac{dz}{dx},$$

$$\lim_{\Delta x \to 0} z \frac{\Delta y}{\Delta x} = z \frac{dy}{dx},$$

But what happens to $\Delta y \, \Delta z / \Delta x$? We can write it as $(\Delta y / \Delta x) \Delta z$ or $(\Delta z / \Delta x) \Delta y$. That is, it's the product of a quantity that tends to a derivative times something that vanishes. It becomes equal to zero. That is the reason Newton was correct to ignore the corner of the rectangle.

Example 6

How can the product rule be used to calculate the derivative of x^3?

Easily. Let's consider x^3 to be the product of two functions, $y(x) = x^2$ and $z(x) = x$. By the product rule we have

$$\frac{d}{dx}(x^3) = y \frac{dz}{dx} + z \frac{dy}{dx}.$$

We already know that

$$\frac{dy}{dx} = 2x \quad \text{and} \quad \frac{dz}{dx} = 1,$$

by Example 4 with $m = 1$ and $b = 0$. Therefore our derivative is

$$\frac{d}{dx}(x^3) = x^2(1) + x(2x) = x^2 + 2x^2,$$

or

$$\frac{d}{dx}(x^3) = 3x^2.$$

Example 7

What is the derivative of $y(x) = x^3 - 2x$?

We realize that we have the sum of two functions, one of which is the product of a constant function and x. To begin, we apply the sum rule:

$$\frac{dy}{dx} = \frac{d}{dx}(x^3) + \frac{d}{dx}(-2x).$$

For the first term, we use our result from Example 6 to write

$$\frac{d}{dx} x^3 = 3x^2.$$

If we consider the second term to be the product of the constant function -2 and x, then the product rule implies that

$$\frac{d}{dx}(-2x) = -2\frac{dx}{dx} + x\frac{d}{dx}(-2).$$

But the derivative of x with respect to x is 1, and the derivative of a constant is 0 (remember Example 3). Consequently, we have

$$\frac{d}{dx}(-2x) = -2.$$

Assembling all the parts together, for our final answer we have

$$\frac{d}{dx}(x^3 - 2x) = 3x^2 - 2.$$

Both Newton and Leibniz applied the product rule to find the derivatives of a very general type of function, the power function $y(x) = x^n$, where n is an integer. What they did was to calculate the derivatives of power functions for $n = 0, 1, 2, 3$, and so on, and in doing so they found a pattern, which they generalized into something known as the *power rule*.

We already know the derivatives of x^n for $n = 0, 1, 2$, and 3, but to see the pattern clearly emerge, let's calculate the derivative of one more function, x^4, by applying the product rule. To do so, we can consider x^4 to be the product of $z(x) = x^2$ and $y(x) = x^2$ (or x and x^3 if you like). Applying Eq. (3.5), we get

$$\frac{d}{dx} yz = y\frac{dz}{dx} + z\frac{dy}{dx}, \tag{3.5}$$

$$\frac{d}{dx} x^4 = x^2 \frac{d}{dx}(x^2) + x^2 \frac{d}{dx}(x^2)$$

But we already know that the derivative of x^2 is $2x$, so we can painlessly write down our result:

$$\frac{d}{dx} x^4 = 4x^3.$$

When we list the derivatives of the power functions that we've calculated, we can see precisely the pattern that Newton and Leibniz found (Table 3.1). To form the derivative of any power function x^n, just bring down the power n and multiply it by x raised to the $n-1$ power; this is the *power rule* for derivatives:

Table 3.1 Derivatives of Power
Functions

$y(x)$	$y'(x)$
x^1	$1x^0 = 1$
x^2	$2x^1$
x^3	$3x^2$
x^4	$4x^3$
$\vdots$	$\vdots$
x^n	nx^{n-1}

$$y'(x) = \frac{dy}{dx} = \frac{dx^n}{dx} = nx^{n-1}. \tag{3.6}$$

By clever reasoning, Newton was able to extend this rule by showing that it holds for negative values of n as well. In fact it holds for *all* values of n – positive or negative, integer or not. Here's how he did it for negative integers:

Writing $x^n \cdot x^{-n} = 1$, he applied the product rule to get

$$x^n \frac{d}{dx}(x^{-n}) + x^{-n}\frac{d}{dx}(x^n) = 0,$$

$$x^n \frac{d}{dx}(x^{-n}) + x^{-n}nx^{n-1} = 0.$$

Solving for the derivative of x^{-n}, Newton found

$$\frac{d}{dx}(x^{-n}) = -nx^{-n-1}.$$

The same method gives a general formula for finding the derivative of the reciprocal of a function. If $z(x) = 1/y(x)$, then $z(x)y(x) = 1$; and if we differentiate this by the product rule, we find

$$z\frac{dy}{dx} + y\frac{dz}{dx} = 0.$$

Solving for dz/dx we have

$$\frac{dz}{dx} = -\frac{z}{y}\frac{dy}{dx} = -\frac{1}{y^2}\frac{dy}{dx}.$$

In other words, here is the rule for the derivative of a reciprocal:

$$\frac{d}{dx}\left(\frac{1}{y}\right) = -\frac{1}{y^2}\frac{dy}{dx}.$$

When $y = x^n$ we get

$$\frac{d}{dx}(x^{-n}) = \frac{-1}{x^{2n}}\, nx^{n-1} = -nx^{-n-1}$$

as above.

Example 8

What is the derivative of $y(x) = 3x^5 + x^{-2}$?

First, we have the sum of two functions, so we can apply the sum rule:

$$\frac{dy}{dx} = \frac{d}{dx}(3x^5) + \frac{d}{dx}(x^{-2}).$$

To calculate the derivative of the first term, we realize that we have the product of a constant (3) and x^5, so we need to use the product rule. But as the derivative of a constant is zero, our result is

$$\frac{d}{dx}(3x^5) = 3\frac{d}{dx}x^5.$$

For future use, it is helpful to remember that the derivative of a constant times some function is always equal to the constant times the derivative of the function.

Next we apply the power law to get

$$\frac{d}{dx}x^5 = 5x^4,$$

and consequently,

$$\frac{d}{dx}(3x^5) = 15x^4.$$

For the second term, we can straightforwardly apply the power rule:

$$\frac{d}{dx}x^{-2} = -2x^{-2-1} = -2x^{-3}.$$

Putting everything back together, our final result is

$$\frac{dy}{dx} = 15x^4 - 2x^{-3}.$$

One other rule of differentiation concerns the derivative of a function that is itself a function of another function. This is actually much simpler than it sounds. Suppose you are told that your car, which gets 25 miles to the gallon, is using fuel at a rate of 2 gal/h.

How fast is it going? Without straining the limits of your mathematical ability, you answer: 50 mi/h.

Now let's ask the question differently. Suppose the distance you go is called y, and the amount of fuel in your tank is called x. The distance you go depends on how much fuel you use, in other words, y is a function of x. As you drive, you consume fuel, so x is changing with time: x is a function of t. Now, given that y is a function of x and x is a function of t, what is your speed, dy/dt?

You knew the answer before we started to confuse the issue. What you said before was

$$25 \, \frac{\text{mi}}{\text{gal}} \times 2 \, \frac{\text{gal}}{\text{h}} = 50 \, \frac{\text{mi}}{\text{h}}$$

or in other words

$$\frac{dy}{dx} \frac{dx}{dt} = \frac{dy}{dt}. \tag{3.7}$$

That's the way it always works; it's called the *chain rule*.

Example 9

Suppose $y = 3x^2$ and $x = 6t$. What is dy/dt?

We have

$$\frac{dy}{dx} = 6x, \qquad \frac{dx}{dt} = 6,$$

$$\frac{dy}{dt} = \frac{dy}{dx} \cdot \frac{dx}{dt} = 36x = 36(6t) = 216t$$

(our result, dy/dt, should be expressed in terms of t, not x). We can check this directly: substituting $6t$ for x, we have

$$y = 3(6t)^2 = 108t^2,$$

$$\frac{dy}{dt} = 216t.$$

The reason the rule works is that dy/dx is the limit as Δx gets small of $\Delta y/\Delta x$, and dx/dt is the limit as Δt gets small of $\Delta x/\Delta t$. Before Δt, and therefore Δx, go to zero, we can form the product of the perfectly ordinary fractions

$$\frac{\Delta y}{\Delta x} \cdot \frac{\Delta x}{\Delta t} = \frac{\Delta y}{\Delta t}.$$

No matter how small Δt (and therefore Δx) become, as long as they are not zero, this remains true, so it remains true in the limit, giving

$$\frac{dy}{dt} = \frac{dy}{dx}\frac{dx}{dt}.$$

(3.7)

One advantage of the Leibniz notation is that you often get the right answer quickly treating dy/dx as if it were an ordinary fraction. Thus we get the last equation by canceling dx top and bottom. Here is another example:

$$\frac{dx}{dy} = \frac{1}{dy/dx}.$$

Example 10

The discussion in the text is a little unrealistic because the fuel consumption of your car (in miles per gallon) depends thoroughly on your speed. A better question might be: The owner's manual says your car gets 20 mi/gal at 70 mi/h. If you are driving at 70 mi/h, how much fuel do you use before the state trooper catches you in 20 min?

We have $dy/dx = 20$ mi/gal and $dy/dt = 70$ mi/h. Because

$$\frac{dy}{dx}\frac{dx}{dt} = \frac{dy}{dt}$$

we find

$$\frac{dx}{dt} = \frac{dy/dt}{dy/dx} = \frac{70}{20} = 3.5\ \frac{\text{gal}}{\text{h}}.$$

In 20 min, you'll use about 1.2 gal.

Example 11

How can you calculate the derivative of $y(x) = x/(x + 1)$?

Here we have the quotient of two functions, but we can write it as the product $x(x + 1)^{-1}$. In this form we can apply the product rule:

$$\frac{dy}{dx} = (x)\frac{d}{dx}(x + 1)^{-1} + (x + 1)^{-1}\frac{d}{dx}(x).$$

The only trouble spot here is the derivative of $(x + 1)^{-1}$, for which we'll invoke the rule for reciprocals:

$$\frac{d}{dx}\left(\frac{1}{z}\right) = \frac{-1}{z^2}\frac{dz}{dx}$$

with $z = 1 + x$. This gives us

$$\frac{d}{dx}\left(\frac{1}{x+1}\right) = \frac{-1}{(x+1)^2}\frac{d}{dx}(x+1) = -(x+1)^{-2}.$$

Consequently, when we substitute back into the product rule, we obtain

$$\frac{dy}{dx} = -x(x+1)^{-2} + (x+1)^{-1}.$$

Example 12

Suppose gas is pumped into a spherical balloon at a rate that makes its volume increase by 250 cm³/s. How fast is the radius of the balloon changing when the radius is 10 cm?

The first step in solving this problem is to decipher all the information given. The volume of a sphere is $V = \frac{4}{3}\pi r^3$, and we are told that the rate of change of the volume with respect to time, which we'll call dV/dt, is 250 cm³/s. What we want to know is the change in radius with respect to time, dr/dt. Here the volume is a function of the radius, which is a function of time. Their derivatives are related by the chain rule:

$$\frac{dV}{dt} = \frac{dV}{dr}\frac{dr}{dt}.$$

Although the equation does not specifically give dr/dt, we can solve for it algebraically once we know dV/dr. And what is dV/dr? By the power rule we have

$$\frac{dV}{dr} = \frac{d}{dr}\left(\frac{4}{3}\pi r^3\right) = 4\pi r^2.$$

Substituting into the chain rule, we get

$$\frac{dV}{dt} = 4\pi r^2\frac{dr}{dt}.$$

Solving for dr/dt, we obtain for the change in radius with respect to time

$$\frac{dr}{dt} = \frac{dV/dt}{4\pi r^2}.$$

To find the change when the radius is equal to 10 cm, we merely substitute for all the factors:

$$\frac{dr}{dt} = \frac{250 \text{ cm}^3/\text{s}}{4\pi(10 \text{ cm})^2},$$

$$\frac{dr}{dt} = 0.20 \text{ cm/s}.$$

Note that we didn't have to express r as a function of time in order to solve the problem. This is one of the advantages of the chain rule.

So far we have encountered three general rules, the sum, product, and chain rules for differentiation, and two special consequences of the product rule: the power rule and the reciprocal rule. By careful application of these rules, a great many functions can be differentiated easily and efficiently; you need not memorize every derivative in the world, but rather you can calculate what you need. We summarize the rules below, which after practice will become second nature:

SUM RULE:

$$\frac{d}{dx}[y(x) + z(x)] = \frac{dy}{dx} + \frac{dz}{dx} \tag{3.4}$$

PRODUCT RULE:

$$\frac{d}{dx} yz = y\frac{dz}{dx} + z\frac{dy}{dx} \tag{3.5}$$

RECIPROCAL RULE:

$$\frac{d}{dx}\frac{1}{y} = \frac{-1}{y^2}\frac{dy}{dx}$$

CHAIN RULE:

$$\frac{dz}{dx} = \frac{dz}{dy}\frac{dy}{dx} \tag{3.7}$$

POWER RULE:

$$\frac{d}{dx} x^n = nx^{n-1} \tag{3.6}$$

Questions

6. Use the product rule to show that

(a) $\dfrac{d}{dx} x^5 = 5x^4$

(b) $\dfrac{d}{dx} x^6 = 6x^5$

7. Using the sum and power rules, evaluate the derivatives with respect to x of the following functions:

(a) $3x^4 - 5x^2 + 2$ (b) $7x^6 - 2x^3 - 8x$

(c) $1/x^4$ (d) $x^{-3} - x^3$.

8. Find the derivatives of the following functions:

(a) $(x - 3)^2$ (b) $(x^3 - x)^5$

(c) $(x - 2x^4)^{-2}$ (d) $(4x^{-3} - x^2)^6$.

9. Determine the derivatives of the following functions:

(a) $x^{-3/2}$ (b) $x^{1/2} - 5/x$

(c) $(x + 1)^{1/2}$ (d) $2x(1 - x)^{-1/2}$.

10. Calculate the derivatives of the following functions:

(a) $(1 - x)/(1 + x)$ (b) $(1 + x^2)(1 - x)$

(c) $4x/(1 - x^4)$ (d) $(1 - 2x + x^2)/(1 - x)$

(e) $5x(2x - 1)/(1 + x)^2$ (f) $3x(3x - x^3)(1 - x)$.

11. Each edge of a cube is expanding at the rate of 1 cm/s. How fast is the volume changing when the length of each edge is

(a) x cm   (b) 2 cm   (c) 5 cm?

3.4 DERIVATIVES OF SPECIAL FUNCTIONS

With our rules for differentiation, we can now find the derivative of even the most complicated function, as long as it is composed of power functions. That is about as far as Newton got, but some progress has been made since then. Phenomena that are cyclic or repeat their motion can conveniently be described by functions that are periodic – the trigonometric functions sine, cosine, and their combinations. Therefore we should learn their derivatives. (Those who need to refresh their trigonometry should see Appendix C for help.)

The derivative of the sine function can be calculated by using the definition of the derivative [Eq. (3.3)], but let's use another method to find the derivative, a geometric one. Our approach is to estimate geometrically the tangent line to the sine curve at a particular angle θ, measure the slope of the tangent line, and then, using that value of the slope, plot it as a function of θ. By repeating this procedure for many points, we shall be able to construct a rough graph of the derivative as a function of θ, that is, $d(\sin \theta)/d\theta$.

This process is illustrated in Fig. 3.7, where the nine numbered points were used. For points 3 and 7, it is easy to see that the slope of the tangent lines is zero (horizontal lines). For the other points, we simply estimate the slopes of the straight lines using our average slope,

$$\text{average slope} = \frac{\Delta y}{\Delta x}. \tag{3.1}$$

All the points and slopes are summarized in Table 3.2.

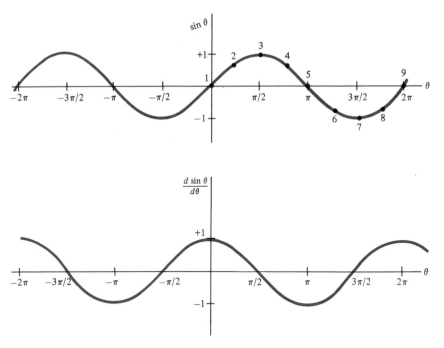

Figure 3.7 Geometric construction of the derivative of sin θ.

Table 3.2 Points Used to Find the Derivative of sin θ

θ (radians)	sin θ	$\dfrac{d}{d\theta} \sin \theta$
0	0	1.0
π/4	0.7	0.7
π/2	1.0	0
3π/4	0.7	−0.7
π	0	−1.0
5π/4	−0.7	−0.7
3π/2	−1.0	0
7π/4	−0.7	0.7
2π	0	1.0

When the points of the derivative function are connected as in Fig. 3.7, an amazing result emerges: the derivative is the same curve shifted by $\pi/2$ radians. But that's the cosine function! Mathematically we summarize this as

$$\frac{d \sin \theta}{d\theta} = \cos \theta. \tag{3.8}$$

If the derivative of sine is cosine, you might suspect that the derivative of cosine is related to the sine function. But how? By repeating our procedure, we can discover how; this is illustrated in Fig. 3.8.

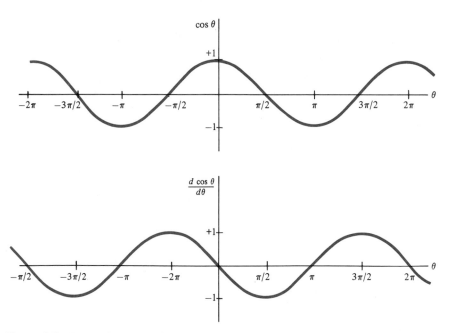

Figure 3.8 Geometric construction of the derivative of $\cos \theta$.

Alternatively, we can use the fact that $\cos \theta = \sin (\pi/2 - \theta)$ and apply the chain rule and Eq. (3.8). The result is that the derivative is not $\sin \theta$ but $-\sin \theta$:

$$\frac{d \cos \theta}{d\theta} = -\sin \theta. \tag{3.9}$$

Example 13
Determine the derivative of $\sin^2 x$.

Let's calculate this derivative by considering the function to be the product of two functions, sin x and sin x. After all, that's what the square means. Once we have broken the function down into a product, we can apply the product rule [Eq. (3.5)]:

$$\frac{d}{dx}\sin^2 x = \sin x \frac{d}{dx}\sin x + \sin x \frac{d}{dx}\sin x,$$

$$\frac{d}{dx}\sin^2 x = 2\sin x \frac{d}{dx}\sin x,$$

$$\frac{d}{dx}\sin^2 x = 2\sin x \cos x.$$

A secret to differentiation of complicated functions is to see them as composed of sums and products of simple functions.

Example 14

What is the derivative of $y(x) = x^3 - \cos x$?

First, we realize that we have a sum of two simple functions and that we can use the sum rule for differentiation. By the power rule, the derivative of the first term is $3x^2$, and we know that the derivative of $\cos x$ is equal to $-\sin x$. Therefore we have

$$y'(x) = 3x^2 + \sin x.$$

Example 15

Find the derivative of $\sin \theta^2$ (here the notation means $\sin(\theta \times \theta)$, not $\sin \theta \times \sin \theta = \sin^2 \theta$).

We can apply the chain rule here, so let's call $y(\theta) = \theta^2$ and $z(y) = \sin y$. Then we have

$$\frac{dy}{d\theta} = 2\theta \qquad \text{(by the power rule)},$$

$$\frac{dz}{dy} = \cos y.$$

The chain rule implies

$$\frac{dz}{d\theta} = \frac{dz}{dy}\frac{dy}{d\theta}, \tag{3.7}$$

or

$$\frac{dz}{d\theta} = 2\theta \cos \theta^2.$$

Trigonometric functions are useful to know because they describe repetitive processes. Another function is also frequently encountered in physics because it describes rapid changes. This function is the solution to the following question: Suppose the human race really did start with Adam and Eve in the Garden of Eden with no predators, plenty of food, and nothing much to do but to reproduce. How fast would the human population grow?

Obviously what we want here is not a detailed history with a lot of names and begats, but rather a rough estimate of how the situation develops once the population gets large enough to treat mathematically.

The key assumption to answering this question is that the rate at which human beings are produced is directly proportional to the number of human beings working to produce them. In other words, if N is the population, then the change in the population with time is given by

$$\frac{dN}{dt} = kN,$$

where k is a constant of proportionality.

This problem is one of a vast class of examples in physics and (as we have seen here) other fields in which something changes at a rate that is proportional to itself. In order to solve the problem in general, let's first make a simple change in variables. We'll define

$$y = kN \qquad \text{and} \qquad x = kt.$$

Then our equation becomes (see Question 14 below)

$$\frac{dy}{dx} = y. \tag{3.10}$$

In this form it represents all problems in which something (in this case y) changes at a rate proportional to itself. It also tells us that we are looking for a function that remains unchanged by differentiation. Geometrically, this means that the solution to Eq. (3.10) is a function whose slope at each point is equal to the function itself. [Of course the function $y(x) = 0$ has this property, but we exclude this trivial solution because it doesn't solve our problem: it would mean $N = 0$, no people.]

We can get a good idea of the shape of the graph of such a function through the geometric interpretation. Let's construct a curve that at each point (x, y) has a slope equal to y. This means that at the point $(x_1, 1)$ the slope is equal to 1, and at $(x_2, 2)$ it is equal to 2, and so on. In order to make our construction as general as possible, let's start with the point $(x_0, y_0) = (0, 1)$ and draw a line of slope 1 joining this point to its neighbor (x_1, y_1). If we make $x_1 = h$, where h is that infamously small number we have frequently encountered, then we can figure out y_1 by using the condition that the slope be equal to 1:

$$\text{slope} = 1 = \frac{y_1 - y_0}{x_1 - x_0} = \frac{y_1 - 1}{h},$$

(3.1)

or

$$y_1 = 1 + h.$$

Figure 3.9 illustrates this construction.

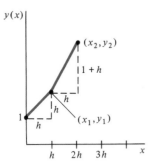

Figure 3.9 Geometric construction of a curve satisfying $y'(x) = y$.

At the point $(h, 1 + h)$ we construct a line whose slope is $1 + h$ and we move along the line to where the x coordinate is $x_2 = h + h = 2h$. We can find y_2, the y coordinate of this point by using the slope. Since the slope of the line is $1 + h$, we have

$$\text{slope} = 1 + h = \frac{y_2 - y_1}{h}.$$

(3.1)

Substituting $y_1 = 1 + h$, we can solve for y_2 to get

$$y_2 = (1 + h) + h(1 + h),$$

which we can factor as

$$y_2 = (1 + h)(1 + h) = (1 + h)^2.$$

To find the next point, we draw a line of slope $(1 + h)^2$, and move along it to $x_3 = 3h$. The new y coordinate y_3 can be found by applying Eq. (3.1) again:

$$\text{slope} = (1 + h)^2 = \frac{y_3 - y_2}{h},$$

or

$$y_3 = y_2 + (1 + h)^2 = (1 + h)^2 + h(1 + h)^2.$$

Factoring out $(1 + h)^2$, we get

$$y_3 = (1 + h)^3.$$

Here you might detect a pattern. If the x coordinate is nh, where n is some integer, then the y coordinate is $y_n = (1 + h)^n$. An interesting property also emerges when we express y_n in terms of x_n. Since $x_n = nh$, then $n = x_n/h$, and our relation becomes

$$y_n = (1 + h)^{x_n/h}.$$

In other words, at each vertex (x, y) of our polygonal curve we have

$$y = u^x,$$

where

$$u = (1 + h)^{1/h}.$$

The smaller h is, the smoother our curve will be. This feature is illustrated in Fig. 3.10, where graphs of $y(x)$ are shown for different values of h.

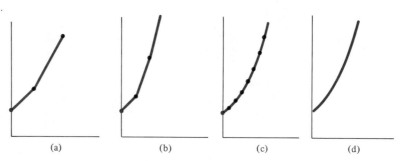

    (a)   (b)   (c)   (d)

Figure 3.10 Polygonal curves for various values of h: (a) $h = 1$, (b) $h = \frac{1}{2}$, (c) $h = 0.1$, (d) $h = 0.001$.

As h takes on various values, $u = (1 + h)^{1/h}$ assumes the values listed in Table 3.3 (try this with a calculator if you have one handy). In the limit as h shrinks to zero, u surprisingly approaches a limiting number. Correct to ten decimal places this number is 2.7182818285; being such a special number it has its own name, the Euler number (pronounced like ''oiler''). Leonhard Euler, who first realized its importance, modestly named it e:

$$e = \lim_{h \to 0} (1 + h)^{1/h} = 2.7182818285. \ldots$$

We now have the answer to our question [the solution of Eq. (3.10)] and it is called the *exponential function*, written

$$y(x) = e^x. \tag{3.11}$$

Most calculators have a key that corresponds to this function and provides the easiest way to calculate values of the function.

The derivative of the exponential function is itself,

$$\frac{d}{dx} e^x = e^x. \tag{3.12}$$

Table 3.3 The Euler Number e as a Limiting Value.

h	$u = (1 + h)^{1/h}$
1	2
0.5	2.25
0.1	2.5937425
0.01	2.7048138
0.001	2.7169239
0.0001	2.7181459
0.00001	2.7182682

Example 16

Find the derivative of $z(x) = e^{-3x^2}$.

Here we wish to take the derivative of a function that itself is a function of another function. In order to clarify matters, let's call $y(x) = -3x^2$. Then $z = e^y$, and by the chain rule we have

$$\frac{dz}{dx} = \frac{dz}{dy}\frac{dy}{dx}.$$

Since $z = e^y$, the first factor is $dz/dy = e^y$, and since $y = -3x^2$, the second factor is $dy/dx = -6x$. Consequently,

$$\frac{dz}{dx} = -6xe^y = -6xe^{-3x^2}.$$

The derivatives of special functions that are frequently used in physics are summarized in Table 3.4. They will soon become routine with practice.

Questions

12. Draw on all of your knowledge of biology, sociology, psychology, and literature to estimate k in the population question. Then, supposing there had been no wars, plagues, floods, etc., estimate how long it should have taken to get from Adam and Eve to the present world population.

Table 3.4 Special Functions and Their Derivatives

$y(x)$	$y'(x) = \dfrac{dy}{dx}$	
c (constant)	0	
x^n	nx^{n-1}	(3.6)
$\sin x$	$\cos x$	(3.8)
$\cos x$	$-\sin x$	(3.9)
e^x	e^x	(3.12)

13. If the equation $dy/dx = y$ represents population growth, what are the units of x and y?

14. Use the chain rule to show that $dy/dx = y$ is the result of changing variables in $dN/dt = kN$, where $y = kN$ and $x = kt$.

15. Construct a graph of e^x for $x = -2$ to $x = 0$.

16. Determine the derivative $y'(3)$ for each of the following functions:

(a) $y(x) = x^3$
(c) $y(x) = x^7 - 5$

(b) $y(x) = e^x$
(d) $y(x) = e^x - 10e^{-x}$.

17. Evaluate the derivatives of the following functions at $x = 90° = \pi/2$ radians:

(a) $\sin x$
(c) $-\sin x + 5$

(b) $\cos x$
(d) $-\cos x - 2$.

18. Use Table 3.4 and the rules for differentiation to evaluate the derivatives with respect to x of the following functions:

(a) $3x^4 - 5x^2 - 1$
(c) $2x \cos x$

(b) $7xe^x$
(d) $\sin^3 x$.

19. Find the derivatives of the following functions:

(a) $\sin x \cos x$
(c) $e^{-x} \sin 2x$

(b) $\sin 5x$
(d) $\cos x \sin^3 x$.

20. Determine the value of x where dy/dx is zero if $y(x)$ is given by

(a) $3x^2 - 27x$
(c) $x^{3/2} - 150x$

(b) $4x^4 - x^2$
(d) $\sin^2 x$.

21. Calculate the derivatives of the following functions:

(a) $\sin\sqrt{x}$
(c) $e^{\alpha x}$

(b) $e^x \cos x$
(d) $3x^6 e^x$.

22. Find the derivatives of the following functions:

(a) $2x/(1 - x^2)$ **(b)** $\sqrt{1 - x^3}$

(c) $\tan x$ **(d)** $(\sin x)/(1 - x)$.

23. Suppose your income y is directly proportional to the number x of hours that you work; then it can be described by $y(x) = cx$, where c is a constant. In addition, let's assume that you're a big spender and that the money z you spend varies with income as $z(y) = be^{y/2}$, where b is another constant. How does the amount of money you spend change with a change in the number of hours worked?

24. (a) Let $y(t) = Ce^{kt}$, where C and k are constants. Show that

$$\frac{dy}{dt} = ky.$$

In other words, the rate of change of y with respect to t is proportional to y, with constant of proportionality k.

(b) Now let $u(t)$ be any function of t with the same property:

$$\frac{du}{dt} = ku,$$

where k is a constant. Prove that $u(t)$ must have the form $u(t) = Ce^{kt}$ for some constant C. *Hint.* Let $f(t) = u(t)e^{-kt}$ and calculate the derivative $f'(t)$, using the product rule.

3.5   A FINAL WORD

In our discussion of functions and derivatives, we have tacitly assumed that functions used to describe nature have no jumps in them, that they are smooth. Mathematicians call these functions *continuous*. We have also assumed that the functions do have derivatives, which means that they are *differentiable*. There are functions that do not have these properties; two examples are shown in Fig. 3.11. The first function depicted there has jumps or discontinuities. The second function shown does not have a unique slope at the peak; at that point the very idea of a derivative has no meaning at all.

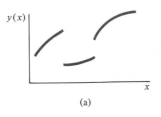
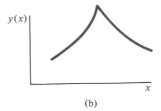

(a)   (b)

Figure 3.11   Examples of (a) a discontinuous function and (b) a nondifferentiable function.

Physicists assume that real discontinuities and nondifferentiable functions seldom occur in nature. On the other hand, mathematicians know that physicists' theories are not nature, but rather idealizations or models of nature in which discontinuities, kinks, and infinities do occur. The mathematician is the guardian of clarity and precision of thought. For the mathematician, every theorem must be stated in a way that is absolutely true, allowing no exceptions, no matter how bizarre or unusual. Physicists tend to ignore the bizarre and unusual because it is rare, but sometimes they can't.

Albert Einstein wrote a letter in 1912 to his friend Arnold Sommerfeld, another famous physicist, in which he said he had "... become imbued with an appreciation of mathematics, the finer points of which I had until now regarded as pure luxury." A few years later, Einstein sent a postcard to the Italian mathematician Tullio Levi-Civita, in which he wrote he had "just finished reading your beautiful new work. It must be nice to ride the horse of true mathematics while the rest of us must make our way laboriously on foot."

Figure 3.12 is a photograph taken in 1931, showing A. A. Michelson, the first American Nobel laureate, Robert A. Millikan, also a Nobel laureate, and Albert Einstein. Behind Einstein is a man named Walther Mayer. He was Albert Einstein's personal mathematician. Just as some notables travel with a retinue that might include a personal physician or hairdresser, Einstein, knowing that he himself was not a good mathematician, brought a mathematician along with him wherever he went.

Figure 3.12 Michelson, Einstein, and Millikan in 1931. (Courtesy of the Archives, California Institute of Technology.)

CHAPTER

INERTIA

We now bring forth a brand new science about an ancient subject. There is perhaps nothing older than the study of motion, and books, neither few nor small have been written about it by philosophers; nevertheless, I have found among its properties many that, although well worth knowing, have never before been observed, much less demonstrated. They point out a few of the more obvious ones, for example that the natural motion of falling bodies accelerates continuously; but in what proportions that acceleration occurs has not been shown. No one, as far as I know, has demonstrated that a body falling from rest traverses, in equal times, distances that have between them the same proportions as the successive odd numbers starting from one. It has been observed that missiles, that is to say, projectiles follow some kind of curved path, but that it is a parabola no one has shown. I will show that it is, together with other things, neither few in number nor less worth knowing, and what I hold to be even more important, they open the doors to a vast and crucial science of which these our researches will constitute the elements; other geniuses more acute than mine will penetrate its hidden recesses.

Galileo Galilei, *Two New Sciences*, Third Day (1638)

4.1 IF THE EARTH MOVES: ARISTOTELIAN OBJECTIONS

In 1543 the revolutionary book appropriately entitled *De Revolutionibus Orbium Coelestium* (*On the Revolutions of the Celestial Spheres*), was published by Nicolaus Copernicus. In its pages Copernicus set the earth spinning on its axis and rotating around the sun. In his attempt to return the heavens to their simple beauty, Copernicus made the sun, not the earth, the center of the universe, and in doing so he tore the heart out of the Aristotelian world. Without the solid, immovable earth at the center of the universe, there could be no Aristotelian laws of motion. And without these laws, there were none at all, for Copernicus had no laws to replace those he destroyed.

For nearly half a century, Copernicus's system was largely ignored. Then, with the rise of Galileo's mechanics and Kepler's developments in astronomy, the Copernican system received serious consideration. The science of mechanics arose as an answer to the challenges of the post-Copernican Aristotelians and completely, irrevocably replaced the old physics of Aristotle.

Aristotle's scheme had described the way everyday objects move: rocks fall, smoke rises, and an apple cart stops when no longer pulled. All motion required a cause, and depending on the nature of its cause was either natural or violent. A heavy rock falling toward its natural place on the earth was natural motion because its cause was the internal nature of the rock, its heaviness. On the other hand, Aristotle knew that a horse was needed to pull a cart, and if the horse stopped, the cart also stopped. This type of motion he classified as violent motion, motion caused by some agent or force outside the object. If the force were removed, violent motion ceased.

But why should a rock fall toward the center of the earth if the earth itself were spinning dizzily through space? If the sun and not the earth is the center of the universe, Aristotle's scheme of natural places and natural motions crumbles. And without the solid stationary earth beneath our feet, what do we have left? Even granting that there is something called gravity that keeps us from flying off into space, how do we begin to understand the motion of other objects?

If the earth rotated from west to east, the Aristotelians argued, then things like clouds not attached to the earth would always seem to move from east to west, just like the sun. Even birds would suffer a similar fate because, with the earth rotating once every 24 h, the speed of the earth at the equator would be 1000 mi/h, and no bird could fly swiftly enough to keep up with it. Therefore birds should always appear to fly from east to west as the earth rapidly revolves underneath.

And the Aristotelians argued further. On a motionless earth, everyone knew that a stone dropped from a tower always lands at the foot of the tower. But if the earth were indeed rotating under a falling stone, then the stone should strike the ground west of the foot of the tower. They made their argument more persuasive by pointing out the size of the effect: If the tower were only 16 ft high, then in its 1-s fall the earth would move 1500 ft!

A falling stone on a spinning earth should behave just like an object dropped on a moving ship. If dropped from the mast, the object lands not at the foot of the mast, but some distance behind it. The distance corresponds exactly to how far the ship moved during the fall. Everyone knew this, so they thought. The Aristotelians never bothered to test their predictions; they preferred words and logic to experiments. Figure 4.1 shows how they compared motions of falling bodies on a moving earth and on a moving ship.

Precisely because birds are seen flying in all directions and stones land directly below the point where they are dropped, seventeenth-century Aristotelians thought they had a sound case against the spinning world of Copernicus. Familiar observations attested to laws of motion of Aristotle and the stationary earth.

Figure 4.1 Aristotelian description of objects in free fall on (a) a
rotating earth and (b) a moving ship.

Questions

1. Using the fact that the circumference of the earth is approximately 25,000 mi (or
   40,234 km), calculate the speed of the rotating earth at the equator. How does
   this compare to the speed of a supersonic airplane, which is about 1200 ft/s (or
   366 m/s)?

2. Substantiate the statement in the text that in 1 s the earth moves 1500 ft.

3. Develop an argument an Aristotelian would present against a rotating earth using the motion of a cannonball fired vertically.

4. Imagine two identical cannons, one aimed horizontally west and the other east. According to an Aristotelian, would one cannon shoot a cannonball further than the other? Explain why or why not.

5. Explain in a sentence or two (before reading the next section) what you understand the word *inertia* to mean.

4.2 THE EARTH MOVES: GALILEO'S LAW OF INERTIA

Galileo was fascinated by the Copernican world view and became a staunch defender of it. Besides, he liked a good fight. By his investigations he hoped to convince the world that the heliocentric system was not merely a fiction that simplified calculations of astronomers, but that it embodied physical truth about the universe. In his work *Dialogue Concerning the Two Chief World Systems*, published in 1632, Galileo summarized his proofs for the Copernican system (Fig. 4.2). He based his arguments in this book on experiments, reasoning, and penetrating insights.

One key experiment involved balls allowed to roll down and then up inclined planes. He purposely chose balls because he wanted to minimize friction and focus purely on

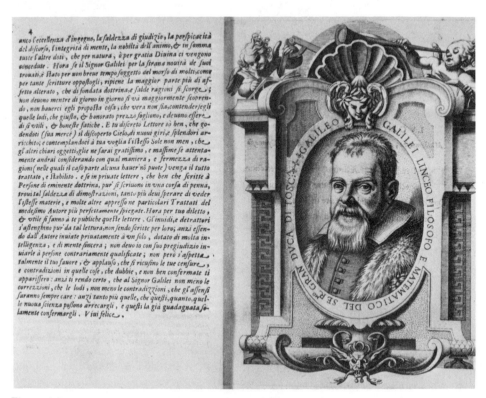

Figure 4.2 Frontispiece from *Dialogue Concerning the Two Chief World Systems*. (Courtesy of the Archives, California Institute of Technology.)

the motion. Figure 4.3 illustrates how Galileo set up his experiment. As he changed the plane on the right from position A to B, he noticed that the ball rolls further along the plane but always to the same height as it was released from.

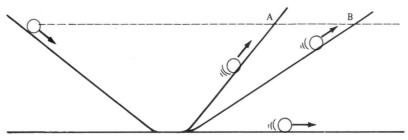

Figure 4.3 Illustration of Galileo's experiment with rolling balls which led him to the idea of inertia.

From this observation Galileo imagined what would happen if the second plane were level, that is, horizontal. If that were the case the ball could never reach its original height; it would keep on going forever with the speed that it had acquired in rolling down the inclined plane. Galileo concluded that any object in motion, if not obstructed, will continue to move with a constant speed along a horizontal line. That was Galileo's version of inertia. Inertia resists changes, from rest, or from motion with a constant speed in the horizontal direction.

Although Galileo's arguments were brilliant, they were not quite right. To him a horizontal surface was one which was everywhere perpendicular to the direction pointing toward the center of the earth. But that describes a sphere with a center at the center of the earth. His law of inertia then means that in the absence of any outside forces an object set into motion would continue moving in a circle around the earth, forever. What he should have said was something different.

Objects set into motion tend to continue moving with a constant speed in a *straight line*, not in a circle. The law of inertia is

> *A body will remain at rest or continue to move with a constant speed in a straight line unless acted on by an outside agent.*

Nonetheless, Galileo was able to take one more step. He realized that the earth is very big and its surface looks flat even though it's really a sphere. Galileo realized that a small portion of a very large circle is very nearly a straight line. Therefore, objects on the earth will seem to move in straight lines. And thinking that this was only approximately true, Galileo perfectly described the law of inertia and worked out many of its consequences. He realized that all arguments against a rotating earth were due to a lack of understanding the law of inertia.

To counter his critics, Galileo shrewdly used precisely the examples his opponents cited. Considering a stone dropped from the crow's nest of a ship, he pointed out that if the ship is at rest, the stone lands at the foot of the mast. No one disagrees with that. But if the ship were moving with a constant speed, Galileo claimed that the stone would still land at the foot of the mast. It makes no difference if the ship is at rest or moving with a constant speed in a straight line; the stone will land in the same place on the ship. But why?

Galileo realized that before the stone is released it is moving along as part of the ship. This means that it has the same speed as the boat along that straight line. When it's released, the stone's inertia keeps it moving with that same speed along a horizontal, straight line. So the stone and ship continue to move together. But in addition, gravity pulls the stone vertically downward. Galileo conjectured that this vertical motion doesn't interfere with the stone's horizontal motion. So as the stone falls, it continues to move horizontally with a constant speed. Because the ship is also moving, the stone lands right at the foot of the mast as illustrated in Fig. 4.4.

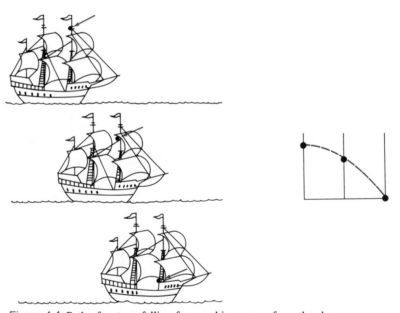

Figure 4.4 Path of a stone falling from a ship as seen from the shore.

Galileo was right on target with his prediction. A sailor on board would see the stone fall in the very same way on a moving ship as on one at rest. It's hard to say whether the Aristotelians would have been convinced of their errors by seeing this demonstrated, but in any case it wasn't until 1640 that the French philosopher Pierre Gassendi carried out the experiment, and showed that Galileo was right.

Using the law of inertia, Galileo was able to explain the natural motion of falling objects on a rotating earth and answer the arguments of his critics. Dropping a stone from the mast of a moving boat is just like dropping a rock from a tower on the moving surface of the earth. Because the rock continues to share the motion of the earth as it falls, the stone lands at the base of the tower. An observer on earth could not tell whether or not the earth were rotating by looking at the stone. Motion, Galileo realized, is relative.

That is to say, the tower, the stone, and the observer, all moving together with the earth, all seem to be at rest.

But he grasped even more. The natural state of motion is motion with a constant speed in a straight line. The tendency of an object was not, as Aristotle thought, to reach a state of rest, but to continue moving without any propelling force. For Galileo, if there were no friction and a horse pulling a cart suddenly stopped and indulgently stepped aside, the cart would continue to move; its inertia would keep it moving.

Questions

6. When a car suddenly brakes to a screeching stop, you lurch forward. Why?

7. Discuss how the law of inertia could explain why a sharp jerk to a dusty coat succeeds in removing the dust.

8. Why is it difficult to stop a speeding hockey player?

9. Must a spacecraft continuously fire its rockets to maintain a constant speed in deep space (where there is practically nothing around to disturb it) once it's out there? Why or why not?

10. Imagine you are Galileo. How would you answer an Aristotelian who said that a cannonball fired horizontally westward would travel further than one fired horizontally eastward, if the earth were rotating?

11. Discuss how your answer to Question 5 is related to the scientific meaning of the word *inertia*.

12. Suppose a supporter of Aristotle were to exclaim that there is no such thing as inertia because when a pebble is released it doesn't remain in midair but falls, whereas the law of inertia states that it should tend to remain at rest. How would you reply?

13. A train moves along a horizontal track with a constant speed of 20 m/s. A passenger drops a ball from the ceiling, which is 2.5 m above the floor of the train.

    (a) According to the passenger, how long does it take the ball to reach the floor?
    (b) According to the passenger, where does the ball hit the floor?
    (c) According to someone standing on the tracks, how far does the train move while the ball is in the air?
    (d) What horizontal distance does a person on the tracks see the ball move while it falls?

4.3 RELATIVE MOTION

By watching a stone fall, you cannot tell whether the earth is at rest or moving. You might wonder if there is any way to tell. Someone on the moon would think he saw the earth moving, but to you it is the man on the moon that's moving. In that sense, all motion is relative. To describe motion you need to specify a frame of reference, that is, the place from which you're describing the motion.

Without a frame of reference, you can't even discuss motion. Imagine a dark, vacant universe that contains nothing but a single blob, a bright red one if you like. Is this object at rest or in motion? That question is nonsense because we have no way of judging its motion; that is, we don't have any reference frame. But now what if a second blob (a blue one) enters the picture? Which blob is at rest? Which one is moving? From the viewpoint of the red one, the blue one might appear to be moving, whereas the blue one could claim that it is the red one that is moving. Here there are two possible reference frames, but the only meaningful description of motion is relative motion: motion of one object in reference to another, which is used as a frame of reference. And because all motion is relative, there is no such thing as absolute rest.

But that is the heart of the law of inertia. If an object is in motion (to some observer), why should the object come to rest? The simplest law one can imagine is that it goes right on moving. That's the law of inertia!

Galileo knew that in the law of inertia he had discovered a fundamental feature of nature. In defending the idea of shifting the center of the universe from the earth to the sun, he found out that any reference frame is suitable to describe motion. To be sure, without leaving their earthly confines, the best minds in medieval Europe could not easily discover that the earth was spinning. The moving surface of the earth, because it is so large, behaves like a reference frame moving with constant speed in a straight line. Any reference frame that is in straight-line motion with a constant speed is a special type of reference frame and comes with a special title: an *inertial reference frame*.

Inertial reference frames have precisely the property that an observer in one has no way of determining whether he is at rest or moving. It is impossible for him to detect his motion by an experiment performed solely within his reference frame. Today this idea is not only important for understanding mechanics; it lies at the very heart of Einstein's theory of relativity.

Questions

14. Imagine you are on the Orient Express, which is chugging along a level track with a constant speed. Can you tell that you are moving without looking out the window or hearing the wheels click? Why?

15. Suppose you rolled a ball along the aisle of the train in problem 14; how would it move? Would it appear to move differently if the train were rounding a curve?

16. If you're driving along a highway at 55 mi/h, what is the reference frame used to determine your speed?

17. Imagine a freighter in deep space cruising along with a constant speed. A mechanical arm reaches into a garbage bay, pulls out a load of garbage, and then releases it. According to a passenger on the ship, describe the motion of the garbage.

18. Suppose you are speeding along at 120 km/h on an interstate highway, and a Ferrari, whose speedometer reads 150 km/h, passes you.

   (a) How fast is the Ferrari moving relative to you?
   (b) How fast are you moving relative to the Ferrari?
   (c) If the Ferrari passed you in the opposite direction, how fast would it be traveling relative to you?

19. Calculate the speed of the earth about the sun, assuming that the earth travels in a circle of radius 150,000,000 km in 1 yr. How does that speed compare to the speed of a point on the equator due to the earth's rotation? (See Question 1.)

20. While a train is stopped at a station, Mike, at the front of a car, tosses a ball back and forth with Ann at the rear of the car. Each one can throw the ball with a speed of 2 m/s.

(a) What is the speed of the ball relative to Mike? to Ann?

(b) If the train is moving at 2 m/s, what then is speed of the ball relative to Mike? to Ann?

(c) While the train is moving as in (b), what is the speed of the ball according to someone on the ground as it goes from Mike to Ann? from Ann to Mike?

4.4 PROJECTILE MOTION: A CONSEQUENCE OF INERTIA

When Galileo was a professor at the University of Padua, he managed to earn a few extra scudi by giving military advice to the Venetian government. He was a valuable consultant because, using his ideas about inertia and falling bodies, he could explain what gunners had known from practical experience, that a cannon shot at a 45° angle to the horizon has the maximum range. After Galileo, no important theoretical advance in gunnery was made until World War I (1914–18), when both sides learned to take high-altitude atmospheric conditions into account when aiming their artillery.

The motion of projectiles had always been a weak spot for Aristotle. He couldn't fit this motion into his scheme of natural and constrained motions, because there obviously is no mover of an arrow once it leaves the bow. So he had to invent one. He suggested that the air through which the projectile moved was the mover. As the projectile was shot into the air, the air in front of it was pushed aside, whereas a void was left behind it. Thinking that nature avoided vacuums, he said that the air rushed in to fill the temporary void and propel the arrow forward. Even Aristotle's devoted medieval followers found this explanation wanting.

In the fourteenth century the idea arose that a projectile was endowed with something called impetus. The bowstring gave the arrow a certain quantity of impetus. According to this theory, an arrow, once shot, should fly along until its impetus ran out, and then fall to the earth. Figure 4.5 is an illustration of this theory from a book published in 1621. Cannons aimed on the basis of this theory tend to do relatively little damage.

Galileo correctly understood that inertia is the key to explaining projectile motion. He saw the motion of a cannonball as a combination of two motions, one horizontal and the other vertical. Moreover, he realized that these motions don't interfere with one another; each motion behaves as if the other is not present. This means that the cannonball falls vertically under the force of gravity while its inertia keeps it moving in the horizontal direction (if no force interferes) with a constant speed. This is the same idea he applied to a stone falling on a moving ship.

Figure 4.6 is a photograph of the paths of two balls taken with a stroboscope which illuminated the balls every $\frac{1}{30}$ s. One ball is projected horizontally at the same instant the other is dropped. The picture reveals that both balls are at the identical vertical position at the same instant; they are both falling vertically according to the law of falling bodies. In addition, the projected ball moves the same horizontal distance between each flash.

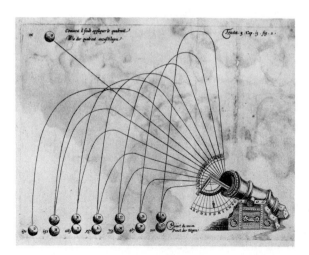

Figure 4.5 A 1621 illustration of the path of a cannonball according
to the impetus theory. (Department of Special Collections, Stanford
University Libraries.)

Remembering that average speed is distance divided by time, we realize that the horizontal
speed of the ball is constant. Galileo was correct!

With a combined knowledge of the law of falling bodies and the principle of inertia,
we can find the shape of a projectile's trajectory. Since different motions take place along
the horizontal and vertical directions, these are the natural directions for coordinate axes.
We'll choose the x axis along the horizontal and the y axis pointing vertically downward,
as shown in Fig. 4.6.

The principle of inertia says that the ball continues to move horizontally with whatever
speed it had when it was fired. This means that the speed along our x axis is constant;
let's call that speed v_x. But if the speed is constant, the distance x in the horizontal
direction must be proportional to the time:

$$x(t) = v_x t. \tag{4.1}$$

Note that this means dx/dt is just the constant speed v_x, which is exactly what we want.

In the vertical direction the ball is falling, and from the law of falling bodies we
know that the vertical distance y it has fallen at any time t is (ignoring air resistance)

$$y(t) = \tfrac{1}{2}gt^2. \tag{4.2}$$

Since the time t is the same time in both of our equations we can solve Eq. (4.1)
for t,

$$t = x/v_x,$$

and substitute it into Eq. (4.2). Our result is

$$y = (g/2v_x^2)x^2. \tag{4.3}$$

But this is the equation of a parabola, $y = ax^2$, where a is a constant. The trajectory of
any projectile (ignoring air resistance) is a parabola still, even today, called a Galilean
parabola.

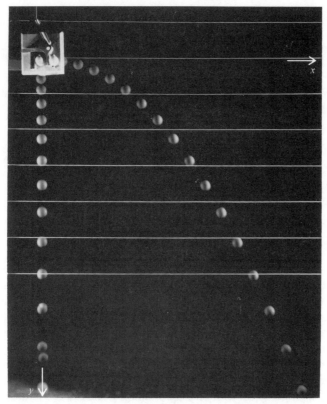

Figure 4.6 Stroboscopic picture of two balls, one projected horizontally at the same instant the other is dropped. The distance between lines was 6 in. and the time between flashes was $\frac{1}{30}$ s. (*PSSC Physics*, 2nd ed., 1965; D.C. Heath and Company with Education Development Center, Inc., Newton, Mass.)

Questions (in all questions ignore air resistance)

21. A bullet is shot horizontally from a rifle and at the same instant another bullet is dropped from the same height. Which of the following statements are false?

    (a) While in the air, the acceleration of both bullets is the same.
    (b) The shot bullet hits the ground first.
    (c) The shot bullet travels a greater distance.
    (d) The speed of both bullets in the vertical direction is the same at any instant while they are in the air.

22. To strike a weapons factory, a bomber should release its bombs

    (a) when it is directly over the target.
    (b) before it is over the target.
    (c) after passing over the target.

23. A plane dropping bales of hay to snow-stranded cattle is in level flight. If the plane continues to move with a constant speed, what is the path the pilot sees for a falling bale? What kind of path does a farmer on the ground see?

24. Suppose the plane in the previous problem is flying horizontally at an altitude of 200 m and with a speed of 65 m/s.

    (a) How long does it take a bale of hay to reach the ground?

    (b) How far does the plane move while the bale is in the air?

    (c) What is the speed of the bale in the horizontal direction just before it strikes the ground?

    (d) How fast is the bale falling just before it hits the ground?

25. A cannon fires a shell horizontally from a cliff with a speed of 80 m/s. The shell impacts 120 m from the base of the cliff.

    (a) How long was the shell in the air?

    (b) What is the height of the cliff?

4.5 A FINAL WORD

Neither Galileo nor artillery gunners would claim that real projectiles move in precisely parabolic trajectories. For example, it has been estimated that air resistance reduces the distance a well-hit baseball travels by as much as 40% of the idealized parabolic trajectory. Artillery gunners must take such effects into account. In Galileo's time they did so simply by watching where the shells landed and adjusting their aim accordingly. In modern times they commonly do so with the aid of computers.

When air resistance is included, the trajectory of a projectile may resemble the impetus theory prediction (Fig. 4.5) more closely than Galileo's parabola. Why then should we prefer Galileo's theory for everyday applications in the earth's atmosphere?

One reason is that even without constructing a vacuum, one can find projectiles of high density, low cross-sectional area, and relatively low speed that follow Galileo's prediction very well. A shot-put follows a much more nearly parabolic orbit than a balloon does.

A second reason concerns the distinction between technology and science. For seventeenth-century technology, unable as it was to compute the effects of air resistance, Galileo's prediction offered no immediate advance in the practice of gunnery. But for science, Galileo's ideas ushered in modern thought and opened the way to Newton's great discoveries.

CHAPTER 5

VECTORS

If I wished to attract the student of any of these sciences to an algebra for vectors, I should tell him that the fundamental notions of this algebra were exactly those with which he was daily conversant . . . In fact, I should tell him that the notions which we use in vector analysis are those which he who reads between the lines will meet on every page of the great masters of analysis, or of those who have probed the deepest secrets of nature.

J. W. Gibbs, in *Nature*, 16 March 1893

5.1 THE RISE OF VECTOR ANALYSIS

Throughout their history, mathematics and physics have been intimately related; a discovery in one field led to an improvement in the other. Early natural philosophers grappling with quantities such as distance, speed, and time used geometry inherited from the Greeks to explore physical problems. But in the tumultuous years of the seventeenth century, physics underwent a transformation – a shift in emphasis from *numerical* quantities, such as distance and speed, to *vector* quantities, such as displacement and velocity, which have direction as well as magnitude.

The transition was neither abrupt nor confined to that century. It was necessary to invent new mathematical objects – *vectors* – and new mathematical machinery for manipulating them – *vector algebra* – to embody the properties of the physical quantities they were to represent. By the nineteenth century the development and acceptance of vectors became a seething pot that boiled over into matters of national pride and individual prestige.

William Rowan Hamilton, an eminent Irish mathematician of the nineteenth century, was a general in the camp that sought to preserve the purity of a geometrical description of space and objects in it. He fathered quaternions, a method designed to represent objects with magnitude and direction without reference to the tainted Cartesian coordinate system advocated on the Continent. For 22 years he labored, nurtured, and fought for his system. Hamilton believed that quaternions held the key to the mathematics of the physical universe. They did not. Instead they exist today as a mathematical curiosity, a harsh reminder that not all paths lead to deeper understanding of nature.

In Germany, working independently and at the same time, an obscure teacher of mathematics, Hermann Grassmann, developed an ingenious method, an algebra, for quantities that have more than one component. With physical applications in mind, he forged mathematical tools to generalize ideas like addition and multiplication. His work was largely ignored (most copies of his first book ended up as waste paper). Abstract, intricate explanations dotted with philosophical ponderings made his book unreadable. Moreover, unlike Hamilton, he had no long list of titles, no reputation to back him. His work could have ushered in vector analysis, but it did not.

Vector analysis emerged in the latter part of the nineteenth century primarily through the efforts of Josiah Willard Gibbs, a Yale physicist, and Oliver Heaviside, a self-taught scientist. Both of these men were influenced by the writings of James Clerk Maxwell on electricity and magnetism. Gibbs noted that many of the features of Hamilton's quaternion analysis were excess baggage in Maxwell's description of electricity and magnetism. He discarded features that obscured the brevity of expression and simplicity of geometrical relations and replaced them with Cartesian geometry. Heaviside independently developed a similar system of symbols and rules to deal with vector quantities. Through his numerous publications in electrical journals, vector methods became widely accepted. In the end, those same elegant methods that best serve the purpose of electricity and magnetism proved best also for mechanics and other branches of physics.

But the birth of vector analysis was not without labor. In the 1890s a bitter struggle took place between the quaternion and vector camps. Through dozens of articles in journals, at numerous scientific meetings, and in secret letters, a battle raged. The fight was over methods, notation, and, most important, what kind of mathematics could best aid physics. The vectors triumphed. But as often happens in war, the losers were not so much vanquished as subsumed and assimilated. Many of Hamilton's ideas survive in modern vector analysis.

5.2 VECTORS – A GEOMETRIC VIEW

It was Hamilton himself who coined the word *vector* (from the Latin for *carrier*) for a new type of mathematical object that has both *magnitude* and *direction* in space. Geometrically, a vector is represented by an arrow whose length corresponds to the magnitude and whose direction, from the tail to the head of the arrow, corresponds to the direction of the vector.

Although a vector has a magnitude and direction, it has no fixed position in space. You can take a vector, like the one shown in Fig. 5.1, and slide it parallel to itself anywhere, and you still have the same vector. Two vectors having the same magnitude and direction are considered equal. They can be translated so that one fits exactly on top of the other.

Figure 5.1 The same vector shown at different positions in space and in different coordinate systems.

Precisely because vectors don't have a definite position, they exist independent of any coordinate system. For this reason, they are useful in mechanics. Why? Before Copernicus, there was only one conceivable reference frame, centered at the center of the earth. The essence of the Copernican Revolution was to move the origin of the coordinate system describing all physics from the center of the earth to the center of the sun. Then, in devising a new mechanics, Galileo discovered through the law of inertia that there is no preferred reference frame. There is nothing unique about coordinate systems; no one is better than any other. Consequently coordinate systems (reference frames) can be anywhere in space, oriented in any way. Similarly, vectors can be anywhere in space. Vectors are natural devices to describe physical quantities with complete generality and independence from particular reference frames.

Because a vector has a magnitude and a direction, it is more than a single number, and as such it should not be represented by the same kind of symbol that we use to represent a single number. Vectors require a special kind of symbol. Physicists often write a vector as a letter with an arrow over it: $\vec{A}$. Others like to use fancier letters, such as $\mathbb{A}$, while still others use the printer's symbol for boldface, a letter with a squiggle under it, as $\underset{\sim}{A}$. You can use whatever notation you wish, but we'll use Gibbs's notation: boldface letters like **A** will denote vectors.

The magnitude or length of a vector **A** is denoted by $|\mathbf{A}|$ or simply by A (in lightface print):

$$\text{magnitude of vector } \mathbf{A} = |\mathbf{A}| \text{ or } A.$$

To emphasize the difference between vectors and ordinary numbers, Hamilton coined the word *scalar* (from the Latin for *stairs* or *scale*) as a name for ordinary numbers like 2, -5, 144, and π. The magnitude of a vector is also a scalar (but it is never negative). Distance and speed are examples of magnitudes of vectors.

In physics, different names are used to distinguish vector quantities from their scalar magnitudes. For example, the velocity $\mathbf{v}$ of a moving object is a vector quantity represented by an arrow pointing in the direction of the motion. The length of the velocity vector, $v = |\mathbf{v}|$, is called the speed; it tells us how fast the object is moving. Another example is displacement versus distance. If a particle moves a distance s from one point to another, the vector $\mathbf{s}$ joining them is called the displacement; its magnitude $s = |\mathbf{s}|$ (the distance moved) is a scalar. Table 5.1 lists common scalars and vectors in physics.

Because vectors are new objects, algebraic operations involving vectors must be defined. We shall define addition, subtraction, and three kinds of multiplication.

The simplest operation is multiplication of a vector by a scalar. If you have a vector $\mathbf{B}$ and multiply it by the scalar 3, then the new vector is three times longer than $\mathbf{B}$, as shown in Fig. 5.2a, and is written $3\mathbf{B}$. More generally, multiplying $\mathbf{B}$ by a scalar c leads to another vector, written $c\mathbf{B}$, possibly longer or shorter, with the same direction if c is positive and with the opposite direction if c is negative. The magnitude of $c\mathbf{B}$ is $|c|\,|\mathbf{B}|$, the absolute value of c times the magnitude of $\mathbf{B}$.

Division of a vector by a scalar presents no new problem because we can always regard division by a scalar c as multiplication by $1/c$ if $c \neq 0$. An example with $c = 2$ is shown in Fig. 5.2b.

A vector multiplied by the scalar -1 is called the negative of the vector. The result is a vector that has the same magnitude but points in the opposite direction, as in Fig. 5.2c.

Multiplication by scalars has a simple application to linear motion with constant speed. If an object moves with constant speed v in a fixed direction, the vector $\mathbf{v}$ in that direction with magnitude v is the velocity of the object. It is a constant vector because both its magnitude and direction do not change. In time t the object undergoes a displacement $\mathbf{s} = t\mathbf{v}$. Thus, for bodies moving with constant velocity, displacement is the result of multiplying the velocity vector by a scalar; in this example the scalar represents time.

Just as zero is an exception among numbers, the zero vector is also an exception

Table 5.1 Common Scalar and Vector Quantities in Physics

Scalars		Vectors	
distance	s	displacement	$\mathbf{s}$
speed	v	velocity	$\mathbf{v}$
acceleration	a	acceleration	$\mathbf{a}$
force	F	force	$\mathbf{F}$
time	t		
mass	m		

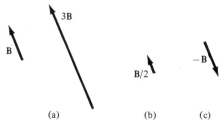

Figure 5.2 (a) Multiplication of the vector **B** by the scalar 3.
(b) Division of the vector **B** by the scalar 2. (c) Multiplication of **B** by
the scalar −1.

among vectors. The zero vector is the result of multiplying any vector by the scalar zero.
It has a magnitude (of zero), but no direction. In vector equations we shall encounter,
the zero vector will simply be written as **0**.

Questions

1. Suppose you wanted to walk from the entrance on the ground floor of a department
store to a point across the building and on the third floor. What is the lowest number of instructions someone could give you that would bring you to that point?

2. Does a car speedometer indicate speed or velocity? If your answer is speed, then
what additional instrument would help you specify your velocity? If your answer is
velocity, explain how.

3. In Table 5.1, time is listed as a scalar, but often we hear about the "arrow of
time," implying that time has a direction, from the past toward the future. Then
shouldn't time be considered a vector?

4. Does the value of a scalar depend on the reference frame in which it is used?

5. For the vector **D** shown below, construct the following vectors:

   (a) $-2\mathbf{D}$ **(b)** $2.5\mathbf{D}$

   (c) $-\frac{1}{4}\mathbf{D}$ **(d)** $\frac{1}{3}\mathbf{D}$.

6. Show that the multiplication of vectors by scalars has the following property:

$$a(b\mathbf{A}) = (ab)\mathbf{A} = b(a\mathbf{A}),$$

where a and b are scalars and **A** is any vector.

In the legendary days of Long John Silver and swashbuckling pirates, when tall ships sailed the high seas, treasure booties were frequently buried on uncharted, out-of-the-way islands. To aid their rum-soaked memories, pirates drew treasure maps, pointing the way to their hidden treasures (Fig. 5.3). These maps usually had a coconut tree or particularly odd-looking rock as a starting point, and consisted of a series of arrows that would steer the lucky map holder to the treasure. Each arrow was labeled with a number of paces to be taken and a compass direction. Unknown to them, pirates were using vectors and vector addition long before Hamilton and Gibbs.

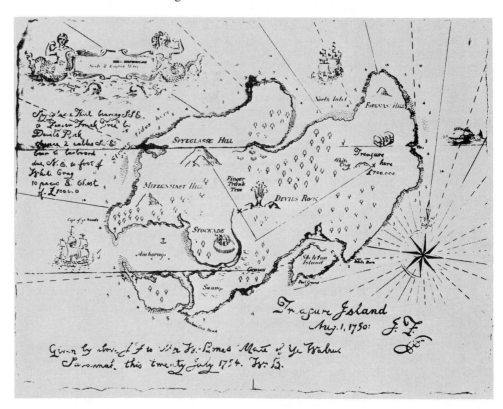

Figure 5.3 Reproduction of a treasure map. (Courtesy Springer/ Bettman Film Archive.)

To define the addition of vectors, let's suppose we have two vectors like those shown in Fig. 5.4a: vector **A** has a length of 3 and a direction of 30° from some reference line (dotted in the figure); vector **B** has a magnitude of 2 at 60°. You can imagine them to be part of a treasure map and wonder, if you followed those two vectors, where would you be from the starting point. That's easy to figure out, because first you would follow **A** and right after that, **B**. If you connect the starting point to where you ended up as in Fig. 5.4b, you have a new vector **C**, which we'll call the sum **A** + **B**; it is "as the crow flies." In vector symbols we write **C** = **A** + **B**. That's *vector addition*.

To add two vectors, begin the second vector where the first one ends. Remember that you can always slide vectors around as long as you keep them pointing in the same

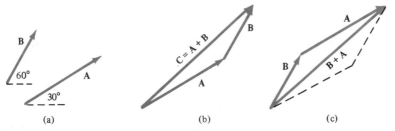

Figure 5.4 (a) Two vectors **A** and **B**. (b) The vector sum **C** = **A** + **B**, found by the tail-to-head method. (c) The vector sum **B** + **A**.

direction, so you can perform this tail-to-head. The *vector sum*, also called the *resultant vector*, points from the beginning of the first vector to the end of the second vector. Once you've constructed the resultant vector, you can use a ruler to measure its length and determine its magnitude. The angle from some reference line can be found easily with a protractor. Then you're done because you have the two things that specify the resultant vector – its magnitude and direction.

Figure 5.4c shows a useful feature of vector addition. In that figure, we started with **B** and added **A** to it; that is, we found the sum **B** + **A**. It turns out to be the same vector **C**. This means that the order in which you add two vectors is unimportant because you'll always get the same result. This property of vector addition is called commutativity.

Commutativity of vector addition:

A + **B** = **B** + **A**.

Note that the sum **A** + **B** is a diagonal of the parallelogram determined by **A** and **B** and shown in Fig. 5.4c. For this reason, vector addition is sometimes called *addition by the parallelogram law*.

The graphical technique of adding vectors can be extended to determine the sum of more than two vectors. For example, suppose you want the sum of three vectors **A**, **B**, and **C**. You can take **A** + **B** and add it to **C**, as shown in Fig. 5.5a, or you can add **A** to **B** + **C**. The two results are the same. This property is known as associativity.

Associativity of vector addition:

A + (**B** + **C**) = (**A** + **B**) + **C**.

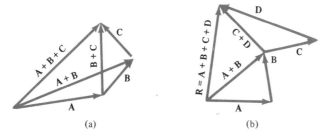

Figure 5.5 (a) Associative law of vector addition. (b) Addition of several vectors.

Because of this, we can write the sum as **A** + **B** + **C** without indicating in which way they are grouped together.

In the same way, if you want the sum of four vectors **A**, **B**, **C**, **D**, just draw them one after another, tail-to-head, in any order until you have all four vectors drawn. The sum or resultant **R** = **A** + **B** + **C** + **D** points from the tail of the first vector **A** to the head of the last vector **D**, as shown in Fig. 5.5b.

The graphical method of vector addition holds not only for vectors that lie in a plane, but also for vectors in space. Figure 5.6 illustrates the addition of three vectors in space. The vectors **A**, **B**, **C** lie along the edges of a room. The vector sum **A** + **B** + **C** is a vector from one corner on the floor to the diagonally opposite corner on the ceiling.

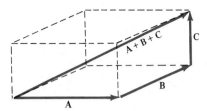

Figure 5.6 Vector addition in space.

Example 1
A ship sails 60 miles in a direction 30° north of east, then 30 miles due east, then 40 miles 30° west of north. Where is the ship with respect to its starting point?

Let's label our three vectors as **A** (60 mi at 30° N of E), **B** (30 mi due E), and **C** (40 mi at 30° W of N). Using a scale of 1 cm = 22.5 mi, we can construct the vector sum **R** = **A** + **B** + **C**, by adding the vectors head-to-tail:

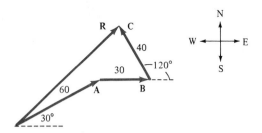

By connecting the starting point to the final point of our construction we obtain the resultant vector **R**. Measuring the length of **R** we find it to be approximately 4 cm, which corresponds to 90 mi. The direction from the starting point (as measured with a protractor) is approximately 46° N of E.

We can also calculate the length of **R** and its direction more accurately using simple trigonometry. By referring to the dotted lines in the following figure, we see that the vector **R** is the hypotenuse of a right triangle whose base and altitude are easily calculated.

The base has length $60(\sqrt{3}/2) + 10$ mi and the altitude is $40(\sqrt{3}/2) + 30$ mi. There-fore, by the Pythagorean theorem we have

$$R^2 = [60(\sqrt{3}/2) + 10]^2 + [40(\sqrt{3}/2) + 30]^2,$$

which implies that $R = 89.5$ mi. The tangent of the angle θ is

$$\tan \theta = (2\sqrt{3} + 3)/(3\sqrt{3} + 1) = 1.043,$$

so $\theta = \tan^{-1}(1.043) = 46.2°$.

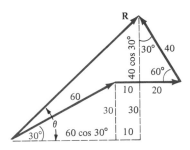

Once we know how to add vectors, subtraction is straightforward because we can always treat subtraction as the addition of a negative vector. As we saw earlier the negative of a vector is a vector with the same magnitude but opposite direction. So if you have two vectors, say **G** and **H**, and you want **G** − **H**, you first form −**H**, then add it tail-to-head to **G**. The vector from the tail of **G** to the head of −**H** is **G** − **H**, as shown in Fig. 5.7b. In summary,

G − **H** = **G** + (−**H**).

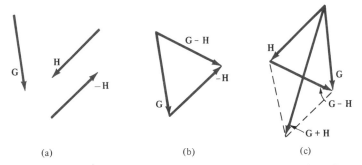

(a)   (b)   (c)

Figure 5.7 (a) Vectors **G**, **H**, and −**H**. (b) Graphical subtraction of vectors to form **G** − **H**. (c) The vector sum **G** + **H** and difference **G** − **H** shown as diagonals of the same parallelogram.

We saw earlier that the sum of two vectors **G** and **H** represents one diagonal of the parallelogram determined by **G** and **H**. The difference **G** − **H** represents the other diagonal from the tip of **H** to the tip of **G**, as shown in Fig. 5.7c. The difference in the other order, **H** − **G**, is, of course, just the negative of this, the diagonal from the tip of **G** to the tip of **H**.

A swimmer in a stream and an airplane in a wind are good physical illustrations of vector addition and subtraction. The swimmer, for example, makes a certain progress with respect to the water he's in, but he's also swept along in some other direction by the current. What is his actual velocity as seen by someone standing on the bank of the stream? It's not hard to see that it is simply the vector sum of the swimmer's velocity in still water plus the velocity of the stream itself. The reason is that each velocity vector is equal to the corresponding displacement vector during one unit of time, and since displacements add vectorially, the same is true for velocities.

Example 2

An excellent swimmer who can swim at 1.7 m/s in still water tries to swim straight across a river. The current in the river causes the swimmer to move instead at an angle 32° from his intended direction with a speed of 2.0 m/s. What is the speed of the river current?

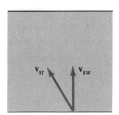

Let's call the swimmer's velocity in still water $\mathbf{v}_{sw}$, the velocity of the river water $\mathbf{v}_{rw}$, and the swimmer's velocity across the river $\mathbf{v}_{sr}$. These velocity vectors are connected by the vector equation

$$\mathbf{v}_{sr} = \mathbf{v}_{sw} + \mathbf{v}_{rw}.$$

From this relation, we can solve for the velocity of the river current,

$$\mathbf{v}_{rw} = \mathbf{v}_{sr} - \mathbf{v}_{sw}.$$

Using a scale of 1 cm = 0.85 m/s, we construct the vector subtraction by the tail-to-head method.

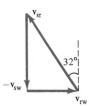

Measuring the length of the vector $\mathbf{v}_{rw}$, we find it to be 1.4 cm, which corresponds to a speed of (1.4)(0.85 m/s) or 1.2 m/s. Therefore, the river current has a velocity of 1.2 m/s downriver.

Example 3

A weekend pilot tries to fly a plane due north with a speed of 120 km/h, but there is a strong wind blowing from the east at 80 km/h. What is the resultant velocity of the plane?

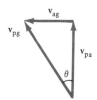

Let's label the velocity of the **plane** with respect to the **air** $\mathbf{v}_{pa}$ and the velocity of the **air** with respect to the **ground** $\mathbf{v}_{ag}$. The velocity of the **plane** with respect to the **ground** $\mathbf{v}_{pg}$ is the vector sum of the other two velocities:

$$\mathbf{v}_{pg} = \mathbf{v}_{pa} + \mathbf{v}_{ag}.$$

Adding the vectors tail-to-head, we have the construction shown. Since here the vectors form a right triangle, let's use the Pythagorean theorem to calculate the magnitude of the resultant vector $\mathbf{v}_{pg}$ (the hypotenuse):

$$v_{pg} = \sqrt{v_{pa}^2 + v_{ag}^2} = \sqrt{(120 \text{ km/h})^2 + (80 \text{ km/h})^2},$$

$$v_{pg} = 144 \text{ km/h}.$$

That's the ground speed – how fast the plane is moving according to someone on the ground. To specify its velocity, we need to indicate its direction. Using the angle θ indicated above, we can use the definition of tangent (or sine or cosine, if you prefer):

$$\tan \theta = \frac{v_{ag}}{v_{pa}} = \frac{80}{120} = 0.67.$$

Consulting a trigonometric table, or using a calculator with an arctangent key, we find that 34° is the angle whose tangent is 0.67. The plane's velocity is 144 km/h at 34° west of north. Note that because the two vectors were perpendicular here, we could figure out everything easily.

Questions

7. What must be true for two vectors to add up to zero?

8. If three vectors add up to zero, what geometric shape do they form when added tail-to-head?

9. Must three vectors be coplanar (lie in the same plane) in order to add up to zero?

10. Can you think of a way in which a person can walk some positive distance and yet have zero displacement?

11. Using the three vectors shown below, construct the following and state the magnitude of the resultant:

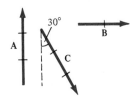

    (a) $C + 2A$ **(b)** $B + \frac{1}{2}A$

    (c) $A - C + B$ **(d)** $C - A - B$.

12. For the vectors given in the preceding problem, construct the following:

    (a) $B + C$ **(b)** $C + B$

    (c) $\frac{1}{2}B - C$ **(d)** $-C + \frac{1}{2}B$.

13. Is vector subtraction commutative? Why or why not?

14. For the vectors given in Question 11, demonstrate associativity of vector addition by constructing $(A + B) + C$ and also $A + (B + C)$.

15. A pilot originally started to fly due north at an air speed of 250 km/h, but a strong wind out of the east results in the plane traveling at 289 km/h 30° west of north. What is the velocity of the wind?

16. If vectors **A** and **B** are consecutive sides of a regular hexagon, determine (in terms of **A** and **B**) vectors forming the other four sides.

17. Given a fixed vector **A**, let **B** be a vector of length 1 that changes its direction. What are the maximum and minimum values of $|A + B|$? Show that, in general, $|A + B| \neq |A| + |B|$.

18. Draw five vectors from the center of a regular pentagon to each of the vertices. Show that the sum of these vectors is the zero vector. Generalize this to a regular polygon of any number of sides.

Thus far we have defined the addition and subtraction of vectors and multiplication of a vector by a scalar. You might wonder if two vectors can be multiplied and divided, and if so, just how this is performed. Hamilton struggled with the problem of multiplication of his quaternions. In the month following his 1843 discovery of quaternions, Hamilton's son would ask him daily at breakfast if he could multiply quaternions. To which Hamilton was always obliged to reply with a sad shake of his head that no, he could only add and subtract them. But weeks later, Hamilton devised two schemes of multiplication, one of which produced a scalar result, the other a vector. We'll examine the scalar product here, deferring the vector product to Section 5.4.

The scalar product is also called the dot product, written **A·B**, with a dot between the vectors. The dot product of **A** and **B** is defined as the scalar

$$\mathbf{A} \cdot \mathbf{B} = AB \cos \theta, \tag{5.1}$$

where θ is the angle between the two vectors (measured so that $0° \leq \theta \leq 180°$). In other words, to compute the scalar product of two vectors, multiply their magnitudes and the cosine of the angle between them.

The scalar quantity

$$b = B \cos \theta$$

is called the component of **B** along **A** and is shown in Fig. 5.8a. The dot product

$$\mathbf{A} \cdot \mathbf{B} = AB \cos \theta = Ab$$

is obtained by multiplying the magnitude A with the component of **B** along **A**. It is also equal to the product of B and the component of **A** along **B**, as shown in Fig. 5.8b. In other words, the scalar product is commutative.

Commutativity of the scalar product:

$$\mathbf{A} \cdot \mathbf{B} = \mathbf{B} \cdot \mathbf{A}.$$

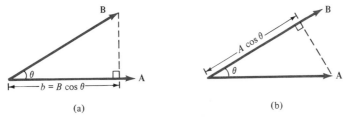

(a) (b)

Figure 5.8 (a) $B \cos \theta$ is the component of **B** along **A**.
(b) $A \cos \theta$ is the component of **A** along **B**.

Example 4

Find the scalar product of the vectors **A** and **B** below. The magnitude of *A* is 10 and the magnitude of *B* is 12.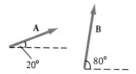

Our first step is to move one of the vectors so that they start from the same point. Of course it doesn't matter which one we move because vectors have no fixed position in space. Once the vectors are drawn from the same point, we can figure out the angle between them. Since one vector makes an angle of 20° with the reference line, and the other vector makes an angle of 80°, the angle between them is 80° − 20° = 60°.

Now we have all the information necessary to compute the scalar product using Eq. (5.1):

$$\mathbf{A \cdot B} = AB \cos \theta, \tag{5.1}$$

$$\mathbf{A \cdot B} = (10)(12) \cos 60°,$$

$$\mathbf{A \cdot B} = 60.$$

Note that the answer is just a number – a scalar.

If **A** and **B** are perpendicular, cos θ is zero, and

$$\mathbf{A \cdot B} = 0.$$

Conversely, if the scalar product is zero, at least one of the vectors is zero or else the two are perpendicular.

If **A** and **B** have the same direction, cos θ = 1 and **A·B** is simply the product of the magnitudes of the two vectors. In particular,

$$\mathbf{A \cdot A} = A^2,$$

the square of the magnitude of **A**.

In Fig. 5.9, *b* and *c* are the components of **B** and **C** along **A**. It is clear that *b* + *c* is the component of **B** + **C** along **A**. But

$$A(b + c) = Ab + Ac,$$

which implies that the dot product is distributive.

Distributivity of the dot product:

$$\mathbf{A}\cdot(\mathbf{B} + \mathbf{C}) = \mathbf{A}\cdot\mathbf{B} + \mathbf{A}\cdot\mathbf{C}.$$

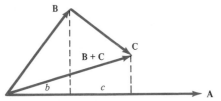

Figure 5.9 Illustration of the distributive law $\mathbf{A}\cdot(\mathbf{B} + \mathbf{C}) = \mathbf{A}\cdot\mathbf{B} + \mathbf{A}\cdot\mathbf{C}$.

Questions

19. Can the scalar product have physical units?

20. Can the scalar product of two vectors be negative? If so, what condition must hold?

21. If $\mathbf{A}\cdot\mathbf{B} = \mathbf{A}\cdot\mathbf{C}$ does it follow that $\mathbf{B} = \mathbf{C}$?

22. Using the vectors shown below, which have magnitudes $A = 3$, $B = 5$, and $C = 6$, calculate the following:

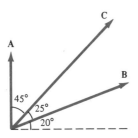

  (a) $\mathbf{A}\cdot\mathbf{B}$
  (b) $\mathbf{A}\cdot\mathbf{C}$
  (c) $(\mathbf{A} + \mathbf{B})\cdot\mathbf{C}$
  (d) Compare $\mathbf{A}\cdot\mathbf{C} + \mathbf{B}\cdot\mathbf{C}$ with $(\mathbf{A} + \mathbf{B})\cdot\mathbf{C}$ and verify the distributive law.

23. If the scalar product of two vectors is equal to the negative of the product of their magnitudes, what is true about their relative directions?

24. If vector $\mathbf{C}$ is perpendicular to each of $\mathbf{A}$ and $\mathbf{B}$, use the dot product to show that $\mathbf{C}$ is also perpendicular to $a\mathbf{A} + b\mathbf{B}$ for all choices of scalars a and b.

5.3 VECTORS: AN ANALYTIC VIEW

In 1637 the French mathematician and philosopher René Descartes introduced a new branch of mathematics known as analytic geometry. His idea was to represent geometric points by numbers. In a plane, like this page, two numbers on a rectangular grid define a point. In space, three numbers are needed to define a point, as shown in Fig. 5.10b. Descartes himself never realized how convenient such rectangular grids would be, but they are nevertheless called Cartesian coordinate systems.

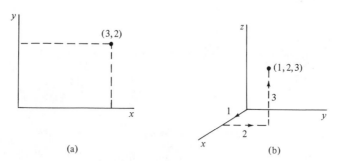

Figure 5.10 Cartesian coordinates in (a) two and (b) three dimensions.

A Cartesian coordinate system in three dimensions starts with three mutually perpendicular axes labeled x, y, and z, like those shown in Fig. 5.10b. Such a coordinate system is called right-handed because of the way the axes are arranged: if you point your right hand in the direction of the positive x axis and curl your fingers toward the positive y axis, your thumb will point in the direction of the positive z axis.

Once a Cartesian coordinate system is introduced, vectors can be described by three numbers, called *components*. They provide instructions telling us how to get from the tail to the head of the arrow. Suppose you have a vector **A** and you want to know its components in a given Cartesian coordinate system. First, you slide the vector so its tail is at the origin. Then the tip will have three rectangular coordinates, which we call the components of **A** and which we denote by (A_x, A_y, A_z). You can think of reaching the head of the arrow by first walking from the origin along the x axis by an amount A_x, then along the y direction by an amount A_y, and last in the z direction an amount A_z, as illustrated in Fig. 5.11.

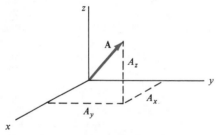

Figure 5.11 Cartesian components (A_x, A_y, A_z) of the vector **A**.

Vectors can be related algebraically to their components with the help of unit vectors. A unit vector is a vector whose length is equal to one – not 1 cm, or 1 m – just 1; it is dimensionless. Following Hamilton's notation, we use the symbol $\hat{\mathbf{i}}$ to represent the unit vector along the positive x axis. A hat (ˆ) over a boldface letter is used to remind us that the vector under the hat is a unit vector. Like a compass needle, $\hat{\mathbf{i}}$ points only in the positive x direction. Similarly, the unit vector $\hat{\mathbf{j}}$ points in the positive y direction, and $\hat{\mathbf{k}}$ points in the positive z direction. Just like the coordinate axes, the unit vectors $\hat{\mathbf{i}}$, $\hat{\mathbf{j}}$, $\hat{\mathbf{k}}$ are mutually perpendicular. Expressed in terms of dot products, this means that

$$\hat{\mathbf{i}} \cdot \hat{\mathbf{j}} = \hat{\mathbf{j}} \cdot \hat{\mathbf{k}} = \hat{\mathbf{k}} \cdot \hat{\mathbf{i}} = 0.$$

The three unit vectors $\hat{\mathbf{i}}$, $\hat{\mathbf{j}}$, $\hat{\mathbf{k}}$ are shown in Fig. 5.12.

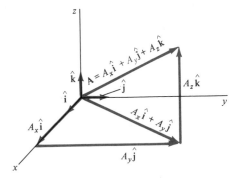

Figure 5.12 A vector **A** expressed in terms of its components.

Using these unit vectors we can express any vector **A** algebraically in terms of its components. The relation, illustrated in Fig. 5.12, is

$$\mathbf{A} = A_x\hat{\mathbf{i}} + A_y\hat{\mathbf{j}} + A_z\hat{\mathbf{k}}.$$

You can think of the displacement from the origin to the tip of **A** as a vector sum of three displacements: $A_x\hat{\mathbf{i}}$ followed by $A_y\hat{\mathbf{j}}$ followed by $A_z\hat{\mathbf{k}}$.

The equation $\mathbf{A} = A_x\hat{\mathbf{i}} + A_y\hat{\mathbf{j}} + A_z\hat{\mathbf{k}}$ also gives us an algebraic way to determine the components. Just take the dot product of each side of this equation with $\hat{\mathbf{i}}$ to get

$$\mathbf{A} \cdot \hat{\mathbf{i}} = A_x\hat{\mathbf{i}} \cdot \hat{\mathbf{i}} + A_y\hat{\mathbf{j}} \cdot \hat{\mathbf{i}} + A_z\hat{\mathbf{k}} \cdot \hat{\mathbf{i}} = A_x,$$

since $\hat{\mathbf{i}} \cdot \hat{\mathbf{i}} = 1$ and $\hat{\mathbf{j}} \cdot \hat{\mathbf{i}} = \hat{\mathbf{k}} \cdot \hat{\mathbf{i}} = 0$. Similarly we find

$$\mathbf{A} \cdot \hat{\mathbf{j}} = A_y \quad \text{and} \quad \mathbf{A} \cdot \hat{\mathbf{k}} = A_z.$$

In other words, the components of **A** are just the dot products of **A** with the unit coordinate vectors. From this it follows that if two vectors are equal then their corresponding components are equal: If $\mathbf{A} = \mathbf{B}$, then $\mathbf{A} \cdot \hat{\mathbf{i}} = \mathbf{B} \cdot \hat{\mathbf{i}}$, $\mathbf{A} \cdot \hat{\mathbf{j}} = \mathbf{B} \cdot \hat{\mathbf{j}}$, and $\mathbf{A} \cdot \hat{\mathbf{k}} = \mathbf{B} \cdot \hat{\mathbf{k}}$, which imply

$$A_x = B_x,$$

$$A_y = B_y,$$

$$A_z = B_z.$$

The simple-looking vector equation $\mathbf{A} = \mathbf{B}$ is shorthand for three scalar equations. This is one of the powerful advantages of vector algebra.

Of course, if the vectors $\mathbf{A}$ and $\mathbf{B}$ are in the xy plane, then the z component of each is zero, and the vector equation $\mathbf{A} = \mathbf{B}$ is equivalent to *two* scalar equations: $A_x = B_x$ and $A_y = B_y$.

Example 5

Construct the vector $\mathbf{B} = 2\hat{\mathbf{i}} - 3\hat{\mathbf{j}}$.

Using the coordinate system shown, we can start at the origin and move $+2$ units along the x direction, which is the vector $2\hat{\mathbf{i}}$. From there we move along the negative y direction 3 units, which is the vector $-3\hat{\mathbf{j}}$. The sum of the two vectors is $\mathbf{B}$.

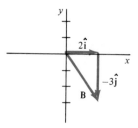

Example 6

What is the negative of the vector $\mathbf{B}$ given in the previous example?

Remembering that we form the negative of a vector by multiplying it by the scalar -1 and that multiplication is distributive, we have

$$-\mathbf{B} = -1(2\hat{\mathbf{i}} - 3\hat{\mathbf{j}}) = -2\hat{\mathbf{i}} - (-3)\hat{\mathbf{j}}$$

$$= -2\hat{\mathbf{i}} + 3\hat{\mathbf{j}}.$$

We can check our result by constructing the vector $-\mathbf{B}$ and seeing if it is in the opposite direction. The result is shown below.

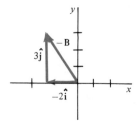

In Chapter 4, we discovered that the trajectory of a projectile is a parabola. In its horizontal motion, following the law of inertia, a cannonball moves with constant speed, given it by the explosion of gunpowder. At the same time and quite independently, in the vertical direction, the ball falls with constant acceleration, obeying the law of falling bodies. What we did when we analyzed this motion was to break the motion down into components.

In Fig. 5.13 the horizontal component of the ball's displacement $\mathbf{s}$ is $s_x = v_x t$, whereas the vertical component (taking the positive y axis to be pointed downward) is, by the law of falling bodies, $s_y = \frac{1}{2}gt^2$. The vector displacement can be written

$$\mathbf{s}(t) = v_x t\,\hat{\mathbf{i}} + \tfrac{1}{2}gt^2\,\hat{\mathbf{j}}.$$

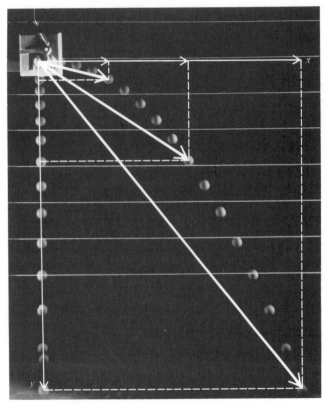

Figure 5.13 Displacement vector of a horizontally projected ball. (*PSSC Physics*, 2nd ed., 1965; D.C. Heath and Company with Education Development Center, Inc., Newton, Mass.)

The arrows in the figure shows this vector at different times.

Since there are two equivalent ways to specify a vector – either by a magnitude and a direction, or by components – there should exist some way of finding components when the magnitude and direction are known, and vice versa. This process of finding components is known as resolving a vector into its components. Many of the applications of vector analysis will involve the resolution of vectors in a plane into their components; therefore, let's examine just how to resolve vectors in a plane (Fig. 5.14).

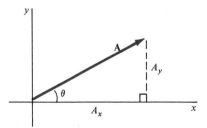

Figure 5.14 Resolution of a vector into its components in a plane.

If we know the magnitude and direction of a vector **A**, we can easily find its components in a plane. From trigonometry (check Appendix C if you're rusty), the components A_x and A_y are related to the magnitude A and angle θ as follows:

$$\sin \theta = \frac{A_y}{A},$$

so the y component can be calculated from

$$A_y = A \sin \theta. \tag{5.2}$$

Similarly, by definition of the cosine,

$$\cos \theta = \frac{A_x}{A},$$

so the x component is

$$A_x = A \cos \theta. \tag{5.3}$$

Equations (5.2) and (5.3) give the prescription for finding the components of a vector when the magnitude and direction are known.

Example 7

On its yearly journey southward, a tern flies at 15 m/s in a direction 57° south of east. What are the components of its velocity along north–south and east–west axes?

Let's set up a coordinate system with the positive x axis to the east and the positive y axis along north. Then for our vector **v** we want to find the components v_x and v_y. Drawing our vector **v** on the axes, we see that the angle specified is measured from the x axis. By constructing a right triangle with sides parallel to the x axis and to the y axis we see the following:

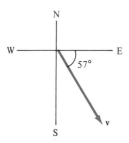

$$v_x = (15 \text{ m/s})\cos 57° = 9 \text{ m/s},$$

$$v_y = -(15 \text{ m/s})\sin 57° = -12 \text{ m/s}.$$

Note that the y component is negative because **v** points downward.

In some cases you might know the components of a vector but want its magnitude and direction. The change from components to magnitude and direction is made by reversing the process we've just described. Suppose you know the components A_x and A_y of a vector as shown in Fig. 5.14. If you draw $\mathbf{A} = A_x\hat{\mathbf{i}} + A_y\hat{\mathbf{j}}$, you automatically have a right triangle. By the Pythagorean theorem, the hypotenuse, which is the magnitude A, is

$$A = \sqrt{A_x^2 + A_y^2}. \tag{5.4}$$

The angle θ as defined in Fig. 5.14 can be calculated in many ways; the easiest, if you're given A_x and A_y, is to use the definition of tangent:

$$\tan \theta = \frac{A_y}{A_x},$$

so that

$$\theta = \tan^{-1}\left(\frac{A_y}{A_x}\right). \tag{5.5}$$

(Most calculators have an inverse tangent key, $\tan^{-1}$, so when you know the components you can easily calculate the angle. Even without a calculator, it's not hard to use trigonometric tables to find an angle when you know its tangent.)

If you've calculated A, then since $A_x = A \cos \theta$, we have $\cos \theta = A_x/A$ and $\sin \theta = A_y/A$.

Example 8

Find the magnitude and direction of the vector $\mathbf{C} = 5\hat{\mathbf{i}} + 12\hat{\mathbf{j}}$.

By the Pythagorean theorem, the magnitude C is

$$C = \sqrt{5^2 + 12^2} = \sqrt{169} = 13.$$

If we measure the angle θ from the x axis, then

$$\theta = \tan^{-1}\frac{12}{5} = 67°.$$

It is usually helpful to draw the vector so that you can see the easiest angle for indicating the direction.

All the algebraic operations on vectors introduced earlier can be expressed entirely in terms of components. For example, if

$$\mathbf{A} = A_x\hat{\mathbf{i}} + A_y\hat{\mathbf{j}} + A_z\hat{\mathbf{k}},$$

and if c is any scalar, then, multiplying both sides of this equation by c, we find

$$c\mathbf{A} = cA_x\hat{\mathbf{i}} + cA_y\hat{\mathbf{j}} + cA_z\hat{\mathbf{k}},$$

so the components of $c\mathbf{A}$ are cA_x, cA_y, cA_z. The components of $c\mathbf{A}$ are obtained by multiplying each of the components of $\mathbf{A}$ by the scalar c.

If we have a second vector $\mathbf{B} = B_x\hat{\mathbf{i}} + B_y\hat{\mathbf{j}} + B_z\hat{\mathbf{k}}$ and add it to $\mathbf{A}$ we find

$$\mathbf{A} + \mathbf{B} = (A_x + B_x)\hat{\mathbf{i}} + (A_y + B_y)\hat{\mathbf{j}} + (A_z + B_z)\hat{\mathbf{k}}.$$

Hence the components of the sum $\mathbf{C} = \mathbf{A} + \mathbf{B}$ are obtained by adding corresponding components of $\mathbf{A}$ and $\mathbf{B}$:

$$C_x = A_x + B_x,$$
$$C_y = A_y + B_y,$$
$$C_z = A_z + B_z.$$

Figure 5.15 illustrates this addition for two vectors in the xy plane.

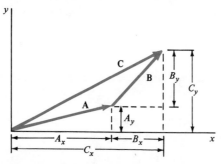

Figure 5.15 Vector addition corresponds to addition of components.

Example 9

Calculate the sum of the following vectors:

$$\mathbf{A} = 3\hat{\mathbf{i}} - 6\hat{\mathbf{j}}, \quad \mathbf{B} = -5\hat{\mathbf{i}} + 9\hat{\mathbf{j}}, \quad \mathbf{C} = 4\hat{\mathbf{i}} - 10\hat{\mathbf{j}}, \quad \mathbf{D} = -\hat{\mathbf{i}} + 9\hat{\mathbf{j}}.$$

Using $R_x = A_x + B_x + C_x + D_x$, we obtain

$$R_x = 3 + -5 + 4 + -1 = 1.$$

The y component of the resultant is $R_y = A_y + B_y + C_y + D_y$,

$$R_y = -6 + 9 + -10 + 9 = 2.$$

Hence, $\mathbf{R} = 1\hat{\mathbf{i}} + 2\hat{\mathbf{j}}$. By the Pythagorean theorem, the magnitude of $\mathbf{R}$ is

$$R = \sqrt{R_x^2 + R_y^2} = 2.2,$$

and by trigonometry,

$$\theta = \tan^{-1}(R_y/R_x) = 63°.$$

Questions

25. Determine the components A_x and A_y if $\mathbf{A}$ is

(a) a velocity of 30 m/s along the x direction.
(b) a displacement of 15 m at an angle of 120° from the x axis.
(c) an acceleration of 20 m/s² directed 90° from the x axis.

26. Find the magnitude and direction of each of the following vectors:

(a) $\mathbf{s} = (10\text{ m})\hat{\mathbf{i}} + (30\text{ m})\hat{\mathbf{j}}$
(b) $\mathbf{B} = -3\hat{\mathbf{i}} - 4\hat{\mathbf{j}}$
(c) $\mathbf{v} = (8\text{ m/s})\hat{\mathbf{i}} - (6\text{ m/s})\hat{\mathbf{j}}.$

27. Using your combined knowledge of the law of falling bodies and the law of inertia, write the velocity vector of the projected ball in Fig. 5.13. What is the acceleration vector?

28. If the sum of two vectors is zero, what must be true about their components?

29. Using the vectors $\mathbf{A} = 7\hat{\mathbf{i}} - 2\hat{\mathbf{j}} + 4\hat{\mathbf{k}}$, $\mathbf{B} = -9\hat{\mathbf{i}} + 3\hat{\mathbf{j}}$, $\mathbf{C} = 2\hat{\mathbf{i}} + 6\hat{\mathbf{j}} - \hat{\mathbf{k}}$, determine the following:

(a) $\mathbf{A} + \mathbf{B} + \mathbf{C}$
(c) $\mathbf{B} - 2\mathbf{A} - \mathbf{C}$
(b) $3\mathbf{C} + \mathbf{B} - \mathbf{A}$
(d) $\frac{1}{3}\mathbf{B} + \mathbf{A}.$

30. If $\mathbf{A} = (8\text{ m})\hat{\mathbf{i}} + (3\text{ m})\hat{\mathbf{j}}$ and $\mathbf{B} = (-5\text{ m})\hat{\mathbf{i}} + (1\text{ m})\hat{\mathbf{j}}$, find the magnitude and direction of

(a) $\mathbf{A} + \mathbf{B}$
(b) $\mathbf{B} - 2\mathbf{A}.$

31. A lost dog runs 20 m due north, then 10 m 60° south of east across a field, then 30 m along a road that runs northwest.

  (a) Choose an appropriate set of coordinate axes and express the dog's three displacements in terms of components.
  (b) What is the dog's resultant displacement?
  (c) How far is the dog from where it started?

32. Rework Example 1 (p. 88) using components.

33. Three vectors **A**, **B**, **C** of magnitudes 1, 2, and 3, respectively, lie along the diagonals of the faces of a cube that meet at a corner.

  (a) Choose a coordinate system with the origin at the corner and express each vector in terms of $\hat{\mathbf{i}}$, $\hat{\mathbf{j}}$, $\hat{\mathbf{k}}$.
  (b) Find the components and the magnitude of the resultant vector **A** + **B** + **C**.

While working on a theory of flows and tides, Grassmann realized that components not only have merit in the addition and subtraction of vectors but in all vector operations. He sought (and found) a way to determine the scalar product of two vectors from components. How to accomplish this feat is no mystery; it follows directly from the definition of the dot product.

For the two vectors

$$\mathbf{A} = A_x\hat{\mathbf{i}} + A_y\hat{\mathbf{j}} + A_z\hat{\mathbf{k}}$$

and

$$\mathbf{B} = B_x\hat{\mathbf{i}} + B_y\hat{\mathbf{j}} + B_z\hat{\mathbf{k}},$$

the dot product **A·B** is given by

$$\mathbf{A}\cdot\mathbf{B} = (A_x\hat{\mathbf{i}} + A_y\hat{\mathbf{j}} + A_z\hat{\mathbf{k}}) \cdot (B_x\hat{\mathbf{i}} + B_y\hat{\mathbf{j}} + B_z\hat{\mathbf{k}}).$$

Next we multiply out the expression on the right, using the distributive law to get

$$\mathbf{A}\cdot\mathbf{B} = A_xB_x\hat{\mathbf{i}}\cdot\hat{\mathbf{i}} + A_xB_y\hat{\mathbf{i}}\cdot\hat{\mathbf{j}} + A_xB_z\hat{\mathbf{i}}\cdot\hat{\mathbf{k}} + A_yB_x\hat{\mathbf{j}}\cdot\hat{\mathbf{i}} + A_yB_y\hat{\mathbf{j}}\cdot\hat{\mathbf{j}}$$

$$+ A_yB_z\hat{\mathbf{j}}\cdot\hat{\mathbf{k}} + A_zB_x\hat{\mathbf{k}}\cdot\hat{\mathbf{i}} + A_zB_y\hat{\mathbf{k}}\cdot\hat{\mathbf{j}} + A_zB_z\hat{\mathbf{k}}\cdot\hat{\mathbf{k}}.$$

This isn't as bad as it looks, because all the terms involving the dot product of a unit vector with a different unit vector are zero. Only the three terms with $\hat{\mathbf{i}}\cdot\hat{\mathbf{i}}$, $\hat{\mathbf{j}}\cdot\hat{\mathbf{j}}$, and $\hat{\mathbf{k}}\cdot\hat{\mathbf{k}}$ survive. So we're left with

$$\mathbf{A}\cdot\mathbf{B} = A_xB_x + A_yB_y + A_zB_z. \qquad (5.6)$$

In other words, to calculate the dot product we simply add the products of corresponding components.

Example 10

Calculate the dot product of $\mathbf{A} = 6\hat{\mathbf{i}} + 4\hat{\mathbf{j}} - 5\hat{\mathbf{k}}$ and $\mathbf{B} = \hat{\mathbf{i}} - 2\hat{\mathbf{j}} + \hat{\mathbf{k}}$.

Using Eq. (5.6) we can easily figure out that

$$\mathbf{A} \cdot \mathbf{B} = (6)(1) + (4)(-2) + (-5)(1) = -7.$$

Example 11

What is the angle between the vectors $\mathbf{A} = 3\hat{\mathbf{i}} + 2\hat{\mathbf{j}}$ and $\mathbf{B} = \hat{\mathbf{i}} - \hat{\mathbf{j}}$?

At first glance, it may appear that we need to draw a graph of the two vectors and then measure the angle between them. But we don't. Instead we can combine our two expressions for the dot product, Eqs. (5.1) and (5.6). By Eq. (5.6) we know that

$$\mathbf{A} \cdot \mathbf{B} = (3)(1) + (2)(-1) = 1.$$

But Eq. (5.1) relates the dot product to the angle between the vectors:

$$\mathbf{A} \cdot \mathbf{B} = AB \cos \theta.$$

So we can calculate the angle θ from

$$\cos \theta = \frac{\mathbf{A} \cdot \mathbf{B}}{AB}.$$

The magnitudes A and B can be found easily by the Pythagorean theorem:

$$A = \sqrt{3^2 + 2^2} = \sqrt{13},$$
$$B = \sqrt{1^2 + (-1)^2} = \sqrt{2}.$$

Putting everything together, we get

$$\cos \theta = \frac{1}{\sqrt{13}\,\sqrt{2}} = 0.20,$$

which tells us that $\theta = 79°$.

When we take the dot product of a vector with itself, we find through Eq. (5.6)

$$\mathbf{A} \cdot \mathbf{A} = A_x^2 + A_y^2 + A_z^2.$$

This says that the square of the length of $\mathbf{A}$ is the sum of the squares of its components. For vectors in the xy plane, this is just the Pythagorean theorem, so we've found an extension of the Pythagorean theorem to three-dimensional space.

Questions

34. Find $\mathbf{A} \cdot \mathbf{B}$ if $\mathbf{A} = 3\hat{\mathbf{i}} + 3\hat{\mathbf{j}} - 2\hat{\mathbf{k}}$ and

(a) $\mathbf{B} = -\hat{\mathbf{i}} + 4\hat{\mathbf{j}} - 2\hat{\mathbf{k}}$ (b) $\mathbf{B} = 4\hat{\mathbf{i}} + \hat{\mathbf{k}}$

(c) $\mathbf{B} = 3\hat{\mathbf{i}} + 2\hat{\mathbf{j}} + \hat{\mathbf{k}}$ (d) $\mathbf{B} = 3\hat{\mathbf{i}} - 3\hat{\mathbf{j}}$.

35. What is the angle between the vectors $\mathbf{A} = 3\hat{\mathbf{j}} - \hat{\mathbf{k}}$ and $\mathbf{B} = 2\hat{\mathbf{i}} + 2\hat{\mathbf{k}}$?

36. Given unit vectors $\mathbf{A}$, $\mathbf{B}$, and $\mathbf{C}$ such that $\mathbf{C}$ makes an angle of $45°$ with $\mathbf{A}$ and an angle of $60°$ with $\mathbf{B}$. Find all scalars a and b such that $\mathbf{C}$ will be perpendicular to $a\mathbf{A} + b\mathbf{B}$.

37. Calculate the value of α so that the vector $\mathbf{A} = \alpha\hat{\mathbf{i}} - 6\hat{\mathbf{j}}$ is perpendicular to the vector $\mathbf{B} = 3\hat{\mathbf{i}} + 2\hat{\mathbf{j}}$.

38. Calculate the angle between the vectors $\mathbf{A} = 2\hat{\mathbf{i}} - 7\hat{\mathbf{j}} + \hat{\mathbf{k}}$ and $\mathbf{B} = 5\hat{\mathbf{i}} + 3\hat{\mathbf{j}} - 2\hat{\mathbf{k}}$. Do you need to know the angles each vector makes with the coordinate axes to find your answer?

39. If $\mathbf{C}$ is the sum of two perpendicular vectors $\mathbf{A}$ and $\mathbf{B}$, use the dot product to show that

$$|\mathbf{C}|^2 = |\mathbf{A}|^2 + |\mathbf{B}|^2.$$

State and prove a corresponding result for the sum of three mutually perpendicular vectors.

40. Let $\mathbf{A}$, $\mathbf{B}$, $\mathbf{C}$ be three mutually perpendicular nonzero vectors in space, and let

$$\mathbf{V} = a\mathbf{A} + b\mathbf{B} + c\mathbf{C},$$

where a, b, c are scalars. (These scalars are called the components of $\mathbf{V}$ relative to the reference frame $\mathbf{A}$, $\mathbf{B}$, $\mathbf{C}$.) Show that

$$a = \mathbf{A} \cdot \mathbf{V} / \mathbf{A} \cdot \mathbf{A}$$

and obtain corresponding formulas for b and c.

5.4 THE CROSS PRODUCT

To describe many phenomena in physics, it is helpful to have a method for constructing a vector perpendicular to each of two given vectors. For this purpose, a special kind of multiplication of two vectors, say $\mathbf{A}$ and $\mathbf{B}$, has been developed, which produces a vector $\mathbf{C}$ perpendicular to both $\mathbf{A}$ and $\mathbf{B}$. It is called the *vector product* or *cross product*, and is written

$$\mathbf{A} \times \mathbf{B} = \mathbf{C},$$

and read "*A* cross *B*."

The direction of $\mathbf{C}$ is related to $\mathbf{A}$ and $\mathbf{B}$ by the right-hand rule. If you place your fingers in the direction of the first vector, $\mathbf{A}$, then curl them through the smallest angle toward $\mathbf{B}$, your thumb points in the direction of $\mathbf{C}$, as shown in Fig. 5.16a. Note that the cross product $\mathbf{B} \times \mathbf{A}$ produces a vector in the opposite direction, as Fig. 5.16b illustrates. Mathematically, the cross product is said to be anticommutative:

$$\mathbf{A} \times \mathbf{B} = -\mathbf{B} \times \mathbf{A}.$$

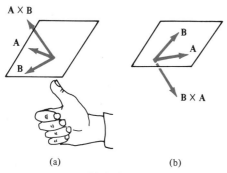

(a)   (b)

Figure 5.16 (a) Right-hand rule used to find the direction of $\mathbf{A} \times \mathbf{B}$. (b) Right-hand rule used to find the direction of $\mathbf{B} \times \mathbf{A}$.

The magnitude of the cross product is defined to be

$$|\mathbf{A} \times \mathbf{B}| = |\mathbf{A}| \, |\mathbf{B}| \sin \theta = AB \sin \theta, \tag{5.7}$$

where θ is the angle between $\mathbf{A}$ and $\mathbf{B}$ measured so that $0 \leq \theta \leq \pi$. As illustrated in Fig. 5.17a, $B \sin \theta$ is the component of $\mathbf{B}$ perpendicular to $\mathbf{A}$, sometimes written as $B_\perp$. In this notation, the magnitude of the cross product is $|\mathbf{A} \times \mathbf{B}| = AB_\perp$. Equivalently, the magnitude can be thought of as B times the component of $\mathbf{A}$ perpendicular to $\mathbf{B}$, which would be called $A_\perp$, so that $|\mathbf{A} \times \mathbf{B}| = A_\perp B$, as Figure 5.17b shows. Geometrically, $|\mathbf{A} \times \mathbf{B}|$ is the area of the parallelogram shown in Fig. 5.17.

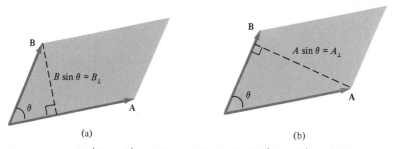

(a)   (b)

Figure 5.17 (a) $|\mathbf{A} \times \mathbf{B}| = AB_\perp = A(B \sin \theta)$. (b) $|\mathbf{A} \times \mathbf{B}| = A_\perp B = (A \sin \theta)B$.

Example 12

Two vectors lie in the xy plane; vector $\mathbf{A}$ has magnitude 1.5 and makes an angle of 30° with the x axis. Vector $\mathbf{B}$ has magnitude 2.0 and makes an angle of 100° with the x axis. Find $\mathbf{A} \times \mathbf{B}$.

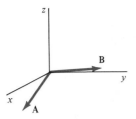

The direction of $\mathbf{A} \times \mathbf{B}$ is found from the right-hand rule: placing the fingers of your right hand in the direction of $\mathbf{A}$ and curling them toward $\mathbf{B}$, you find that your thumb points along the positive z direction. To find the magnitude, we need to know the angle between $\mathbf{A}$ and $\mathbf{B}$; the angle $\theta = 100° - 30° = 70°$. Therefore, the magnitude of the cross product is

$$|\mathbf{A} \times \mathbf{B}| = AB \sin \theta = 1.5(2.0) \sin 70° = 2.8.$$

Using the definition of the magnitude of the cross product, Eq. (5.7), we can see that the cross product of two parallel vectors is zero (because $\sin 0 = 0$ and $\sin \pi = 0$). For the units vectors $\hat{\mathbf{i}}, \hat{\mathbf{j}}, \hat{\mathbf{k}}$, we find that

$$\hat{\mathbf{i}} \times \hat{\mathbf{i}} = \hat{\mathbf{j}} \times \hat{\mathbf{j}} = \hat{\mathbf{k}} \times \hat{\mathbf{k}} = 0,$$

whereas by the right-hand rule

$$\hat{\mathbf{i}} \times \hat{\mathbf{j}} = \hat{\mathbf{k}},$$

$$\hat{\mathbf{j}} \times \hat{\mathbf{k}} = \hat{\mathbf{i}},$$

$$\hat{\mathbf{k}} \times \hat{\mathbf{i}} = \hat{\mathbf{j}}.$$

For two vectors expressed in component form,

$$\mathbf{A} = A_x\hat{\mathbf{i}} + A_y\hat{\mathbf{j}} + A_z\hat{\mathbf{k}},$$

$$\mathbf{B} = B_x\hat{\mathbf{i}} + B_y\hat{\mathbf{j}} + B_z\hat{\mathbf{k}},$$

the cross product is

$$\mathbf{A} \times \mathbf{B} = (A_x\hat{\mathbf{i}} + A_y\hat{\mathbf{j}} + A_z\hat{\mathbf{k}}) \times (B_x\hat{\mathbf{i}} + B_y\hat{\mathbf{j}} + B_z\hat{\mathbf{k}}).$$

Multiplying out the terms and using the properties of the cross product between unit vectors, we find

$$\mathbf{A} \times \mathbf{B} = (A_yB_z - A_zB_y)\hat{\mathbf{i}} + (A_zB_x - A_xB_z)\hat{\mathbf{j}} + (A_xB_y - A_yB_x)\hat{\mathbf{k}}. \tag{5.8}$$

Equation (5.8) provides a useful way to calculate the cross product of two vectors when the components are known.

If you are familiar with determinants, there is a handy mnemonic device for remembering the cross product (think about the "determinant" as being expanded along the first row):

$$\mathbf{A} \times \mathbf{B} = \begin{vmatrix} \hat{\mathbf{i}} & \hat{\mathbf{j}} & \hat{\mathbf{k}} \\ A_x & A_y & A_z \\ B_x & B_y & B_z \end{vmatrix}. \tag{5.9}$$

Example 13

Find the cross product and angle between the vectors $\mathbf{A} = \hat{\mathbf{i}} - \hat{\mathbf{j}} + 3\hat{\mathbf{k}}$ and $\mathbf{B} = \hat{\mathbf{i}} - 5\hat{\mathbf{j}} - 2\hat{\mathbf{k}}$.

Using Eq. (5.8), we calculate the cross product to be

$$\mathbf{A} \times \mathbf{B} = [-1(-2) - 3(-5)]\hat{\mathbf{i}} + [3(1) - 1(-2)]\hat{\mathbf{j}} + [1(-5) - (-1)1]\hat{\mathbf{k}},$$

$$\mathbf{A} \times \mathbf{B} = 17\hat{\mathbf{i}} + 5\hat{\mathbf{j}} - 4\hat{\mathbf{k}}.$$

To find the angle between $\mathbf{A}$ and $\mathbf{B}$, let's make use of the magnitude of the cross product, Eq. (5.7):

$$|\mathbf{A} \times \mathbf{B}| = AB \sin \theta. \tag{5.7}$$

If we know all the magnitudes, we can use this relation to find $\sin \theta$ and consequently θ. We easily calculate the following:

$$A = \sqrt{1^2 + (-1)^2 + 3^2} = \sqrt{11},$$

$$B = \sqrt{1^2 + (-5)^2 + (-2)^2} = \sqrt{30},$$

$$|\mathbf{A} \times \mathbf{B}| = \sqrt{(17)^2 + 5^2 + (-4)^2} = \sqrt{330},$$

Therefore, we find

$$\sin \theta = \frac{|\mathbf{A} \times \mathbf{B}|}{AB} = \frac{\sqrt{330}}{\sqrt{11}\ \sqrt{30}} = 1,$$

which tells us that the angle between $\mathbf{A}$ and $\mathbf{B}$ is 90°. This could also be seen by taking the dot product, $\mathbf{A} \cdot \mathbf{B} = 0$.

Questions

41. Use Eq. (5.8) for the cross product to show this multiplication obeys the distributive law:

$$\mathbf{A} \times (\mathbf{B} + \mathbf{C}) = \mathbf{A} \times \mathbf{B} + \mathbf{A} \times \mathbf{C}.$$

42. Two vectors, $\mathbf{A}$ and $\mathbf{B}$, lie in the xy plane and have magnitudes 2.8 and 3.2, respectively, whereas their directions are 210° and 45° measured from the x axis. Find $\mathbf{A} \times \mathbf{B}$.

43. Two vectors are given by $\mathbf{A} = 3\hat{\mathbf{i}} + 5\hat{\mathbf{j}}$ and $\mathbf{B} = -1\hat{\mathbf{i}} + 2\hat{\mathbf{j}} - 3\hat{\mathbf{k}}$. Find $\mathbf{A} \times \mathbf{B}$ and the angle between the vectors.

44. For any scalar c, show that $c(\mathbf{A} \times \mathbf{B}) = (c\mathbf{A}) \times \mathbf{B} = \mathbf{A} \times c\mathbf{B}$.

45. For any vectors **A** and **B**, what does **A**·(**A** × **B**) equal? Why?

46. Supply the steps leading to Eq. (5.8).

47. The two vectors shown have magnitudes $A = 3$, $B = 4$. Find

    (a) $A_\perp$ **(b)** $B_\perp$ **(c)** **A** × **B**.

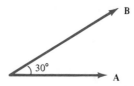

48. For the vectors $\mathbf{A} = 3\hat{\mathbf{i}} - 2\hat{\mathbf{k}}$, $\mathbf{B} = 2\hat{\mathbf{i}} + 3\hat{\mathbf{k}}$, and $\mathbf{C} = 6\hat{\mathbf{i}} - 4\hat{\mathbf{k}}$, find

    (a) **A** × **B**   **(b)** |**A** × **B**| **(c)** **A** × **C**.

49. The dot product **A**· (**B** × **C**) is called the *scalar triple product* of **A**, **B**, and **C**.

    (a) Prove that its absolute value represents the volume of the parallelepiped spanned by **A, B, C,** as shown below
    (b) For the vectors shown in the figure, show by geometrical arguments that
        $\mathbf{A} \cdot (\mathbf{B} \times \mathbf{C}) = \mathbf{B} \cdot (\mathbf{C} \times \mathbf{A}) = \mathbf{C} \cdot (\mathbf{A} \times \mathbf{B}).$

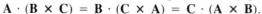

5.5 A FINAL WORD

Before Copernicus, the center of the earth was the center of the universe. That was the starting point for describing something's location in the Aristotelian world. Any other idea about place had no meaning. Then Copernicus came along, and went through what amounts to a routine mathematical exercise. He described the motion of the planets in a different coordinate system, and changed the world forever. That's very significant, but somewhat misleading. It suggests that the really important thing in science is to have the right coordinate system. But the truth is exactly the opposite. What we finally learned is that all coordinate systems, if they work, *are equally good*. Copernicus said the origin is in the sun; the United States Coast Guard, pinpointing the position of a distressed craft, says that the origin is in an airbase – and *they're both right!* That's a very valuable lesson, and as a statement, it's even more profound. What it really means is that the laws of physics are the same everywhere in the universe. The laws that Newton gave us work as well in the Crab Nebula as in Kansas City. Because we believe that's true, we need a mathematical device for expressing those laws in a way that's the same in all coordinate systems.

    That device is the vector. The idea of a vector is a little disconcerting, because it has a size and a direction, but not a place, unless we find it convenient to give it one. But that makes it a perfect tool for expressing laws that work equally well everywhere. And that is just what Newton's laws do. That's why the next thing we'll study is a vector equation that lies at the heart of our understanding of the world.

CHAPTER 6

NEWTON'S LAWS

Then from these forces, by other propositions which are also mathematical, I deduce the motions of the planets, the comets, the moon, and the sea. I wish we could derive the rest of the phenomena of Nature by the same kind of reasoning from mechanical principles, for I am induced by many reasons to suspect that they may all depend upon certain forces by which the particles of bodies, by some cause hitherto unknown, are either mutually impelled towards one another, and cohere in regular figures, or are repelled and recede from one another. These forces being unknown, philosophers have hitherto attempted the search of Nature in vain, but I hope the principles here laid down will afford some light either to this or some truer method of philosophy.

Isaac Newton, *Principia*, 1687

6.1 THE END OF THE CONFUSION

In 1543, Copernicus published his book, and a tremor rocked the foundations of the Aristotelian world. A century later the Aristotelian world lay in ruins, but nothing had arisen to replace it. Galileo and Kepler had made mighty discoveries, but there was no central principle that could organize the world. The unified harmony of the Aristotelian view had been replaced by buzzing confusion.

Galileo was concerned not with the causes of motion, but instead with its description. The branch of mechanics he reared is known as kinematics; it is a mathematically descriptive account of motion without concern for the causes. Central to Galileo's arguments was the law of inertia, which we discussed in Chapter 4. Armed with this principle, Galileo could neutralize Aristotelian arguments against a moving earth, but the reconstruction of a new mechanics that he promised in his final book had scarcely begun.

The law of inertia created confusion in the minds of seventeenth-century scholars because it overturned a centuries-old teaching that a body needs a force to keep it in motion. Not fully grasping the ideas of the new mechanics, these scholars failed to define accurately the numerous scientific terms they employed. The confusion was aggravated even further when the term *force* was used in different ways. Originally force was considered chiefly as the cause of movement. But some thinkers considered it to be a result of motion. The term *centrifugal force*, for example, was invented for the tendency of a revolving body to fly away from the center because of its motion. Traditionally gravity had been considered to be a property of a body, its heaviness, which causes it to fall. Some physicists now considered it to be the result of the motion of a swirling fluid of matter.

In 1642, Galileo – nearly 80 years old, blind, and imprisoned – died. In the same year, Isaac Newton was born. Newton's work was the culmination of the Copernican Revolution. It was his task to create order in the chaos of terms, notions, and misconceptions that permeated seventeenth-century physics.

Newton was not a man of half-hearted pursuits. When he thought on something, he thought on it continually, to the neglect of food, sleep, and human society. The period from the autumn of 1684 to the spring of 1686 is a virtual blank in his life except for his *Principia*. Once he adopted the principle of inertia and developed the central concept of force, his dynamics quickly fell into place. He had seized on the essence of his second law of motion 20 years before and had never altered it as he wrestled with the first law, which was a statement of the law of inertia. He realized that an impressed force alters a body's motion and that the corresponding change in motion is proportional to it. By this understanding, he capped and completed Galileo's kinematics with a dynamics – a theory of the *causes* of motion. Few periods of history have been so intense or held greater consequences for science than the six months in the autumn and winter of 1684–5, when Isaac Newton created the modern science of dynamics.

6.2 NEWTON'S LAWS OF MOTION

The *Principia* was published in 1687, when Newton was 44, and details all his work on the motion of bodies. The style of the *Principia* is reflective of its author: cold and rigid; its pages are laden with diagrams and geometric proofs. Although Newton undeniably arrived at his results by using his newly developed calculus, in the *Principia* he presented geometric proofs – the language of physics in the seventeenth century. At Newton's own Cambridge University, a stately institution not given to undue haste, the *Principia* was used as a textbook right into the twentieth century.

To remove the confusion in terminology, Newton began by saying carefully what he meant by terms such as inertia, mass, and force. He used these terms in his axioms of motion, which he presented as statements that need no proof. These axioms are the basis from which the nature of all types of motion can be deduced.

Newton inherited from Galileo and Descartes the essential idea that motion along a straight line with a constant speed was the natural state of any body, needing no further explanation. This is Newton's first law, the law of inertia. Stated in his own words,

First Law: *Every body continues in its state of rest, or of uniform motion in a straight line, unless it is compelled to change that state by forces impressed upon it.*

Example 1

What would be the path of the planets if there were no force acting on them?

The first law tells us that in the absence of any force, a body will continue to move in a straight line. Therefore, if there were no force on the planets, they would travel in straight lines, not in nearly circular orbits about the sun.

Newton, like Galileo before him, realized that an object's inertia was somehow connected to its mass. He defined mass as the quantity of matter that arises conjointly from an object's density and size. The greater the mass of an object, the more difficult it is to prevent it from continuing in motion with a constant velocity. This idea led to his second law. In the *Principia* it is modestly stated as

Second Law: *The change of motion is proportional to the force impressed; and is made in the direction of the straight line in which the force is impressed.*

What Newton meant by ''motion'' involved not only a body's velocity, but also its mass. It is the quantity we call *momentum*, the product of mass m and velocity $\mathbf{v}$. Stated as an equation, the second law is

$$\mathbf{F} = \frac{d}{dt}(m\mathbf{v}) . \tag{6.1}$$

The d/dt is a mathematical symbol meaning ''the instantaneous rate of change of.'' When we understand that, the equation expresses Newton's second law almost as we would in an English sentence in which $\mathbf{F}$ is the subject and '' $=$ '' is the verb. The mathematical sentence says, ''Force is equal to the rate of change of momentum of an object.''

If the second law is applied to a body for which the mass m is a constant, then by a rule of differentiation this means that $d(m\mathbf{v})/dt = m\,d\mathbf{v}/dt$. But we know that the instantaneous rate of change of velocity is acceleration, $\mathbf{a} = d\mathbf{v}/dt$. So for an object (or collection of objects) whose mass doesn't change, Newton's second law tells us that acceleration is caused by forces. It is usually written

$$\boxed{\text{Second Law:}\quad \mathbf{F} = m\mathbf{a}.} \tag{6.2}$$

This form was first presented by the Swiss mathematician Leonhard Euler 65 years after the publication of the *Principia*. It is probably the most useful equation in all of physics.

To understand Newton's second law, we need to understand the concept of force. In everyday language, force is associated with a push or a pull. When you push on something, you can feel yourself exerting a force. Once armed with that sensation, you look around and find countless examples of things exerting forces on other things. Pushes, pulls, gravity, tension in a string, and friction are all examples of forces that enter Newton's second law. But these forces must originate *outside* the object whose motion we're trying to describe. In other words, only *external* forces acting on an object can change its motion.

The force **F** need not be just one force acting on the body. It is the *vector sum* of *all* external forces acting on the object. Even though vectors hadn't been invented yet, Newton knew that forces have both a magnitude and a direction. Whenever we write **F** = *m***a**, **F** symbolizes the *vector sum* of external forces acting on a body. The acceleration of an object is the result of the total force acting on it. Some physicists use $\mathbf{F}_{net}$ to remind them that it is the net or total force that enters the equation. Others write Σ **F** to symbolize the vector sum. The symbol Σ is the Greek letter sigma and means sum.

The second law, being a vector equation, is shorthand for three equations involving Cartesian components:

$$\Sigma F_x = ma_x, \tag{6.3a}$$

$$\Sigma F_y = ma_y, \tag{6.3b}$$

$$\Sigma F_z = ma_z. \tag{6.3c}$$

It is this form that is more useful for solving problems. Once the respective components of the external forces acting on a body are known, the components of the acceleration are determined by the equations, and the motion of the body can be deduced.

Example 2

A Rolls Royce Silver Shadow and a Volkswagen Rabbit are traveling at the same speed along a level road. How do the forces required to stop them compare?

A Rolls Royce is not only more expensive, but also has a greater mass than a Volkswagen. By Newton's second law, if the acceleration of each car is the same, as they are here because they are to be brought to rest from the same speed, then the more massive car requires a greater force. The ratio of the forces is equal to the ratio of the masses, because from the relations $\mathbf{F}_{RR} = m_{RR}\mathbf{a}$ and $\mathbf{F}_{VW} = m_{VW}\mathbf{a}$, we get

$$\frac{F_{RR}}{F_{VW}} = \frac{m_{RR}}{m_{VW}}.$$

Newton needed one additional law to express what happens when several bodies interact with each other. His third law is

Third Law: *To every action there is always opposed an equal reaction: or, the mutual actions of two bodies upon each other are always equal, and directed to contrary parts.*

When you push on anything – a door, a pencil – it pushes back on you with a force equal in magnitude but in the opposite direction. In other words, you can't touch without being touched. That's the essence of the third law – a law of interactions. As illustrated in Figure 6.1, if Body 1 exerts a force $\mathbf{F}_{12}$ on Body 2, then Body 2 exerts a force $\mathbf{F}_{21}$ on Body 1 such that $\mathbf{F}_{12} = -\mathbf{F}_{21}$.

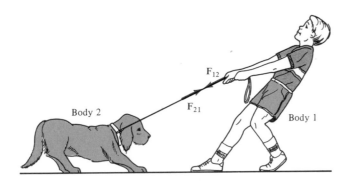

Figure 6.1 An illustration of Newton's third law.

Sometimes it is difficult to isolate the action–reaction pairs of forces in Newton's third law. As a guide, remember that they always act on *different* bodies, never on the same body. If you know one force – for example, you pull on a rope – you can find the reaction force by turning around the sentence: the rope pulls on you. As with the second law, the third law is best understood through applications.

Example 3

Identify the action–reaction pair of forces for each of the following cases:

(a) a child pulling a dog's leash,
(b) a bird flapping its wings,
(c) raindrops hitting a roof.

Let's form two categories: Action Force and Reaction Force.

Action Force	Reaction Force
(a) child pulls on leash	(a) leash pulls on child
(b) wings push down on air	(b) air pushes up on wings
(c) raindrops push down on roof	(c) roof pushes up on raindrops

If we wanted to understand the motion of the child, the bird, or the raindrops, we would use the forces exerted on these objects as listed in the second column. Those are the external forces that enter the second law.

Questions

1. Suppose you have two identical cans, one filled with lead and the other empty, in an orbiting spacecraft where everything is weightless. How can you tell which can is empty without looking inside?

2. What physical principles are behind the reasoning for making wrecking cranes with massive weights at the end of a cable?

3. Without seatbelts, you would hit the windshield during a quick stop. Why? Why would you be in danger of whiplash if your car had no headrests and you suffered a collision from behind?

4. What kind of motion does a constant force produce?

5. A train consisting of an engine and three boxcars moves down the track with a constant acceleration. Between what two cars is the tension in the coupling the greatest? the least? Why?

6. Often when parents spank a child they say, "This hurts me as much as it does you." Is there any physical basis for this statement?

7. While you are driving along the freeway, a bug splatters on your windshield. Which experiences the greater force, the bug or the windshield?

8. Discuss whether the following pairs of forces are action–reaction forces:

    (a) An athlete standing on a scale pushes down on it; the scale pushes up on the athlete.
    (b) The earth attracts a stone; the stone attracts the earth.
    (c) The tires of a car push on the road; the earth pulls down on the tires.
    (d) A chair pushes down on the floor; gravity pulls down on the chair.

9. A farmer urges an Aristotelian horse to pull his wagon, but the horse refuses to try. In his defense, the horse cites Newton's third law and claims, "If I pull on the wagon, the wagon pulls equally back on me. I can never exert a greater force on the wagon than it exerts on me, so I could never start it moving." What advice would you give the farmer to counter this argument?

10. A fan is mounted on a cart as shown below. If the fan is turned on, does the cart move? If so, in which direction?

    Suppose a sail were added to the cart. What would be the motion of the cart if the fan were now turned on?

6.3 UNITS OF MASS, MOMENTUM, AND FORCE

Mass, length, and time form the basic physical quantities used in mechanics. The unit of mass in the metric system (SI) is the kilogram, abbreviated kg. The standard kilogram is a platinum–iridium cylinder kept in a vault at the International Bureau of Weights and Measures in Sèvres, France. It is the only SI unit still defined by such an artifact. Secondary standards are housed all over the world. With an equal-arm balance, these standards can determine the mass of objects to a precision of two parts in 100,000. The kilogram is also equal to 1000 grams (1000 g); the gram is the unit of mass in the cgs (centimeter, gram, second) system. Other SI prefixes can be used with the gram, such as milligram (1 mg $= 10^{-3}$ g) and microgram (1 μg $= 10^{-6}$ g). The unit of mass in the British engineering system is the slug; one slug is equal to 14.58 kg.

Momentum, being a product of mass and velocity, is a derived physical quantity. In the metric system, the basic unit is kg m/s. It has been suggested that this unit be called a descartes, after the French mathematician-philosopher, or a clout, which is perhaps more descriptive. In the British system, momentum comes in units of slug ft/s.

Through Newton's second law, force should have the same units as mass times acceleration. Therefore, the SI unit of force is the kg m/s^2, called a newton and abbreviated N. One newton is the force required to accelerate a 1-kg mass at 1 m/s^2:

$$1 \text{ N} = 1 \text{ kg} \frac{\text{m}}{\text{s}^2}.$$

The unit of force in the cgs system is called a dyne and is the force that will accelerate a 1-g mass at an acceleration of 1 cm/s^2. Using 1 kg $= 10^3$ g and 1 m/s$^2 = 10^2$ cm/s^2, you can show that 1 N $= 10^5$ dyn.

In the SI and cgs systems, mass, length, and time are the fundamental quantities. Force is a derived unit. In the British system, however, the standard quantities are force, length, and time. The unit of force in the British system is the pound, abbreviated lb, which is 1 slug ft/s^2. By working out the conversion of units, you can show that 1 lb $=$ 4.45 N.

When an object is in free fall, gravity accelerates it downward with a constant acceleration g. Newton's second law tells us that the force must be

$$\mathbf{F} = m\mathbf{g},$$

where the direction of the force is vertically downward, toward the center of the earth. This is what we mean by the weight of an object; weight is the force of gravity acting on an object (whether it is falling or not). Being a force, weight is a vector. If we call the magnitude of the vector W, then $W = mg$ near the surface of the earth.

In countries that still use the British system, people are often confused between kilograms and pounds. These units refer to different physical quantities. Yet labels list the weight of an item in pounds along with its mass in kilograms and do not specify that one is weight and the other is mass. Unlike the mass of a body, which is an intrinsic property of a body, the weight of a body depends on its location. If you know the mass of an object, you can find its weight if you also know the acceleration of gravity at that location. Moving an object around on the surface of the earth doesn't change its weight very much, but moving it to the moon does change its weight considerably, without changing its mass. In Chapter 8 we'll find out why weight varies with location, when we discuss Newton's universal law of gravity.

Table 6.1 summarizes the units and conversions between the three systems of units.

Table 6.1 Units of Mass and Force in the SI, cgs, and British Systems

	Units of Mass		
	kilogram	gram	slug
1 kg	1	10^3	0.0685
1 g	10^{-3}	1	6.85×10^{-5}
1 slug	14.58	0.0146	1
	Units of Force		
	newton	dyne	pound
1 N	1	10^5	0.2248
1 dyn	10^{-5}	1	2.248×10^{-6}
1 lb	4.448	4.448×10^5	1

Questions

11. Determine whether the following combinations of units are units of mass, momentum, or force:

    (a) N s,   (b) dyn s²/cm,
    (c) slug ft/s,   (d) lb s.

12. How many newtons does a typical 160-lb man weigh?

13. What is the mass of a 0.75-lb can of beans?

14. What does it mean for an object to be weightless?

15. A rock weighs 60 N on the moon, where the acceleration due to gravity is one-sixth that on earth. What is the mass of this rock on the earth?

16. A jar of lightning bugs is tightly capped. Does it weigh more, less, or the same when the bugs are flying around compared to when they are at rest?

17. Suppose you hand an object to each of two people and ask them to guess its weight. One holds it still and guesses; the other hoists it up and down before guessing. One is estimating the mass, and the other the weight. Which is which? Explain.

6.4 PROJECTILE MOTION: AN APPLICATION OF NEWTON'S SECOND LAW

Galileo was the first to describe the motion of a projectile correctly. Using his laws of inertia and of falling bodies, he showed that projectiles follow parabolic trajectories and was able to deduce many other properties of their motion. Let's reexamine how a projectile moves through space by starting with Newton's second law.

If a cannonball has mass m, and you know the net force $\mathbf{F}$ acting on it, you can predict the cannonball's motion using $\mathbf{F} = m\mathbf{a}$. What force acts on a cannonball? Initially there is the force from the explosion of gunpowder. That force acts on the cannonball only momentarily and determines the initial speed of the cannonball. What we need to know is the force acting on the cannonball after it has left the cannon, while it is in the air. And that's simple; the only force acting on it is gravity (neglecting air resistance).

We already deduced what that force is for any body near the surface of the earth. Since force equals mass times acceleration, the force on the cannonball must be mg, its weight. The direction of this force is vertically downward. If we choose a coordinate system like that in Fig. 6.2, then gravity is directed in the negative z direction. Using $\hat{\mathbf{k}}$ as the unit vector in the positive z direction, we have

$$\mathbf{F} = -mg\hat{\mathbf{k}}. \tag{6.4}$$

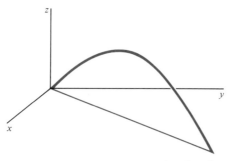

Figure 6.2 Coordinate system used to describe a projectile's motion.

Substituting this force into the second law, we get

$$-mg\hat{\mathbf{k}} = m\mathbf{a}.$$

The mass is the same on both sides of the equation and cancels out, so we are left with

$$-g\hat{\mathbf{k}} = \mathbf{a}.$$

This tells us that the flight of a projectile doesn't depend on its mass. Galileo knew that! He told us about it in Chapter 2.

Our equation looks quite simple, but remember it is a vector equation – it stands for three equations, not just one. Let's write out the equations for each of the x, y, and z components of the vectors. The acceleration vector can have three components, which we'll call a_x, a_y, and a_z. The left-hand side of the equation has only one component, which is in the z direction. In other words, its components are 0, 0, and $-g$. In terms of components, our equation is

$$a_x = 0, \tag{6.5a}$$

$$a_y = 0, \tag{6.5b}$$

$$a_z = -g. \tag{6.5c}$$

We expect these three equations to tell us everything we want to know about the cannonball's motion: how fast it is going, in what direction, and how far it will travel. To begin to answer these questions, let's find the velocity, which is related to the acceleration by the derivative

$$\mathbf{a} = \frac{d\mathbf{v}}{dt}.$$

This stands for three equations relating the components:

$$a_x = \frac{dv_x}{dt}, \tag{6.6a}$$

$$a_y = \frac{dv_y}{dt}, \tag{6.6b}$$

$$a_z = \frac{dv_z}{dt}. \tag{6.6c}$$

The situation now looks complicated because we have six equations, but we can quickly combine the last three with the first three.

·Let's take Eq. (6.5a), which tells us that $a_x = 0$, and substitute that information into Eq. (6.6a). The result is

$$\frac{dv_x}{dt} = 0. \tag{6.6d}$$

This, like Eqs. (6.5) and (6.6), is an example of a differential equation. A differential equation is any equation involving derivatives of an unknown function. Many problems in physics require solving differential equations because physical laws often involve rates of change. For example, the differential equation in (6.6d) tells us that the derivative of v_x, the x component of the velocity, is zero. Solving the equation means finding v_x. What can v_x be if it isn't changing? That's easy: if it isn't changing, then it must be a constant, so all solutions are

$$v_x = \text{constant}.$$

Thus, Eq. (6.6d) has many solutions, one for each value of the constant. This makes sense physically because the solution v_x will depend on how the cannonball was initially aimed and how much gunpowder was used.

The flight of one cannonball differs from another, not because the physical law is different, but because the things v_x depends on are different. Let's call one particular value v_{x0} – the x component of the initial velocity. That's how we can solve the differential equation for a specific case; we use the initial value.

Our result reflects the law of inertia. It tells us that the velocity, or more precisely, its component in the x direction, is constant. The reason is that in that direction there is no force (the axes are oriented so that gravity is in the z direction). If there's no force acting on a body, the velocity doesn't change – that's the principle of inertia.

The next question is, what happens to v_y? Looking at Eqs. (6.5b) and (6.6b), we see that the differential equation for v_y is the same as that for v_x. Therefore, we can immediately write down the answer. The y component of the velocity remains equal to whatever value it started out as. We'll call that initial value v_{y0}.

Example 4

A cannonball is fired with a speed v_0 of 200 m/s at an angle of 30° from the horizontal. What are the horizontal and vertical components of the velocity?

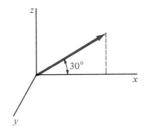

We are given the magnitude and direction of the initial velocity and are asked for the components. Resolving the vector **v** into components as we did in Chapter 5, and letting x be the horizontal direction in which the cannon is fired, we have

$$v_{y0} = 0,$$

$$v_{x0} = v_0 \cos \theta = (200 \text{ m/s}) \cos 30° = 173 \text{ m/s},$$

$$v_{z0} = v_0 \sin \theta = (200 \text{ m/s}) \sin 30° = 100 \text{ m/s}.$$

Combining Eqs. (6.5c) and (6.6c) we get an equation that tells us what happens to v_z:

$$\frac{dv_z}{dt} = -g.$$

Here we have a slightly different equation to solve. This equation tells us that the derivative of v_z with respect to time is a constant, not zero. So we ask, what function of time can we take the derivative of and end up with a constant? One answer, as you can easily verify by taking the derivative, is

$$v_z = -gt.$$

But this is not the only answer. If we add a constant to the above function, giving $-gt$ + constant, the derivative will still be $-g$ because the derivative of a constant is zero.

The constant will be the value v_z started with, in other words, the value of v_z at time $t = 0$. We'll call that initial value v_{z0}. Then the solution to our differential equation for v_z is

$$v_z = v_{z0} - gt. \tag{6.7}$$

Figure 6.3 illustrates the trajectory of a projectile and the velocity vector at subsequent times.

We now have the velocity vector of the projectile at any instant of time in terms of its components. What we've discovered is the process of solving the very simplest types of differential equations. The cases we've examined are so simple that we could find all solutions by guessing. For more complicated equations, which we'll encounter in later chapters, we shall learn other ways to find solutions besides just guessing.

There's always a sure-fire method to find out whether the solution we've guessed is correct or not. Just take the derivative and see if it satisfies the differential equation. For our last result, we would ask, is it true that if $v_z = v_{z0} - gt$, then $dv_z/dt = -g$? If you perform the differentiation you can see that it is true.

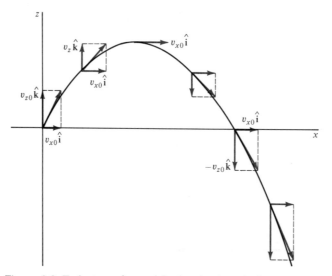

Figure 6.3 Trajectory of a particle showing its velocity.

Example 5

When does the cannonball of Example 4 reach its highest point?

First, we ask what specifies the highest point of a trajectory. At the highest point, the cannonball stops rising vertically; that is, the z component of its velocity is zero. Using the result of Eq. (6.7), we know that at any time t,

$$v_z(t) = v_{z0} - gt.$$

We can solve for the instant of time at which $v_z = 0$ from this equation, then substitute for v_{z0} from Example 4. The result is

$$t = v_{z0}/g = (100 \text{ m/s})/(9.8 \text{ m/s}^2) = 10.2 \text{ s}.$$

Is the x component of the velocity also zero at this instant? No. Our analysis showed that the x component remains constant, and here the constant is 173 m/s.

If you care at all about a cannnonball, what you probably care about most is where it is going to land. We can find out if we know its trajectory – where it is in space as a function of time. We need $x(t)$, $y(t)$, and $z(t)$, but we know instead how fast those positions are changing:

$$v_x = \frac{dx}{dt},$$

$$v_y = \frac{dy}{dt},$$

$$v_z = \frac{dz}{dt}.$$

Again we have differential equations, so let's try to guess the solution for each one. Our solutions will have to be consistent with the solutions for v_x, v_y, and v_z we found earlier. Let's start with $v_x = dx/dt$, remembering that $v_x = v_{x0}$ is constant. We ask ourselves what function when differentiated will give a constant [this is just like the procedure we followed for finding $v_z(t)$]. The answer, as you can readily verify by differentiation, is

$$x(t) = x_0 + v_{x0}t, \tag{6.8}$$

where x_0 is the initial position of the cannonball in the x direction.

Just as for x, we can immediately write down $y(t)$ because we have the same differential equation. The result is

$$y(t) = y_0 + v_{y0}t. \tag{6.9}$$

The remaining equation is a bit more complicated. Using Eq. (6.7),

$$v_z(t) = v_{z0} - gt,$$

we have

$$\frac{dz}{dt} = v_{z0} - gt.$$

Since we have two terms we expect that the solution $z(t)$ will consist of the sum of two terms. (Remember that the derivative of a sum is the sum of the derivatives.) If we differentiate $v_{z0}t$, we get v_{z0}, so that will be the first term in our solution.

What function can we differentiate that will yield $-gt$? By the power rule, whenever we take d/dt of t^n, the exponent is decreased by one and we have

$$\frac{d}{dt}(t^n) = nt^{n-1}. \tag{3.6}$$

So if the derivative is t^1, then the function we took the derivative of should involve t^2. But that's not all. We know that $(d/dt)/(-gt^2) = -2gt$. But we want the derivative to be $-gt$. Therefore, we need to divide by 2; in other words, the function we're looking for is $-gt^2/2$.

Now we almost have our solution for $z(t)$. It will be the sum of the two terms we just guessed plus a constant, z_0. We include this constant term because the derivative of a constant is zero. Here z_0 is the initial position of the cannonball in the z direction. Adding all the terms together, we have

$$z(t) = z_0 + v_{z0}t - \tfrac{1}{2}gt^2. \tag{6.10}$$

Example 6
If the cannonball of Example 3 starts out at $z_0 = 0$, what is the maximum height it reaches?

We already know when the cannonball is at its highest point; we found that out in Example 5, where it turned out to be 10.2 s. Equation (6.10) will tell us its position at that time. Substituting, we obtain

$$z(10.2 \text{ s}) = 0 + 100(10.2) - \tfrac{1}{2}(9.8 \text{ m/s}^2)(10.2 \text{ s})^2 = 510 \text{ m}.$$

Example 7
A football is kicked at an angle of 37° with a speed of 10 m/s. How long is it in the air? How far does it travel?

We can find out how long the football is airborne by realizing that it starts at $z_0 = 0$ and ends up after some time t at $z(t) = 0$. (This is a Pee-wee league punt, so we can ignore air resistance and the fact that it actually starts a short distance off the ground.)

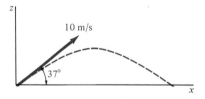

In the z direction the initial velocity is $v_{z0} = (10 \text{ m/s}) \sin 37°$. Using Eq. (6.10) and setting $z_0 = 0$, $z(t) = 0$, we can solve for t and get

$$0 = 0 + v_{z0}t - \tfrac{1}{2}gt^2,$$

$$t = 2v_{z0}/g = 2(10 \text{ m/s}) \sin 37°/(9.8 \text{ m/s}^2),$$

$$t = 1.2 \text{ s}.$$

Once we've found how long the ball is in the air, we can find out how far it travels, $x(t)$, by using Eq. (6.8):

$$x(t) = x_0 + v_{x0}t,$$

$$x(1.2 \text{ s}) = 0 + (10 \text{ m/s}) \cos 37°(1.2 \text{ s}),$$

$$x(1.2 \text{ s}) = 8.8 \text{ m}.$$

The laws of physics are expressed as differential equations whenever they involve rates of change. Newton's second law is a particularly important example. If you understand what these equations say and how to find solutions, then you can apply the second law to *any* situation. As we've seen in our application, solving Newton's second law led to trajectories of all projectiles. Once the solutions are in hand [that is, Eqs. (6.8), (6.9), and (6.10)], specific trajectories are determined by the initial position (x_0, y_0, z_0) and the initial velocity (v_{x0}, v_{y0}, v_{z0}). The particular solutions need not be memorized; understanding how they were arrived at is more important. The famous twentieth-century physicist Enrico Fermi once remarked that if he had had a good memory, he would have been a biologist. In physics you need not memorize equations; if you understand the laws and how to solve them, then you have solutions for particular cases at your disposal.

Questions (Ignore air resistance in 18–25)

18. Suppose you made a movie of an arrow flying through the air, and then played it backward. The arrow would be seen to move in reverse. What would be the direction of the arrow's acceleration? Is it also reversed?

19. When air resistance is ignored, does a projectile's trajectory always lie in a plane?

20. A toy dart is shot vertically upward from the ground and reaches a maximum height of 15 m.

    (a) What was the initial speed of the dart?
    (b) How long did it take to reach its maximum height?
    (c) Is the total time it is in the air equal to twice the time in (b)? Explain why.
    (d) What was the speed of the dart just before it hit the ground?

21. An arrow is shot with a velocity of 15 m/s 53° above the horizontal.

    (a) How long is it in the air?
    (b) What horizontal distance does it travel?

22. A monkey hunter sits on the ground armed with a tranquilizer dart gun. He points the gun directly at a monkey hanging from a tree as shown. Startled by the noise of the gun, the monkey lets go of the branch at the same instant the dart leaves the gun. Explain why the dart will strike the monkey as he falls to the ground. How does the initial speed of the dart affect where, along his line of fall, the monkey will be hit?

23. A cannonball is fired horizontally with a speed of 100 m/s from a tower 40 m above the ground.

(a) How long is the cannonball in the air?

(b) How far from the base of the tower does it strike the ground?

24. Suppose that instead of finding $v_x = v_{x0}$, Newton's second law had led to $v_x = \alpha t^2$, where α is a constant. By repeating the procedure presented in the text, try to guess what $x(t)$ would be.

25. Two identical cannons, A and B, are aimed at each other as shown. Cannonballs from each are fired simultaneously and at the same speeds.

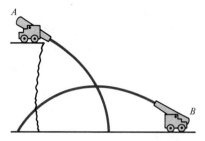

Discuss which of the following statements about their subsequent motion is correct:

(a) Cannonball A hits the ground first.

(b) Cannonball A is in the air longer.

(c) The two cannonballs collide in midair.

(d) They reach the ground at the same time.

26. Suppose air resistance is a force in the direction opposite to the body's motion, and proportional to the body's speed. How can this be expressed using Newton's equations? Can you write three differential equations for the components of the motions? (You don't have to solve the equations.)

6.5 A FINAL WORD

$\mathbf{F} = m\mathbf{a}$ is probably the most useful equation in all physics. But what does it mean? To understand Newton's second law, we need to answer the questions: What is force? What is mass? Both of these quantities appear in the equation along with acceleration, which we already understand. In the years since Newton, philosophers have debated these questions at great length.

Even though Newton explained what he meant by mass, some people think that mass is a quantity that is defined by the equation $\mathbf{F} = m\mathbf{a}$. If you know the magnitude of the force acting on an object, measure the magnitude of the resulting acceleration; then the mass of the object is $m = F/a$. Other people think that the second law is a definition of force. If you know the mass of an object and measure its acceleration, then the applied force can be found. Now, if the second law is a definition of either mass or force, it is not a deep discovery in physics; it is simply a definition. But if the equation is used to define *two* of the three quantities that appear in it, then it has no meaning at all. Can it be that Newton's famous law is meaningless?

The way to understand what $\mathbf{F} = m\mathbf{a}$ means is to make use of it. Through applications to specific problems, such as those given throughout this chapter, we can see how it works and understand how it organizes the world. Sometimes it is used to determine a mass. Once determined, the response of that mass to various forces can be studied. Sometimes $\mathbf{F} = m\mathbf{a}$ is used to determine a force, whose effect on various different masses is then studied. In many cases m and $\mathbf{F}$ are known independently and $\mathbf{F} = m\mathbf{a}$ successfully predicts the motion. Though the precise logical status of Newton's second law may pose a philosphical question, in practice, simple, workable determinations of $\mathbf{F}$ and m exist that lead to consistent results over the vast range of phenomena described by classical physics.

CHAPTER 7

INTEGRATION

"Also I do not at all believe that the judgment wch is given can be taken for a final judgment of the [Royal] Society. Yet Mr Newton has caused it to be published to the world by a book printed expressly for discrediting me, and sent it into Germany, into France and into Italy in the name of the Society . . .

As for me I have always carried myself with the greatest respect that could be towards Mr Newton. And tho it appears now that there is great room to doubt whether he knew my invention before he had it from me; yet I have spoken as if he had of himself found something like my method; but being abused by some flatterers ill advised, he has taken the liberty to attaque me in a manner very sensible. Judge now S^r, from what side that should principally come wch is requisite to terminate this controversy."

Leibniz's reply to the Royal Society on the priority claim, 28 April 1714

7.1 ANTIDIFFERENTIATION, THE REVERSE OF DIFFERENTIATION

We saw by examples in earlier chapters that laws of physics are often expressed as equations about the rate at which things change – that is, equations about derivatives.

In discussing falling bodies in Chapter 2, we started with a knowledge of the distance function (how far a body falls in a given time), then took its derivative to find its speed (how fast it is falling), and then took the derivative of the speed to find its acceleration (how fast it was getting faster).

In Chapter 6 we worked in reverse. From a knowledge of the acceleration of a cannonball, we determined its velocity, and from the velocity we found its trajectory.

This last example is typical of many problems in physics where we try to determine something from its rate of change. The process of finding a function or quantity whose derivative is known is called *antidifferentiation*. Antidifferentiation is also called *integration*, a term which is also used for an entirely different concept, the process of calculating areas. The connection between the two processes is developed in Section 7.2.

Although we did not call it by that name, we have already done some antidifferentiation in Chapter 6. Let's recap what we did in treating the vertical component of a body's motion near Earth because this will teach us something about the general process of antidifferentiation. Here are the basic equations:

acceleration $a = -g,$

speed $v = v_0 - gt,$

position $z = z_0 + v_0 t - \frac{1}{2} g t^2.$

To obtain these equations, we started with the first one, $a = -g$, a fact we learned about falling bodies in Chapter 2, and then we used our additional knowledge that

$$a = \frac{dv}{dt}.$$

Finding v was therefore simply a matter of reversing dv/dt, or in other words, finding a function v whose derivative is equal to the constant $-g$.

We know one such function, namely, $v = -gt$. Another is $v = 3 - gt$, and yet another is $v = -5 - gt$. In fact, if we take $v = C - gt$ for any constant C, then $dv/dt = -g$, so we see that there are many such functions, one for each value of C.

This is typical of the process of antidifferentiation. A function is not uniquely determined by its derivative, because there can be many functions having the same derivative. But it is easy to see that any two of them can differ only by a constant. In fact, if two functions $g(t)$ and $f(t)$ have the same derivative, $g'(t) = f'(t)$, their difference $g(t) - f(t)$ has derivative 0, which means the difference doesn't change, so the difference $g(t) - f(t) = C$, where C is a constant. Hence $g(t) = f(t) + C$. In other words, if $f(t)$ is one antiderivative of $f'(t)$, then *all* antiderivatives are $f(t) + C$, where C is an arbitrary constant.

In our velocity problem, the constant C has a specific physical meaning. It represents the velocity when $t = 0$, that is, the initial velocity v_0. Thus, among all possible functions $v(t) = C - gt$ with derivative $-g$, we choose that one for which $C = v_0$ and get

$$v = v_0 - gt.$$

Now we repeat the process. Knowing that $v = v_0 - gt$ and that

$$v = \frac{dz}{dt},$$

we want to find z, an antiderivative of v. Again, we know (or can guess) one such antiderivative, namely $z = v_0 t - \frac{1}{2} g t^2$ because its derivative is $v_0 - gt$. But we also know that all antiderivatives must be equal to this one plus some constant C, so

$$z = C + v_0 t - \tfrac{1}{2} g t^2.$$

In this case, the constant C represents the initial vertical displacement z_0, and thus we get the required position function,

$$z = z_0 + v_0 t - \tfrac{1}{2} g t^2.$$

In Chapter 3, on p. 66, there is a table of derivatives of some functions which are required in physics. By reversing the table and adding constants, we obtain the following table of antiderivatives:

Function	Antiderivative
nx^{n-1}	$x^n + C$
$\cos x$	$\sin x + C$
$-\sin x$	$\cos x + C$
e^x	$e^x + C$

Even more useful is Table 7.1. It involves the same types of functions, but has extra constant factors a thrown in.

Table 7.1 Antiderivatives Used in Physics

Function	Antiderivative	
ax^n	$a\,\dfrac{x^{n+1}}{n+1} + C$	$(n \neq -1)$
$\cos ax$	$\dfrac{1}{a} \sin ax + C$	$(a \neq 0)$
$\sin ax$	$-\dfrac{1}{a} \cos ax + C$	$(a \neq 0)$
e^{ax}	$\dfrac{e^{ax}}{a} + C$	$(a \neq 0)$

Note: Sometimes the word *primitive* is used instead of antiderivative. Thus we say $P(x)$ is a primitive of $f(x)$ if $P'(x) = f(x)$. If $P(x)$ is one primitive of $f(x)$, then all primitives are $P(x) + C$, where C is an arbitrary constant.

Example 1

Determine all antiderivatives $P(x)$ of the function $f(x) = 2x^3$.

This function is of the form $f(x) = ax^n$, with $a = 2$ and $n = 3$, so from the first entry in our table we get

$$P(x) = 2\frac{x^4}{4} + C = \frac{x^4}{2} + C.$$

Of course, this is easily checked by differentiation:

$$\frac{d}{dx}\left(\frac{x^4}{2} + C\right) = 2x^3.$$

Example 2

Determine all primitives $P(x)$ of the function

$$f(x) = 3x^2 + \cos 2x.$$

Since the derivative of the sum of two functions is the sum of their derivatives, the same is true for primitives. The function x^3 is a primitive of $3x^2$ and the function $\frac{1}{2}\sin 2x$ is a primitive of $\cos 2x$, so $x^3 + \frac{1}{2}\sin 2x$ is one primitive of $f(x)$. Hence all the primitives are

$$P(x) = x^3 + \frac{1}{2}\sin 2x + C.$$

Again, the result is easily checked by differentiation:

$$\frac{d}{dx}(x^3 + \frac{1}{2}\sin 2x + C) \doteq 3x^2 + \cos 2x.$$

Questions

1. Verify each entry in Table 7.1 by differentiating each antiderivative.

2. Determine all antiderivatives $P(x)$ for each of the following functions $f(x)$:

   (a) $f(x) = 3x^4$ (b) $f(x) = 6/x^2$ $(x \neq 0)$

   (c) $f(x) = -3\cos x$ (d) $f(x) = 2\sin 4x$

   (e) $f(x) = x^{1/2}$ $(x \geq 0)$ (f) $f(x) = 3e^{2x}$.

3. Determine all antiderivatives $P(x)$ for each of the following functions $f(x)$:

   (a) $f(x) = 3x^2 - x + 5$ (b) $f(x) = 2x^3 + \cos\frac{1}{2}x$

   (c) $f(x) = 3e^{2x} + 2e^{3x}$ (d) $f(x) = \frac{1}{2}(e^x + e^{-x})$.

4. For each of the examples in Question 3, find that particular antiderivative $P(x)$ such that $P(0) = 1$.

7.2 ANTIDIFFERENTIATION AND QUADRATURE

A famous problem from antiquity that challenged the world's best minds for nearly 2000 years was that of *quadrature*: Given a region with curved boundaries, find a square having the same area. One of the most important events in the history of mathematics was the discovery of Newton and Leibniz that the ancient problem of quadrature could be solved with the help of antidifferentiation.

It is not at all obvious that quadrature and differentiation are related. Quadrature deals with area, while differentiation deals with rate of change. We can discover the relationship between the two by looking at some simple examples.

Figure 7.1 shows the graph of the constant acceleration of a falling body: $a = g$.

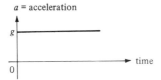

Figure 7.1 Graph of constant acceleration.

Let's calculate the area of the region between this graph and the time axis. The area depends, of course, on where we start and where we stop. Suppose we start at time 0 and stop at time t. Then the region is the shaded rectangle in Fig. 7.2, and its area is simply gt, the product of the base times height.

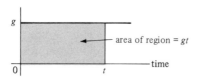

Figure 7.2 Area of the region under the graph from 0 to t.

If we think of t not as a fixed number but as a variable, then the area gt is a function of t, which we can call the area function. This function has derivative g, so it is an antiderivative of the constant function a that we started with. But an antiderivative of acceleration is speed, hence the area function here is equal to the speed of a body falling from rest:

$$v(t) = gt.$$

Now let's draw the graph of the speed function, a line of slope g shown in Fig. 7.3a, and let's calculate the area of the region under its graph from time 0 to any time t. This region, shown shaded in Fig. 7.3b, is a triangle of base t and height gt, so *its* area (half the product of base and height) is $\frac{1}{2}gt^2$, a new function of t.

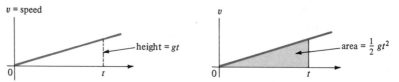

Figure 7.3 Speed curve of a body falling from rest: (a) graph of
speed function $v(t) = gt$, (b) area of region under its graph.

Again, we see that this area function is an antiderivative of the curve we started
with. The area $\frac{1}{2}gt^2$ is equal to the distance fallen in time t.

The relationship between area and antidifferentiation revealed in these two examples
is not merely an accident. It is the underlying idea behind the stunning discovery made
by Newton and Leibniz.

To explore this idea further, let's try to calculate the area of the parabolic segment
shown in Fig. 7.4. The curve (part of a parabola) is the graph of the function $y = x^2$,
and we want to find the area of the region between the curve and the x axis, from $x =
0$ to $x = t$. This region (shaded in Fig. 7.4) is called a parabolic segment of base t and
altitude t^2.

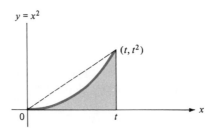

Figure 7.4 Parabolic segment of base t and altitude t^2.

From Fig. 7.4 it is clear that the area of the parabolic segment is less than that of a
triangle with the same base and the same altitude. This triangle has area $\frac{1}{2}t(t^2) = \frac{1}{2}t^3$,
so the area of the parabolic segment is smaller than $\frac{1}{2}t^3$. How much smaller is it?

By an ingenious geometric argument, Archimedes (287–212 B.C.) showed that the
area is exactly $\frac{1}{3}t^3$. He was the first to solve the quadrature problem for a parabolic
segment. We shall now solve the same problem by antidifferentiation.

Let $A(t)$ denote the area of the parabolic segment. This is the function we are trying
to determine by antidifferentiation, so let's try to find its derivative dA/dt. Recall that
the derivative dA/dt is the limit of the quotient

$$\frac{A(t + h) - A(t)}{h}$$

as h shrinks to 0. The numerator of this quotient is the difference of two areas, the area
of a parabolic segment of base $t + h$ and the area of a parabolic segment of base t. This
difference is shown in Fig. 7.5c.

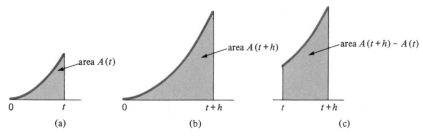

Figure 7.5 $A(t + h) - A(t)$ is the difference of areas of two parabolic segments.

In Fig. 7.6 we have constructed a rectangle with the same area as the region in Fig. 7.5c. It has base h and height x^2 for some (unknown) value of x between t and $t + h$.

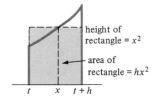

Figure 7.6 A rectangle having the same area as $A(t + h) - A(t)$.

The area of the rectangle is hx^2, hence we have

$$A(t + h) - A(t) = hx^2$$

or, dividing by h,

$$\frac{A(t + h) - A(t)}{h} = x^2$$

for some x between t and $t + h$. Now we let h shrink to 0 and see what happens to this equation. The left-hand side becomes dA/dt, and x^2 becomes t^2. In other words, we have shown that

$$\frac{dA}{dt} = t^2,$$

the derivative of $A(t)$ is t^2. Therefore $A(t)$ must be an antiderivative of t^2. But we know all antiderivatives of t^2 have the form $\frac{1}{3}t^3 + C$ for some constant C, so

$$A(t) = \tfrac{1}{3}t^3 + C.$$

But when $t = 0$ the area is 0 hence $C = 0$, and we find $A(t) = \frac{1}{3}t^3$. We have obtained Archimedes's quadrature formula using antidifferentiation!

The reason this discovery is important is not that Newton and Leibniz solved the quadrature problem for a parabolic segment. After all, Archimedes had already done it nearly 2000 years earlier. The importance of their discovery is that the same method is applicable when the parabola is replaced by any smooth curve. The argument for any smooth curve, which follows closely the one we have just given for a parabolic segment, is given in Appendix 1. Here are the main results:

Take a function $f(x)$ whose graph lies above the x axis as shown in Fig. 7.7, and let $A(t)$ denote the area of the shaded region from $x = a$ to $x = t$. Then the area $A(t)$ is related to $f(t)$ by the equation

$$\frac{dA}{dt} = f(t). \tag{7.1}$$

Consequently, if $P(t)$ is an antiderivative of $f(t)$, then

$$A(t) = P(t) + C \tag{7.2}$$

for some constant C. But because the area $A(t)$ is zero when $t = a$ we find $C = -P(a)$ and hence (7.2) becomes

$$A(t) = P(t) - P(a). \tag{7.3}$$

This formula gives us a straightforward recipe for finding areas. If you want to find the area of the region under a curve $y = f(x)$ from $x = a$ to $x = t$, first find an antiderivative of the given function, any function $P(x)$ whose derivative is $f(x)$. Then the required area is simply $P(t)$ minus $P(a)$.

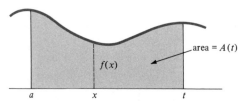

Figure 7.7 Area $A(t)$ of the region under the curve $y = f(x)$ from x ' $= a$ to $x = t$.

Example 3

Calculate the area of the region under one arch of the curve $y = \sin x$ (shown shaded in the figure).

Let $A(t)$ denote the area from $x = 0$ to $x = t$. We want $A(\pi)$, but it is no more effort to find $A(t)$ for all t between 0 and π. All we need is an antiderivative of $\sin x$. The function $P(x) = -\cos x$ is an antiderivative, because $P'(x) = \sin x$. Therefore, by formula (7.3), we have

$$A(t) = P(t) - P(0) = -\cos t + \cos 0 = 1 - \cos t.$$

When $t = \pi$ we get

$$A(\pi) = 1 - \cos \pi = 1 - (-1) = 2.$$

Example 4

Find the area of the region under the graph of $y = x^n$ from $x = a$ to $x = t$, where n is a positive integer and $0 \le a \le t$. (The graph is sometimes called a *generalized parabola*.)

Again, all we need is an antiderivative of x^n. One such antiderivative is $P(x) = x^{n+1}/(n + 1)$, so the area $A(t)$ from $x = a$ to $x = t$ is equal to

$$A(t) = P(t) - P(a) = \frac{t^{n+1} - a^{n+1}}{n + 1}.$$

This quadrature formula for the generalized parabola had been obtained prior to Newton and Leibniz by Cavalieri, Fermat, Pascal, and Roberval, but they did not use antiderivatives.

Questions

5. Use the method of antiderivatives to calculate the area of the region under the graph of $f(x) = 2x^2$ from $x = 1$ to $x = 4$.

6. Calculate the area of the region under the graph of $f(x) = x^{1/2}$ from $x = 0$ to $x = 3$.

7. The line $y = \frac{1}{2}x$ intersects the curve $y = x(2 - x)$ at the origin and at another point Q.

(a) Find the coordinates of Q.
(b) Find the area of the shaded region between the two curves.

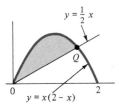

7.3   THE LEIBNIZ INTEGRAL NOTATION

Leibniz introduced a special notation for representing the area $A(t)$ of the region under the graph of a function $f(x)$ from $x = a$ to $x = t$ (the region shown in Fig. 7.7). He denoted this area symbolically as

$$A(t) = \int_a^t f(x) \, dx.$$

This is read "$A(t)$ is the integral from a to t of $f(x)\ dx$." The symbol $\int$ (an elongated S) is called an *integral sign*. You can see it today in copies of the original Declaration of Independence or the Constitution. Leibniz introduced it to mathematics in 1675.

The function $f(x)$ under the integral sign is called the *integrand*, and the interval from a to t is called the *interval of integration*, with the numbers a and t being the *limits of integration*.

Liebniz's symbol for the integral was readily accepted by many early mathematicians because they liked to think of the region under the graph as being composed of many thin rectangles of height $f(x)$ and base dx, as suggested in Fig. 7.8. The symbol $\int_a^t f(x)\ dx$ represented the process of summing together the areas of all these thin rectangles.

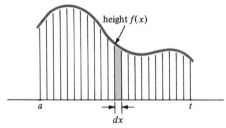

Figure 7.8 The region under the curve conceived as being filled with thin rectangles of height $f(x)$ and base dx.

All the results we obtained earlier concerning the area function $A(t)$ can now be expressed in Leibniz's integral notation. For example, the formula for quadrature of the parabolic segment becomes

$$\int_0^t x^2\ dx = \tfrac{1}{3}t^3.$$

Equation (7.1), which tells us that the derivative of the area function $A(t)$ is the function $f(t)$ we started with, now becomes

$$\frac{d}{dt}\int_a^t f(x)\ dx = f(t). \tag{7.4}$$

This charming result, which relates the derivative and the integral, is known as the *first fundamental theorem of calculus*. It says that the derivative of an integral $\int_a^t f(x)\ dx$ with respect to the upper limit is equal to the integrand evaluated at t.

The first to notice this connection between derivatives and integrals was Isaac Barrow, a Cambridge professor of Greek and later of mathematics and theology. He was also Newton's teacher, and later resigned his prestigious chair in favor of Newton. But Barrow never realized the importance of his discovery. Both Newton and Leibniz, however, appreciated the significance of the result and exploited it to develop their powerful technique for solving quadrature problems by antidifferentiation, as described above by

Eq. (7.3). In integral notation, Eq. (7.3) can be stated as follows:
If $P(x)$ is any antiderivative of $f(x)$, then

$$\int_a^t f(x)\,dx = P(t) - P(a). \tag{7.5}$$

This is known as the *second fundamental theorem of calculus*. It can also be written entirely in terms of P. Since $P'(x) = f(x)$, it states that

$$\int_a^t P'(x)\,dx = P(t) - P(a). \tag{7.6}$$

In other words, if you integrate the derivative $P'(x)$ of some function $P(x)$ from $x = a$ to $x = t$, the value of the integral is $P(t) - P(a)$, the difference of the function values at the end points.

Sometimes the special symbol

$$P(x)\,\Big|_a^t$$

is used to designate the operation of evaluating $P(x)$ first for $x = t$ and then for $x = a$ and subtracting. With this symbol, the second fundamental theorem, as stated in Eq. (7.5), can be written as follows:

$$\int_a^t f(x)\,dx = P(x)\,\Big|_a^t = P(t) - P(a).$$

Note that the value of the integral, $P(t) - P(a)$, depends only on the endpoints a and t and not on the running variable that varies from a to t. Because the result doesn't depend on x, we call x a *dummy variable*.

In any integral $\int_a^t f(x)\,dx$ we can replace the dummy variable x by any other convenient symbol, for example, y, u, or τ (the Greek letter tau). Thus we write

$$\int_a^t f(x)\,dx = \int_a^t f(y)\,dy = \int_a^t f(u)\,du = \int_a^t f(\tau)\,d\tau.$$

In choosing a dummy variable it is best to avoid letters that are already used for other purposes. Thus it's not a good idea to write $\int_a^t f(t)\,dt$, since the t attached to the integral sign is supposed to represent an endpoint of the interval, whereas the dummy variable is supposed to run through all values in the interval.

Example 5
Show that it doesn't matter which antiderivative is chosen when you apply the second fundamental theorem.

If $P(x)$ is one antiderivative of $f(x)$, all others have the form $P(x) + C$ for some constant C. If we use $P(x) + C$ instead of $P(x)$ in Eq. (7.5), we get the same result because the constant C cancels in the subtraction:

$$[P(x) + C]\Big|_a^t = [P(t) + C] - [P(a) + C] = P(t) - P(a).$$

Example 6

Calculate the integral $\int_1^t x^4 \, dx$.

An antiderivative of x^4 is $P(x) = \frac{1}{5}x^5$, so by the second fundamental theorem we have

$$\int_1^t x^4 \, dx = P(t) - P(1) = \frac{1}{5}t^5 - \frac{1}{5}.$$

Example 7

What is the value of $\int_0^1 e^{2x} \, dx$?

Table 7.1 shows that $\frac{1}{2}e^{2x}$ is an antiderivative of e^{2x}. Using the second fundamental theorem we get

$$\int_0^1 e^{2x} \, dx = \frac{1}{2}e^{2x}\Big|_0^1 = \frac{1}{2}e^2 - \frac{1}{2}e^0 = \frac{1}{2}(e^2 - 1).$$

Using a calculator with an e^x key, we obtain a value of 3.19 for this integral.

Leibniz adapted his integral notation to introduce a special symbol for antiderivatives. He wrote

$$\int f(x) \, dx$$

without any limits attached to the integral sign to denote any antiderivative of $f(x)$. In this notation, an equation like

$$\int f(x) \, dx = P(x) + C$$

simply means that $P'(x) = f(x)$. For example, since $d(\sin x)/dx = \cos x$ we can write

$$\int \cos x \, dx = \sin x + C.$$

The symbol C represents an arbitrary constant. Thus we could write

$$\int \cos x \, dx = \sin x + 5 \quad \text{or} \quad \int \cos x \, dx = \sin x - 7.$$

All of these are correct because the derivative of each right-hand side is $\cos x$.

Despite similarity in appearance, the symbol $\int f(x) \, dx$ is conceptually distinct from the symbol $\int_a^x f(t) \, dt$. The symbols originate from two different processes – the first from antidifferentiation, the second from quadrature. But they are related to each other because of the first and second fundamental theorems. Each represents a function whose derivative is $f(x)$. Therefore they differ only by a constant, so we can write

$$\int f(x) \, dx = \int_a^x f(t) \, dt + C \tag{7.7}$$

for some constant C.

Because of long historical usage, the symbol $\int f(x) \, dx$ is often referred to as an *indefinite integral* rather than as an antiderivative. By contrast, $\int_a^x f(t)$ is called a *definite integral*. This is justified, in part, by Eq. (7.7) which tells us that $\int f(x) \, dx$ is an integral from some unspecified point a to x, plus some unspecified constant C. Handbooks of mathematical tables often contain extensive lists of formulas labeled *tables of indefinite integrals* which, in reality, are *tables of antiderivatives*. Our skill in calculating integrals depends on our ability to find antiderivatives, so such tables are very useful. Any systematic method for finding antiderivatives is called a *technique of integration*. When one is asked to "integrate $\int f(x) \, dx$" what is really wanted is the most general antiderivative of $f(x)$.

Table 7.2 is how our list of antiderivatives in Table 7.1 would appear in the Leibniz notation.

Table 7.2 Indefinite Integrals (Antiderivatives) Used in Physics

$$\int ax^n \, dx = a \, \frac{x^{n+1}}{n+1} + C \qquad (n \neq -1)$$

$$\int \cos ax \, dx = \frac{1}{a} \sin ax + C \qquad (a \neq 0)$$

$$\int \sin ax \, dx = -\frac{1}{a} \cos ax + C \qquad (a \neq 0)$$

$$\int e^{ax} \, dx = \frac{e^{ax}}{a} + C \qquad (a \neq 0)$$

Note that in the important formula for the indefinite integral of a power, $\int ax^n \, dx$, the case $n = -1$ is specifically excluded (for obvious reasons). It must be stated separately:

$$\int \frac{a}{x} \, dx = a \log_e x + C \qquad (x > 0) \tag{7.8}$$

where $\log_e x$ (also written $\ln x$) is called the natural logarithm. It is related to the Euler number $e = 2.718 \ldots$ (see Chap. 3) because

$$\log_e e = \ln e = 1$$

or, more generally

$$\ln e^x = x.$$

Another way to express Eq. (7.8) is

$$\int_1^t \frac{a}{x} \, dx = a \log_e t \qquad (t > 0).$$

These results are worked out in detail in Appendix 2.

Example 8

A certain type of bacteria breed uninhibitedly such that at time t, the number $N(t)$ of bacteria is given by

$$N(t) = N_0 e^{kt},$$

where N_0 is the initial number of bacteria, and $k = \frac{1}{5} s^{-1}$. If originally there were 5 bacteria, how long would it take for there to be 1000?

Our equation for $N(t)$ implies that $N/N_0 = e^{kt}$, so

$$kt = \ln(N/N_0),$$

$$t = (1/k)\ln(N/N_0) = (1/k)\ln(200).$$

Substituting the numerical value of k and either looking up the logarithm in a table or using a calculator, we find $t = 26.5$ s. In less than half a minute they increase their numbers 200-fold!

In defining the integral in terms of area, we assumed that the integrand $f(x)$ was nonnegative, so its graph never went below the x axis. If $f(x)$ takes both positive and negative values, as shown in Fig. 7.9, we define the integral to be the algebraic sum of the areas of the regions above the axis, minus the sum of the areas of the regions below the axis. Areas above the axis are added, whereas those under the axis are subtracted. Under this extended definition the second fundamental theorem is still valid.

One more remark about the definition of the integral. In writing $\int_a^t f(x) \, dx$ we have always assumed that the lower limit a is less than the upper limit t. If $a < t$ we also define

$$\int_t^a f(x) \, dx = -\int_a^t f(x) \, dx.$$

In other words, switching the limits around changes the sign of the integral. Finally, if

$a = t$ we define $\int_a^a f(x)\, dx = 0$. This is consistent with the previous equation when $t = a$.

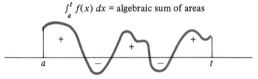

$\int_a^t f(x)\, dx$ = algebraic sum of areas

Figure 7.9 The integral is the sum of the areas of the regions above the axis minus the sum of those below.

Example 9
Evaluate the integral $\int_0^{\pi/2} (\sin 2x + \cos x)\, dx$.
From Table 7.2 we have

$$\int \sin 2x\, dx = -\tfrac{1}{2} \cos 2x \qquad \text{and} \qquad \int \cos x\, dx = \sin x.$$

Since the antiderivative of a sum is the sum of antiderivatives we have

$$\int (\sin 2x + \cos x)\, dx = -\tfrac{1}{2} \cos 2x + \sin x.$$

Using the second fundamental theorem we find

$$\int_0^{\pi/2} (\sin 2x + \cos x)\, dx = \left. (-\tfrac{1}{2} \cos 2x + \sin x) \right|_0^{\pi/2}$$

$$= (-\tfrac{1}{2} \cos \pi + \sin \pi/2) - (-\tfrac{1}{2} \cos 0 + \sin 0)$$

$$= (\tfrac{1}{2} + 1) - (-\tfrac{1}{2} + 0) = 2.$$

The same result could have been obtained by writing

$$\int_0^{\pi/2} (\sin 2x + \cos x)\, dx = \int_0^{\pi/2} \sin 2x\, dx + \int_0^{\pi/2} \cos x\, dx$$

and evaluating each integral on the right separately.

Questions
8. Determine each of the following indefinite integrals:

(a) $\int x^6\, dx$

(b) $\int x^{-2}\, dx \quad (x > 0)$

(c) $\int 5x\, dx$

(d) $\int x^{1/2}\, dx \quad (x \geq 0)$.

9. Verify each of the following formulas for indefinite integrals:

(a) $\int 3e^{-x}\,dx = -3e^{-x} + C$

(b) $\int (4x^3 - x^4)\,dx = x^4 - \frac{1}{5}x^5 + C$

(c) $\int xe^x\,dx = xe^x - e^x + C$

(d) $\int (\sin x - \cos x)\,dx = -\sin x - \cos x + C$.

10. Evaluate the following integrals by using the second fundamental theorem:

(a) $\int_0^\pi \sin(x/2)\,dx$

(b) $\int_0^{\pi/2} 2\cos(2x)\,dx$

(c) $\int_{-\pi}^\pi \sin x\,dx$.

11. Perform the following integrations:

(a) $\int_0^1 3e^{-3x}\,dx$

(b) $\int_0^1 x^{3/2}\,dx$

(c) $\int_{-1}^1 x^5\,dx$.

12. Evaluate $\int_a^b f(t)\,dt$ for the following functions $f(t)$, given that $0 < a < b$:

(a) $f(t) = 5e^{-t}$

(b) $f(t) = t^{-n}$ $(n \neq 1)$

(c) $f(t) = 2t^{-3}$.

13. Evaluate the following integrals:

(a) $\int_0^1 (1 + t + t^2)\,dt$

(b) $\int_{-\pi}^\pi (\cos t + \frac{1}{2})\,dt$

(c) $\int_{-\pi}^\pi (1 - e^{-x})\,dx$.

14. Calculate the integral $\int_0^1 f(x)\,dx$ for each of the following functions:

(a) $f(x) = x^3 - x$

(b) $f(x) = 2\sin x + \cos 2\pi x$

(c) $f(x) = e^x - e^{-x}$.

15. Perform the following integrations:

(a) $\int_0^\pi (x^2 + \sin x)\,dx$

(b) $\int_1^2 x(x + 5)\,dx$

(c) $\int_0^1 x^{1/2}(x + 1)\,dx$.

16. Evaluate the following integrals:

(a) $\displaystyle\int_1^2 \frac{1}{2x}\,dx$

(b) $\displaystyle\int_1^4 \frac{1}{x}\,dx$

(c) $\displaystyle\int_2^5 \frac{1}{x + 1}\,dx$ (*Hint:* Interpret the integral as an area.)

17. The radioactive isotope ^{11}C (carbon-11) decays in such a way that the number of atoms at any instant, $N(t)$, is given by $N(t) = N_0 e^{-kt}$, where N_0 is the initial number of atoms and $k = 0.035$ min^{-1}.

  (a) What fraction of radioactive ^{11}C atoms remain after 14 min?
  (b) How many minutes are needed for there to be one-tenth of the original amount of ^{11}C atoms?

18. For the bacteria of Example 8, how long will it take to double the original number of bacteria?

19. Using the functional equation for the exponential [Eq. (7.19) in Appendix 2], show that

$$t^n = e^{n \log_e t}$$

for any integer n.

20. Is there any value of x that can satisfy the equation $\ln x = e^x$?

7.4 APPLICATIONS OF THE SECOND FUNDAMENTAL THEOREM TO PHYSICS

The second fundamental theorem of calculus states that

$$\int_a^t P'(x)\, dx = P(t) - P(a),$$

or

$$P(t) = P(a) + \int_a^t P'(x)\, dx. \tag{7.9}$$

This formula tells us how to recover a function $P(t)$ from its derivative: Integrate the derivative $P'(x)$ from $x = a$ to $x = t$ and then add $P(a)$. We now have a mathematical prescription for what we did in Chapter 6 when we solved Newton's second law for projectiles. Let's cast those results into our new language of integrals.

We can recover the velocity of a body at any instant, $v(t)$, from a knowledge of the acceleration function $a(\tau)$ by applying (7.9):

$$v(t) = v(0) + \int_0^t a(\tau)\, d\tau \tag{7.10}$$

(we're using τ for the dummy variable since τ suggests time). Here $v(0)$ is the initial velocity which we've previously called v_0. The power of this approach is that $a(\tau)$ can be *any* acceleration function. If we know the acceleration function and can find its integral, then we also know the velocity of the object at any instant. The speed at any instant t is equal to the initial velocity plus the integral of the acceleration from the initial time to the instant t.

Rewriting this in a slightly different way helps us show why the Leibniz notation is so popular. Since $a(\tau) = dv(\tau)/d\tau$, we can substitute $dv(\tau)/d\tau$ for $a(\tau)$ in (7.10) to get

$$v(t) = v(0) + \int_0^t \frac{dv(\tau)}{d\tau} d\tau.$$

Now, just as we have done before, we treat the differentials $d\tau$ as if they were ordinary numbers, canceling $d\tau$ from top and bottom, to obtain

$$v(t) = v(0) + \int_0^t dv(\tau).$$

We can interpret the symbol $\int_0^t dv(\tau)$ to mean "the total change in $v(\tau)$ from time $\tau = 0$ to time $\tau = t$." We read the equation as "v at time t is equal to the value it has at time zero, plus the total change between zero and t."

Equation (7.10) also applies to any component of the velocity, say v_z. In that case we have

$$v_z(t) = v_z(0) + \int_0^t a_z(\tau) \, d\tau,$$

where we use the corresponding component a_z to find the velocity component v_z.

Example 10

As a racecar starts along a course, its acceleration is described by $a(t) = ct^3$, where $c = 0.20$ m/s^5 and t is in seconds. If the car started from rest, how fast is it traveling after 5 s?

We are given $v_0 = 0$, $a(\tau) = c\tau^3$, and want $v(5)$. According to Eq. (7.10), the velocity at time $t = 5$ s is

$$v(5) = 0 + c \int_0^5 \tau^3 \, d\tau = \left. \tfrac{1}{4}c\tau^4 \right|_0^5,$$

where we used Table 7.1 for the antiderivative of τ^3. Substituting for the value of c, we obtain $v(5) = 31.25$ m/s.

Similarly, the distance function $s(t)$ can be recovered from the speed function by integration:

$$s(t) = s(0) + \int_0^t v(\tau) \, d\tau. \tag{7.11}$$

If we know the speed function at all times, we can find the distance the object has traveled after any time t. The initial distance is denoted by $s(0)$. Equation (7.11) is a mathematical summary of the process we followed in Chapter 6.

Of course, Eq. (7.11) holds for any component of the displacement. To find $z(t)$ when we know $v_z(t)$, we would just replace $s(t)$ by $z(t)$ and $v(\tau)$ by $v_z(\tau)$.

Example 11

Show that if a body has a constant acceleration of $-g$ in the negative z direction, then $z(t) = z_0 + v_{z0}t - \frac{1}{2}gt^2$.

From Eq. (7.10), we know that if the acceleration is constant, then the z component of the velocity at time t is

$$v_z(t) = v_{z0} + \int_0^t (-g)\, d\tau = v_{z0} - gt.$$

Now we substitute into Eq. (7.11) and get

$$z(t) = z_0 + \int_0^t (v_{z0} - g\tau)\, d\tau.$$

The integral is equal to the difference of two integrals, namely,

$$\int_0^t (v_{z0} - g\tau)\, d\tau = v_{z0} \int_0^t d\tau - g \int_0^t \tau\, d\tau.$$

Knowing that one antiderivative of 1 is τ, and one of τ is $\frac{1}{2}\tau^2$, we substitute and obtain

$$z(t) = z_0 + v_{z0}t - \frac{1}{2}gt^2,$$

which is the same result we got in Chapter 6. The advantage of the procedure described by Eq. (7.11) is that it allows us to find the distance *any* object travels if we can describe its acceleration.

Example 12

For the car of Example 10, find the distance traveled after any time t.

From Example 10, we know that the speed of the car at any instant is $v(t) = \frac{1}{4}ct^4$, where the initial speed was zero. Substituting this into Eq. (7.11), we have

$$s(t) = s(0) + \frac{c}{4} \int_0^t \tau^4\, d\tau.$$

Using Table 7.1 again, we know an antiderivative of τ^4 is $\frac{1}{5}\tau^5$. Therefore, if the car starts at $s(0) = 0$, after t seconds it has traveled (in meters)

$$s(t) = \frac{1}{20}ct^5.$$

Example 13

The following questions refer to the speed versus time curve shown below for a car in motion:

(a) When is the car at rest?

(b) When does the car have a constant positive acceleration?

(c) How far has the car traveled in the first 30 min?

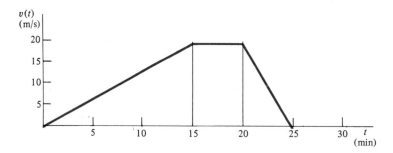

(a) The car is at rest when the speed is zero. From the graph we see that this occurs at $t = 0$ and at $t \geq 25$ min.

(b) Acceleration is the derivative of speed, so the acceleration is a positive constant when the graph is a straight line with positive slope. This occurs from $t = 0$ to $t = 15$ min.

(c) The distance traveled is equal to the area under the curve from $t = 0$ to $t = 30$ min. Breaking the shaded region in the figure into two triangles and a rectangle and converting time to seconds, we find the area to be

$$s = \tfrac{1}{2}(15 \text{ min})(60 \text{ s/min})(20 \text{ m/s})$$

$$+ \ (5 \text{ min})(60 \text{ s/min})(20 \text{ m/s})$$

$$+ \ \tfrac{1}{2}(5 \text{ min})(60 \text{ s/min})(20 \text{ m/s}) = 12{,}600 \text{ m}.$$

Questions

21. The speed versus time curve for a run of the monorail at Disneyland is shown below. The following questions refer to this graph.

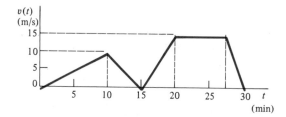

(a) When does the monorail have zero acceleration?
(b) When is the acceleration constant but negative?
(c) Sketch the acceleration versus time graph.
(d) Calculate how far the monorail has traveled during its trip.

22. A toy car has the velocity versus time graph shown below:

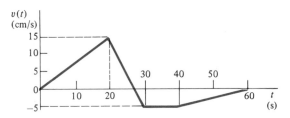

(a) Sketch the acceleration versus time graph.
(b) Calculate the distance traveled in the first 30 s.
(c) Find the distance traveled in the interval 30 s $< t <$ 60 s.

23. An object has an acceleration given by $a(t) = a_0 e^{-kt}$ where a_0 and k are constants. If the object starts from rest when $t = 0$, find its speed at any time t.

24. A body has an acceleration $a(t) = a_0 \sin \omega t$, where a_0 and ω are constants. Find the velocity $v(t)$ of the object, and the displacement $s(t)$, assuming it started with $s(0) = 0$ and $v(0) = 0$.

25. A car is waiting at a stoplight, and when the light turns green the car accelerates uniformly for 6 s at 2 m/s² and then moves with constant speed. At the instant the car started, a truck moving in the same direction with a constant speed of 10 m/s passed it.

(a) Sketch graphs for the car and the truck on the same axes.
(b) When will the car catch up with the truck?
(c) How far will the car have traveled before it catches the truck?

26. An object starts from rest and moves for 10 s with an increasing acceleration $a(t) = (2 \text{ m/s}^3)t$, where t is in seconds.

(a) What is the speed of the object at the end of 10 s?
(b) How far does the object travel in 10 s?

27. A stone is dropped from a diving board which is 5 m above the surface. After hitting the water, it moves with the speed it had at the surface as it sinks 3 m to the bottom of the pool.

(a) How long is the stone in the air (ignoring air resistance)?
(b) What is the speed of the stone when it hits the water?
(c) How long does it take the stone to reach the bottom of the pool from the point it is dropped?

28. A body has a velocity described by $v(t) = v_0 e^{-kt}$, where v_0 is the initial velocity and is equal to 20 m/s, and $k = 0.01 \text{ s}^{-1}$.

(a) What is the acceleration of the body at any instant?

(b) How far does the body travel in 10 s?

7.5 A FINAL WORD

More than 2000 years ago Greek mathematicians developed a method to determine the areas of various geometric figures with curved boundaries. The method of exhaustion, as it is known, consisted of inscribing polygons with an increasing number of edges into a region whose area is to be determined. Archimedes (287–212 B.C.) successfully used this method to calculate the area of a circle and some other special figures.

By the seventeenth century, when algebraic symbols and manipulations had become standard techniques, the method of exhaustion was slowly transformed into what is today known as the process of *integration*, a systematic method for calculating areas and volumes. Many mathematicians in various countries had developed special techniques to treat special problems. In Germany, Kepler found formulas for the volumes of barrels. In Italy, Cavalieri formulated a comparison principle to determine when two solids cut by parallel planes have equal volumes. In France, Descartes, Fermat, and Pascal calculated areas of special regions, as did Wallis in England and Guldin in Switzerland. The air of seventeenth-century Europe was swarming with ideas of differential and integral calculus.

Yet Newton and Leibniz are considered the founders of calculus. Why? They provided an important missing ingredient. Newton and Leibniz recognized that *differentiation and integration are inverse processes*. A few mathematicians had come close to this knowledge, but it was Newton and Leibniz who independently realized the significance of this discovery. Moreover, they exploited this relation to create a systematic and satisfactory calculus which treated whole classes of problems by routine operations, to be executed strictly by rules, without using geometric arguments.

At the ripe age of 23, during the plague year of 1665–6, Isaac Newton undertook an incredible program of study executed in solitude at his family residence in Lincolnshire. One result of his concentration was the genesis of differential and integral calculus. But Newton wasn't much of an extrovert. Being a reserved, uncommunicative individual, he delayed publication of his work for 20 years. As a result, Leibniz's publications of his calculus preceded those of Newton. Each had his own notation, but the ideas and methods were the same. Both of these lofty scholars were conscious of the profound power of calculus to such an extent that they waged a bitter battle over the priority of discovery. Newton's followers charged Leibniz with plagiarism, suggesting that during an exchange of letters with Newton he had learned crucial ideas and later used them in his published work without giving credit to Newton.

In an attempt to settle the dispute, Leibniz appealed to the Royal Society of London, of which he was a member. This was an unfortunate step. Newton was president of the Society. Embittered over the controversy, Newton stage-managed the final report, which was backed by the investigating committee but unsigned. It ruled that Leibniz was essentially guilty as charged.

The controversy was based on nationalistic rivalry rather than scholarship. Today, historians realize that the important discoveries of Newton and Leibniz were made independently, and both men deserve to be recognized as the founders of calculus.

Ironically, the temporary English victory had a deplorable effect on British mathematics for over a century. Blinded by patriotic loyalty to Newton, English mathematicians refused to adopt Leibniz's superior notation and cut themselves off from the spectacular advances in eighteenth-century mathematics and physics that came from continental mathematicians using Leibniz's symbols rather than Newton's.

APPENDIX 1:   THE FUNDAMENTAL THEOREMS OF CALCULUS

Suppose we have a function $y = f(x)$ whose graph lies above the x axis, as shown in Fig. 7.10. We'll denote the area of the shaded region bounded by the graph, a fixed vertical line $x = a$, and a moving line $x = t$ by $A(t)$. The area $A(t)$ is a number which depends on the

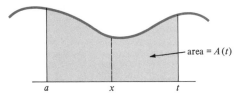

Figure 7.10 Area $A(t)$ of the region under the curve $y = f(x)$ from $x = a$ to $x = t$.

position of the line $x = t$, so it is a function of t, which we'll call the area function. Let's proceed as we did for the parabolic segment and try to find the derivative of this function. As before, we look at the quotient

$$\frac{A(t + h) - A(t)}{h}.$$

The numerator of this quotient is the difference of two areas, as shown in Fig. 7.11a. In Fig. 7.11b we have constructed a rectangle having the same area as the shaded region

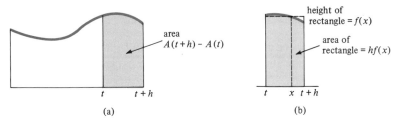

Figure 7.11 (a) A region with area $A(t + h) - A(t)$. (b) A rectangle with the same area.

in Fig. 7.11a. Its base is h and its height is $f(x)$ for some x between t and $t + h$, so its area is $hf(x)$. Therefore

$$A(t + h) - A(t) = hf(x)$$

or

$$\frac{A(t + h) - A(t)}{h} = f(x)$$

for some x between t and $t + h$. But in the limit as h shrinks to zero, the left-hand side of our equation becomes the derivative dA/dt, and the right-hand side $f(x)$ goes to $f(t)$. Thus we have

$$\frac{dA}{dt} = f(t). \tag{7.12}$$

In other words, the derivative of the area function $A(t)$ is the function $f(t)$ we started with. Therefore $A(t)$ itself is an antiderivative of $f(t)$. This shows that the quadrature problem in this general setting is really a problem of finding an antiderivative.

Now suppose that $P(t)$ is some antiderivative of $f(t)$. Then the area function $A(t)$ must be equal to $P(t)$ plus some constant,

$$A(t) = P(t) + C. \tag{7.13}$$

To determine the constant C we use the fact that when $t = a$ (the starting point of our region in Fig. 7.10) the area $A(a) = 0$. But because

$$A(a) = P(a) + C$$

it follows that $C = A(a) - P(a) = -P(a)$. Therefore Eq. (7.13) becomes

$$A(t) = P(t) - P(a). \tag{7.14}$$

Equation (7.12) is now called the first fundamental theorem of calculus, whereas Eq. (7.14) is called the second fundamental theorem.

APPENDIX 2: QUADRATURE OF A HYPERBOLIC SEGMENT. THE LOGARITHM FUNCTION

Figure 7.12 shows the graph of the function $y = 1/x$ for $x > 0$. The curve is called a rectangular hyperbola, and the shaded region between the hyperbola and the x axis from $x = 1$ to

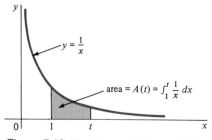

Figure 7.12 Quadrature of a hyperbolic segment.

$x = t$ is called a hyperbolic segment. The area $A(t)$ of this segment is denoted by the integral

$$A(t) = \int_1^t \frac{1}{x}\, dx. \tag{7.15}$$

To calculate this integral by the second fundamental theorem we need to know an antiderivative of the function $1/x$. We turn to our trusty Table 7.2 and find the entry

$$\int x^n\, dx = \frac{x^{n+1}}{n+1} + C \qquad (n \neq -1).$$

But the exponent we need, $n = -1$, is excluded because $n + 1$ appears in the denominator on the right. Since the table doesn't list any antiderivative for x^{-1}, how can we calculate the area function $A(t)$?

Remarkably enough, we can find out everything we want to know about $A(t)$ from the fact that it is the area of a hyperbolic segment. For example, let's see how to determine the shape of its graph.

When $t = 1$, $A(1) = 0$ so the graph crosses the t axis at $t = 1$. As t increases the region gets larger so its area $A(t)$ also increases. To find how fast $A(t)$ increases we calculate its derivative. Since $A(t)$ is an integral, the first fundamental theorem tells us that its derivative is the value of the integrand at the upper limit, so

$$\frac{dA}{dt} = \frac{1}{t}.$$

Therefore we know the slope of the graph of $A(t)$ for each t. When $t = 1$ the slope is 1, and as t increases the slope decreases and gets closer to slope 0 as t gets larger. Thus the graph look something like this for $t \geq 1$:

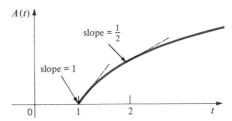

Figure 7.13 Graph of $A(t)$ for $t \geq 1$.

What about values of t smaller than 1? If $0 < t < 1$, we use the same formula (7.15) to define $A(t)$ but we rewrite it as follows:

$$A(t) = -\int_t^1 \frac{1}{x}\, dx,$$

the minus sign arising because we switched the limits. But now $A(t)$ is a negative number, since the integral $\int_t^1 x^{-1}\, dx$ represents the area of the shaded region in Fig. 7.14, and this area is a positive quantity.

As t moves closer to 0 the shaded region in Fig. 7.14 gets larger so $A(t)$ gets more and more negative. Also, the slope $1/t$ gets larger as t approaches 0, so now we can fill in the rest of the graph in Fig. 7.13 and we get a curve like that in Fig. 7.15.

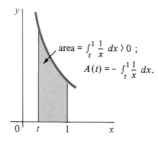

Figure 7.14 $A(t)$ is negative if $0 < t < 1$.

The function $A(t)$ has a name. It's called the *natural logarithm* function, and its values are denoted by $\ln t$. Thus, by definition,

$$\ln t = \int_1^t \frac{1}{x}\, dx \qquad (t > 0).$$

The name is not important. What is important is what we can say about this function. We know its derivative and the general shape of its graph. But its most important property

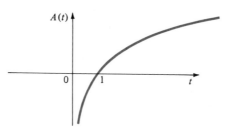

Figure 7.15 The graph of $A(t)$ for $t > 0$.

is about to be revealed: The logarithm of a product is the sum of the logarithms of each of the factors. In symbols, this can be expressed as

$$\ln(at) = \ln a + \ln t$$

for every $a > 0$ and $t > 0$. This relation is called the functional equation for the logarithm.

We can show it is true from our knowledge of the derivative of $\ln t$, together with the chain rule. Let $A(t) = \ln t$ and let $B(t) = A(at) = \ln (at)$, where the factor a is kept fixed. We know that $dA/dt = 1/t$ so by the chain rule for derivatives we find

$$\frac{dB}{dt} = a\frac{1}{at} = \frac{1}{t}.$$

Since $A(t)$ and $B(t)$ have the same derivative, they differ only by a constant, so we can write

$$B(t) = A(t) + C.$$

When $t = 1$ we find $A(1) = 0$ and $B(1) = \ln a$ so the constant in this equation is $C = \ln a$. Therefore the last equation becomes

$$\ln(at) = \ln a + \ln t,$$

which is the functional equation.

Applying this relation repeatedly, you can show two useful properties for every positive integer n:

$$\ln(x^n) = n \ln x,$$

and

$$\ln(x^{-n}) = -n \ln x.$$

Next we show that there is exactly one number whose natural logarithm is equal to 1. This number, like π, occurs often in physics and in mathematics. In fact, we already encountered it in Chapter 3. It is the famous Euler number e,

$$e = \lim_{h \to 0} (1 + h)^{1/h} = 2.718. \ldots \tag{7.16}$$

To see why $\ln e = 1$, or in other words, that

$$\int_1^e \frac{1}{x} \, dx = 1,$$

we introduce a family of curves that are close to the hyperbola, namely, $y = x^{h-1}$ where $h > 0$, and we calculate the integral

$$\int_1^a x^{h-1} \, dx,$$

using the second fundamental theorem. If $h = 0$ this is the integral for $\ln a$, but if $h \neq 0$, the function x^{h-1} is the derivative of x^h/h so we can use the second fundamental theorem to obtain

$$\int_1^a x^{h-1} \, dx = \frac{x^h}{h} \bigg|_1^a = \frac{a^h - 1}{h}. \tag{7.17}$$

Let's find the value of a that makes this integral equal to 1. This requires

$$\frac{a^h - 1}{h} = 1$$

so $a^h = 1 + h$ and hence

$$a = (1 + h)^{1/h}. \tag{7.18}$$

Now we let h shrink to zero. Our family of curves $y = x^{h-1}$ approaches the curve $y = x^{-1}$ in Fig. 7.12, and the number a in (7.18) approaches the Euler number e. This is why $\ln e = 1$.

We can use the same argument to find the value of a that makes $\ln a = b$ for any real b. All we have to do is choose a in (7.17) so that

$$\frac{a^h - 1}{h} = b$$

before we let h shrink to zero. This gives us $a^h = 1 + bh$, so

$$a = (1 + bh)^{1/h} = (1 + k)^{b/k} = [(1 + k)^{1/k}]^b$$

where $k = bh$. Now we let h shrink to zero as before. Then k also goes to 0 and we find

$$\lim_{k \to 0}(1 + k)^{b/k} = e^b$$

because of (7.16). In other words, if $a = e^b$ then $\ln a = b$.

Thus the equation $y = e^x$ means the same as $\ln y = x$. The function $y = e^x$ is the exponential function we found in Chapter 3 as the solution of a special differential equation: it is its own derivative. Because of its connection to the logarithm function, the exponential function is sometimes called the antilogarithm. The logarithm and exponential functions are inverses of each other. If, for example, you know the value of an exponential function and you want to know the power of e that will yield that number, the answer is the natural logarithm of that number. The graphs of these functions are shown in Fig. 7.16; they are reflections of each other through the line $y = x$.

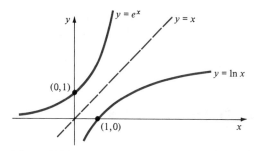

Figure 7.16 Graphs of the functions $y = \ln x$ and $y = e^x$.

Each property of the logarithm can be translated into a property of the exponential. For example, the two equations

$$\ln 1 = 0 \quad \text{and} \quad \ln e = 1$$

become

$$e^0 = 1 \quad \text{and} \quad e^1 = e.$$

The functional equation for the logarithm,

$$\ln(at) = \ln a + \ln t,$$

implies a corresponding functional equation for the exponential,

$$e^u e^v = e^{u+v},\tag{7.19}$$

which is valid for all real u and v. To prove (7.19) let $a = e^u$, $t = e^v$. Then $u = \ln a$, $v = \ln t$, and $u + v = \ln a + \ln t = \ln(at)$, so $at = e^{u+v}$. In other words, $e^u e^v = e^{u+v}$, which is (7.19).

Questions

29. If $a > 0$, $a \neq 1$, and $x > 0$, we define a^x by the equation $a^x = e^{x \ln a}$.
Use this definition to deduce the following properties:

(a) $a^x a^y = a^{x+y}$
(b) $(ab)^x = a^x b^x$.

30. Refer to Question 29. Show that $y = 10^x$ if and only if $x = \ln y / \ln 10$. The quotient $\ln y / \ln 10$ is called the logarithm of y to the base 10 and is often denoted $\log_{10} y$.

CHAPTER 8

THE APPLE AND THE MOON

Hitherto we have explained the phenomena of the heavens and of our sea by the power of gravity, but have not yet assigned the cause of this power. This is certain, that it must proceed from a cause that penetrates to the very centres of the sun and planets, without suffering the least diminution of its force; that operates not according to the quantity of the surfaces of the particles upon which it acts (as mechanical causes used to do), but according to the quantity of the solid matter which they contain, and propagates its virtue on all sides to immense distances, decreasing always as the inverse square of the distances.

Isaac Newton, *Principia* (1686)

8.1 THE GENESIS OF AN IDEA

The year was 1665, the month was August, and Cambridge, England, was besieged by bubonic plague. Isaac Newton, then a 23-year-old university student, retired to the solitude of his family's farm in Lincolnshire until the plague subsided and the university reopened. Not taking kindly to inactivity, Newton composed 22 questions for himself to tackle, ranging from geometric constructions to Galileo's new mechanics to Kepler's planetary laws. During the next 18 months, he immersed himself in the search for answers and along the way discovered calculus, the laws of motion, and the universal law of gravity.

There is a myth in physics that Isaac Newton was inspired one of those plague days in his Lincolnshire orchard, by the fall of an apple, to consider whether gravity was responsible for the motion of the moon as well. Newton himself never wrote about that day in the orchard, but he did reminisce about it to friends some 50 years later. He must have had something in mind when he compared the way the moon "falls" with gravity on Earth, and there is every reason to believe that the fall of an apple gave rise to it.

But the story of the apple reduces one of the greatest discoveries of mankind to a simple bright idea – a flash of insight. The universal law of gravitation did not yield even to the great Newton at his first effort. He battled with it and struggled with questions on the behavior of gravity. In what way must gravity decrease to account for Kepler's third law, which relates the period and radius of a planet's orbit? What other physical quantities could this force depend on? And how is Galileo's law of falling bodies – gravity on the earth – related to gravity in the heavens?

The secret of the heavens was Newton's for nearly 20 years. In 1684 he amazed a trusted friend, Edmund Halley, when he calmly stated that a force law that decreases inversely as the square of the distance leads to orbits that are conic sections (ellipses, circles, parabolas, and hyperbolas). At Halley's entreaty, Newton wrote a nine-page paper, "On the Motion of Bodies in Orbit," which divulged his secret of universal gravitation to the world and later grew into the *Principia*. Halley recognized that Newton's short paper embodied an immense step forward: the physics of the earth became the same as the physics of the heavens.

8.2 THE LAW OF UNIVERSAL GRAVITATION

Newton had been struggling to find an explanation for the basic rules of planetary motion, which had been laid down by Johannes Kepler half a century earlier. What he perhaps realized that day in Lincolnshire was that the explanation of Kepler's orbits would also explain why an apple falls to the earth. But the answer, if he could find it, would also have to resolve the riddle of why all bodies fall at the same rate regardless of their mass.

From his study of Kepler's orbits, Newton already had an inkling of what he needed. The force between any two bodies in the universe would have to diminish as the bodies moved farther apart. The force, he said, would be inversely proportional to the square of the distance between the two bodies. This relationship would satisfy Kepler's empirical law relating the radius of an orbit to its period.

To complete his law of universal gravitation, Newton said that the force of gravity is proportional to the mass of each of the two bodies involved. If m_1 and m_2 are two masses and r is the distance between them, Newton's law of universal gravitation may be expressed as

$$F = G \frac{m_1 m_2}{r^2}.$$

The constant G is a universal constant, having the same value for any two bodies in the universe. This constant should not be confused with g, the acceleration of a body on the surface of the earth due to the earth pulling on it. The constant G must be found from an experiment, but once it is determined, it is the same for *any* two bodies; that's why it is a *universal* constant. In Chapter 10 we'll find out how it was first measured; the value of G is 6.67×10^{-11} N m^2/kg^2.

Force is a vector quantity, but the relation above states only its magnitude. Since gravity tends to pull objects directly toward one another, the direction of the gravitational force between any two bodies is along the line that joins them. We signify the direction by using the unit vector $\hat{\mathbf{r}}$, which, as shown in Fig. 8.1, points in the direction from one

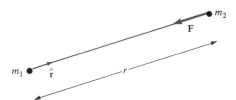

Figure 8.1 Vector quantities in Newton's universal law of gravity.

object to the body on which you want to know the force. The force, being attractive, is in the direction opposite to $\hat{\mathbf{r}}$ Therefore the vector equation for the universal law of gravity is

$$\mathbf{F} = -G\frac{m_1 m_2}{r^2}\,\hat{\mathbf{r}}. \qquad (8.1)$$

Equation (8.1) tells us the force between two *point masses*, which are idealized objects having all their mass concentrated at one point. If there are several masses, like m_1, m_2, and m_3 as shown in Fig. 8.2, how would we calculate the gravitational force on

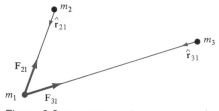

Figure 8.2 Gravitational force on one point mass due to two other point masses.

one of them, say m_1? If only m_1 and m_2 were present, the force from m_2 on m_1 would be

$$\mathbf{F}_{21} = -G\,\frac{m_2 m_1}{r_{21}^2}\,\hat{\mathbf{r}}_{21},$$

where r_{21} is the distance between m_1 and m_2 and $\hat{r}_{21}$ is the unit vector pointing from m_2 to m_1, as shown in Fig. 8.2. Similarly, if only m_1 and m_3 were present, the force from m_3 on m_1 would be

$$\mathbf{F}_{31} = -G\,\frac{m_3 m_1}{r_{31}^2}\,\hat{r}_{31}.$$

How do we calculate the force when both m_2 and m_3 are attracting m_1? A surprising feature of gravity is that the total force on m_1 is the *vector sum* of the forces $\mathbf{F}_{21}$ and $\mathbf{F}_{31}$:

$$\mathbf{F}_1 = \mathbf{F}_{21} + \mathbf{F}_{31},$$

$$\mathbf{F}_1 = -G\,\frac{m_2 m_1}{r_{21}^2}\,\hat{r}_{21} + -G\,\frac{m_3 m_1}{r_{31}^2}\,\hat{r}_{31}.$$

The gravitational force obeys the *superposition principle – the resultant force on a mass is the vector sum of the individual forces*. The superposition principle is a powerful idea which allows us to calculate the gravitational force on an object from any number of bodies.

Example 1

Locate the point between two fixed masses $m_3 = 50.0$ kg and $m_2 = 80.0$ kg, which are separated by 1.0 m, where a third mass $m_1 = 10.0$ kg, feels no force.

Let's call x the distance from m_1 to m_2; then the distance from m_1 to m_3 is $1 - x$, as shown below.

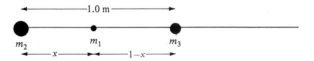

The forces $\mathbf{F}_{21}$ and $\mathbf{F}_{31}$ are in opposite directions, so the magnitude of the total force on m_1 is

$$F_1 = G\,\frac{m_3 m_1}{(1 - x)^2} - G\,\frac{m_2 m_1}{x^2}.$$

Setting this force equal to zero and solving for x, we obtain (after a dose of algebra)

$$x = \frac{\sqrt{m_2/m_3}}{1 \pm \sqrt{m_2/m_3}}.$$

The $\pm$ comes from taking square roots. Since the distance x must be positive and less than 1 m, the physically acceptable solution is the one with the $+$ sign. Therefore, our answer is

$$x = (\sqrt{80/50})/(1 + \sqrt{80/50}) = 0.56 \text{ m}.$$

Newton accounted for the difference in the motion of the moon and of a falling apple by assuming that the force of gravity is inversely proportional to the square of the distance from the earth. But what is "the distance from the earth?" To use the universal law of gravity, you first need to know what r is, and in what direction $\mathbf{r}$ is.

Suppose you have an apple of mass m plummeting to the earth. As illustrated in Fig. 8.3, we can draw many different vectors from some point in the apple to some point in the earth. The gravitational law says that each bit of the apple is attracted by each bit of the earth; the forces $\mathbf{F}_1$ and $\mathbf{F}_2$ in the figure are two such forces. To find the total force on the apple, we would use the superposition principle: apply the law an infinite number of times to cover every bit of the apple and every bit of the earth and vectorially add up all the forces. In essence, the calculation requires integration.

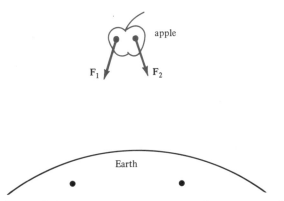

Figure 8.3 Calculating the gravitational force on an apple.

Newton apparently understood how to solve this problem immediately, but perhaps he had difficulty in proving it mathematically. He was a man of rigor and would not consider exposing his ideas in print before he had completely satisfied himself of their soundness. Whether it finally took him years to prove his solution or not, Newton, with the power of his integral calculus, showed that when two spherical objects are not touching, each acts as if all its mass were concentrated at its center. This important realization allows us to consider the earth as a point mass having all its mass concentrated at its center.

Newton proved that not only do spherical objects behave as point masses, but so do spherical shells: the gravitational force from a spherical shell is the same as that from a point mass located at the center of the shell if you're at a point outside of the shell. Spherical shells, he realized, have an additional intriguing property: the force of gravity from a spherical shell itself is zero anywhere inside the shell. To prove this property, Newton unsheathed his integral calculus; his reasoning, though, was based on his physical insight into universal gravitation.

Newton imagined a uniformly dense spherical shell. Since the density is the same everywhere in the shell, the mass contained in a small section of the shell is the same for any part of it. To find the gravitational force on a small mass m located at point P inside the shell, he constructed a narrow double cone with apex at P intersecting the areas A_1 and A_2 on the shell as shown in Fig. 8.4. Area A_1 is a distance r_1 from P and A_2 is a distance r_2 from P. The mass contained in each of these areas pulls on m; in fact, they pull in opposite directions.

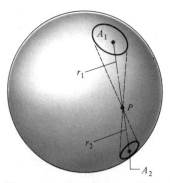

Figure 8.4 Determination of the force on a point mass inside a spherical shell.

Next Newton needed to know how much mass was contained in each area. That's simple. Because the density of the shell is constant, the mass m_1 in the area A_1 is the product of the density ρ, the area A_1, and the thickness t. In other words, $m_1 = \rho A_1 t$. But the area A_1 depends on r_1^2, so m_1 is proportional to r_1^2, $m_1 \propto r_1^2$. Similarly, we have $m_2 \propto r_2^2$. Now we can write the magnitude of the force on m:

$$F = Gm_1 m/r_1^2 - Gm_2 m/r_2^2.$$

Using the proportionality between m_1 and r_1 and between m_2 and r_2, we have

$$F \propto Gm(r_1^2/r_1^2 - r_2^2/r_2^2) = 0.$$

This can be repeated with double cones covering all of the mass in the shell. Thus, because the gravitational force decreases inversely with distance, the force on a mass inside any spherical shell is zero.

Questions

1. If gravity is a result of the earth pulling on, say, an apple, what is the reaction force (of Newton's third law)?

2. If the distance between two objects is tripled, how does the gravitational force between them change?

3. Assuming that gravity caused the earth to condense from interstellar gas, use the universal law of gravity to explain why it is approximately spherical.

4. According to Newton's universal law of gravity, if there were nothing in space except two objects, they would attract each other. Suppose that all space were filled with water, except for two bubbles (which we'll assume contain nothing). Would these bubbles attract, repel, or not affect each other? Why?

5. Imagine yourself to be in a cave deep below the surface of the earth, say about 1000 km down. In the cave there is

   (a) just as much gravity as on the surface of the earth;
   (b) less gravity than on the surface of the earth;
   (c) more gravity than on the surface of the earth.
   Explain your answer.

6. Using the astronomical data of Appendix D, calculate the gravitational force between (a) the earth and the moon and (b) the earth and the sun. Obtain the ratio of these forces.

7. Three point masses, each of mass m, are fixed at the corners of an equilateral triangle of side a, as shown below. Calculate the gravitational force on any one of the masses.

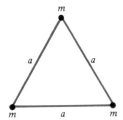

8.3 ACCELERATION OF GRAVITY ON THE EARTH

When you apply the universal law of gravity to the earth, you can think of the earth as having all its mass concentrated at its center. Let's apply this law to an apple falling near the surface of the earth. By Eq. (8.1), if m is the mass of the object and M_E is the mass of the earth, the magnitude of the force of gravity on the object is

$$F = G\frac{mM_E}{r^2}.$$

The direction of this force is along the line from the center of the apple to the center of the earth and r is the distance from the center of the apple to the center of the earth. Suppose, as shown in Fig. 8.5, r is equal to the radius of the earth, R_E, plus the height

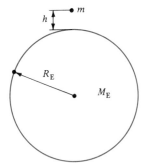

Figure 8.5 Quantities pertaining to gravity on the surface of the earth.

of the apple above the surface of the earth, h. Then the force of gravity can be expressed as

$$F = G \frac{mM_E}{(R_E + h)^2}.$$

We can gain greater insight into this result by an approximation. The radius of the earth is about 6000 km or 4000 mi. The tallest building is about 350 m high. So if we dropped something from the tallest building, the distance from the center of the earth to the object changes from 6000 km + 350 m to 6000 km. The fractional change in height is (350 m)/(6 × 10⁶ m) or about 6/100,000 m. In other words, to extremely high accuracy, the distance between the center of the earth and an object falling near its surface hardly changes at all. Therefore, a very good and very safe approximation is to ignore h completely in the force, and write

$$F = G \frac{mM_E}{R_E^2}.$$

According to Newton's second law, $\mathbf{F} = m\mathbf{a}$, we can calculate the acceleration a of a falling body by substituting F from the gravitational law. Doing just this, we write

$$G \frac{mM_E}{R_E^2} = ma.$$

The mass of the object appears on both sides of the equation, which means that it cancels out, leaving

$$a = G \frac{M_E}{R_E^2}.$$

This result tells us something amazing. Everything on the right-hand side of the equation is a constant on the earth, yet the left-hand side refers to the acceleration of an object falling near the surface of the earth. We've found the reason for Galileo's law of falling bodies. All objects fall to the earth with the same constant acceleration. This happens because the force of gravity depends on the mass of the falling object. A more massive object feels a stronger force, so that by $a = F/m$, the acceleration is constant. That constant acceleration we called g, and now we see how it is related to the characteristics of the earth (or any other planet):

$$g = G \frac{M_E}{R_E^2}. \tag{8.2}$$

Example 2

Knowing that the mass of Mars is one-tenth the mass of the earth and that its radius is half that of the earth, what is the acceleration due to gravity on the surface of Mars?

From Eq. (8.2) we can set up a ratio of g_M to g on the earth:

$$\frac{g_M}{g} = \frac{GM_M/R_M^2}{GM_E/R_E^2} = \frac{M_M}{M_E}\left(\frac{R_E}{R_M}\right)^2,$$

and inserting values of the ratios, we have

$$g_M = g(0.1)(1/0.5)^2 = 0.4g = 3.9 \text{ m/s}^2.$$

Taking the earth to be a sphere of uniform density, but nevertheless imagining that a particle of mass could move from the center of the earth outward to infinity, we see that the force of gravity on the particle would be greatest at the earth's surface. Moving outward from the surface, the force decreases as $1/r^2$. Moving inward, $1/r^2$ increases, but the mass of that part of the earth affecting the particle decreases faster, as we will see in Example 3. Figure 8.6 shows a graph of the magnitude of the gravitational force as a function of distance from the center of the earth.

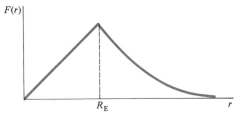

Figure 8.6 Graph of gravitational force versus distance from the earth's center.

Example 3
Suppose that a tunnel could be drilled completely through the earth, passing through its center. What is the gravitational force on a mass m inside the tunnel?

We know that the gravitational force on a body located a distance r from the center of the earth is due entirely to the amount of matter within a sphere of radius r. The shell of matter outside of the object exerts no force on it. Let's assume that the earth's density is ρ. Then the mass M inside a sphere of radius r is the product of ρ and the volume of the sphere, $\frac{4}{3}\pi r^3$, that is, $M = \frac{4}{3}\pi\rho r^3$. This mass can be treated as if it were concentrated at the center of the earth. Therefore, the force on the mass m is

$$F = -G\frac{Mm}{r^2} = -G\frac{\frac{4}{3}\rho\pi r^3 m}{r^2} = -kr,$$

where $k = \frac{4}{3}G\rho\pi m$. The direction of this force is toward the center of the earth. This proves the assertion in the text and explains the origin of Fig. 8.6.

Questions

8. Using the value of g, G, and the radius of the earth (6.4×10^6 m), calculate the mass of the earth.

9. Suppose that an apple is 100 km above the surface of the earth and that the earth somehow uniformly expanded its radius by 100 km while the mass remained constant. Would the force of gravity on the apple be more, less, or the same after the expansion as before? Why?

10. Knowing that the mass of the moon is 7.34×10^{22} kg and its radius is 1.74×10^6 m, calculate the acceleration of gravity on the surface of the moon.

11. The acceleration of gravity on the surface of the earth is known to vary with latitude. For example, $g = 9.78$ m/s^2 at the equator and 9.83 m/s^2 at the poles. In light of Eq. (8.2), can you offer any explanation for this variation?

12. Because a spherical, uniformly dense object can be considered to have all its mass concentrated at its center when calculating the effects of gravity, a weight hung on the end of a string (forming a plumb line) should point toward the center of the earth. Yet a plumb line near the Himalaya Mountains exhibits a small deviation and points slightly toward the mountains. Can you explain why?

13. At what height above the earth's surface would the acceleration of gravity be 4.9 m/s^2? (The mass of the earth is 6.0×10^{24} kg and its radius is 6.4×10^6 m.)

14. At what distance from the center of the earth is the acceleration of gravity half of the value at the surface.

8.4 WHY THE MOON DOESN'T FALL TO THE EARTH

Although Newton at first didn't name the force explicitly, he knew that something had to attract the moon if it was to remain in orbit. The law of inertia stated that the moon would tend to travel in a straight line unless some force acted on it. He coined the word *centripetal force* for any force that is directed inward, toward the center of an object's motion (centripetal means *center seeking*). Gravity is the centripetal force that holds the moon in its orbit.

Before he compared the force of gravity on a falling apple and on the moon, Newton realized that any satellite (the moon for example) is a projectile. He considered horizontally projecting a stone from a high mountain. Figure 8.7 is an illustration of this idea taken from the *Principia*. If the stone is given a small initial speed, it doesn't travel far horizontally before hitting the earth, following the path from V to D in the figure. As we discussed in Chapter 4, its inertia keeps it moving horizontally with a constant speed while at the same time it is falling under the influence of gravity. Now if the stone is projected with a greater speed, it travels farther before it is at last brought to the ground as path VE illustrates.

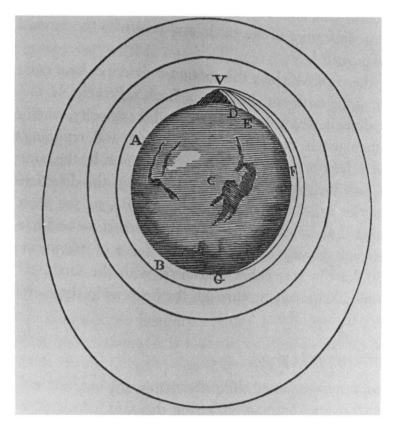

Figure 8.7 Paths of a stone horizontally projected with different speeds from a tall mountain leading to orbital motion (from the *Principia*).

The greater the initial speed of the projectile, the farther it goes before it falls to the earth. But we must also remember that as the projectile falls, the earth curves away from under it. In the absence of air resistance, the stone could be projected so fast that it would follow path VBA and return to the mountaintop. The stone is continually falling, but because of the curvature of the earth it never reaches the ground. Instead, the stone orbits the earth. The moon, Newton realized, has just the right speed to orbit the earth as it does; it is always falling toward the earth, but never reaching it. This, too, is how satellites and space shuttles orbit the earth.

Once Newton understood that the moon is falling, he could determine how far it should fall in 1 s in comparison to an apple falling near the surface of the earth. We already know that the distance s_a an apple falls is described by

$$s_a = \tfrac{1}{2}gt^2. \tag{2.23}$$

In 1 s this turns out to be 16 ft. Using Eq. (8.2) we can express g as

$$g = G\frac{M_E}{R_E^2}, \tag{8.2}$$

and write

$$s_a = \tfrac{1}{2}G(M_E/R_E^2)t^2. \tag{8.3}$$

The moon is not close to the surface of the earth, and consequently it falls with an acceleration different than that given by Eq. (8.2). What is its acceleration? We can find out by using the universal law of gravity as we did earlier to calculate the acceleration on the surface of the earth. The answer is

$$a = G\frac{M_E}{r_m^2},$$

where r_m is the distance from the center of the earth to the center of the moon. The moon, therefore, should fall according to

$$s_m = \tfrac{1}{2}at^2,$$

or

$$s_m = \tfrac{1}{2}G(M_E/r_m^2)t^2. \tag{8.4}$$

On the earth, an apple falls 16 ft (4.9 m) in 1 s. How far, Newton asked, does the moon fall? In other words, what is the ratio of s_m to s_a? Using Eqs. (8.3) and (8.4), we find

$$\frac{s_m}{s_a} = \frac{\tfrac{1}{2}at^2}{\tfrac{1}{2}gt^2} = \frac{a}{g} = \frac{GM_E/r_m^2}{GM_E/R_E^2} = \frac{R_E^2}{r_m^2}.$$

The ratio of the distance the moon falls in 1 s to the distance an apple falls is equal to the square of the ratio of the radius of the earth to the distance to the moon.

Ancient Greek mathematicians had figured out that the distance to the moon is about 60 times the radius of the earth. In his *Principia*, Newton cited values of this distance from Ptolemy, Kepler, Tycho, and Copernicus and used the value of 60 earth radii. Thus the ratio is

$$\frac{s_m}{s_a} = \frac{1}{60^2} = \frac{1}{3600}.$$

This means that if an apple falls 16 ft in 1 s, the moon in 1 s falls

$$s_m = \frac{16 \text{ ft}}{3600} = 0.05 \text{ in.}$$

Each second the moon is falling $\frac{1}{20}$ in.

Now we have a prediction. How can we check it? In other words, how do we calculate how much the moon actually falls in 1 s? We know that the moon goes around the earth in a nearly circular orbit, taking about one month to complete a revolution. According to the law of inertia, the moon does not want to travel in a circle, but rather fly tangentially out and keep moving in a straight line. Gravity from the earth always makes it fall, thereby causing to move in a circle.

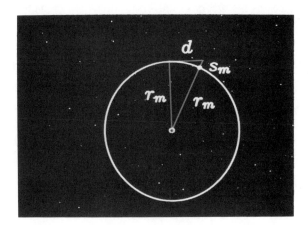

Figure 8.8 Geometry to determine the distance the moon falls in 1 s.

The distance the moon falls in 1 s, s_m, is shown in Fig. 8.8. The distance d is the horizontal distance the moon moves in the same amount of time. From the theorem of Pythagoras we have $(r_m + s_m)^2 = r_m^2 + d^2$, so

$$r_m^2 + 2r_m s_m + s_m^2 = r_m^2 + d^2.$$

Canceling r_m^2, we get $2r_m s_m + s_m^2 = d^2$. Now s_m^2 is a quantity so much smaller than r_m and d that we can safely neglect it to get

$$2r_m s_m = d^2$$

or

$$s_m = \frac{d^2}{2r_m}. \tag{8.5}$$

If we know d, then we know how far the moon falls in 1 s. Because d is so much smaller than r_m, we can approximate d as part of the circle constituting the moon's orbit. Then the ratio of d to the circumference of the circle ($2\pi r_m$) is equal to the ratio of time it takes the moon to travel the distance d, which is 1 s, to the time for the moon to travel once around the circle, which is its period – one month. This means that

$$\frac{d}{2\pi r_{\mathrm{m}}} = \frac{1\ \mathrm{s}}{1\ \mathrm{month}} = 4.2 \times 10^{-7}.$$

Therefore, we know that

$$d = 2\pi(240{,}000\ \mathrm{mi})(4.2 \times 10^{-7}) = 0.63\ \mathrm{mi}.$$

Substituting this value into our expression for s_{m} we obtain

$$s_{\mathrm{m}} = (0.63\ \mathrm{mi})^2/(2 \times 240{,}000\ \mathrm{mi}) = 8.3 \times 10^{-7}\ \mathrm{mi} = 0.05\ \mathrm{in}.$$

Aristotelian physics had been crumbling for more than a century. But all through that time, no one imagined that an experiment done on the earth could reveal the laws of the heavens. That moment arrived when Isaac Newton realized that the moon falls $\frac{1}{20}$ in. every second, just as his theory of universal gravitation predicted. The physics of the heavens and physics on earth became united in one coherent science.

Questions

15. Using the astronomical data of Appendix D, calculate how much the earth falls in 1 s in its orbit around the sun.

16. When an astronaut is orbiting the earth, is his weight, as defined in Chapter 6, really zero? Does this suggest a connection between the term *weightlessness* and free fall? If so, how?

17. Which planet falls toward the sun more in 1 s, Mercury or Earth? Explain your reasoning.

8.5 A FINAL WORD

That $\frac{1}{20}$ in. would have been enough of an accomplishment for any ordinary lifetime. But for Newton, it was barely the beginning. The list of his scientific and mathematical discoveries leaves us breathless.

But not everything he did was scientifically respectable. He spent years of his life immersed in alchemy, Biblical chronology, and other arcane pursuits. In his view, these studies were part and parcel of his search for a system of the world. He was also more than an amateur politician. He was twice elected to Parliament, and, in the year 1705, Queen Anne knighted him, making him Sir Isaac Newton. He was also given a sinecure – a lifetime position as Warden of the Mint. In that capacity, he was responsible for the coin of the realm – and for capturing and interrogating forgers. In 1693, he suffered a nervous breakdown. Some people today think he was suffering from mercury poisoning, possibly contracted during his experiments in alchemy. Some evidence for that has been found by chemical analysis of hairs from his head. Not all historians agree. In any case, he recovered from his illness and went on to become president of the Royal Society, a position he held from 1703 to his death in 1727. Compared to his accomplishments, however, the personal details of Newton's life hardly matter.

Newton gave us not only a series of scientific discoveries, but also a coherent view of how and why the universe works. That view has dominated all aspects of Western thought from his time right down to our very own. Isaac Newton was a human being with faults and flaws – maybe even more than his share of them. But he was also a giant, almost unparalleled in our history.

In *Don Juan*, Lord Byron wrote

> When Newton saw an apple fall, he found
>   In that slight startle from his contemplation –
> 'Tis *said* (for I'll not answer above ground
>   For any sage's creed or calculation) –
> A mode of proving that the earth turn'd round
>   In a most natural whirl, called ''gravitation;''
> And this is the sole mortal who could grapple,
> Since Adam, with a fall, or with an apple.
>
> Man fell with apples, and with apples rose,
>   If this be true; for we must deem the mode
> In which Sir Isaac Newton could disclose
>   Through the then unpaved stars the turnpike road,
> A thing to counterbalance human woes:
>   For ever since immortal man hath glow'd
> With all kinds of mechanics, and full soon
> Steam engines will conduct him to the moon.

CHAPTER

MOVING IN CIRCLES

How I came to dare to conceive such motion of the Earth, contrary to the received opinion of the Mathematicians and indeed contrary to the impression of the senses, is what your Holiness [the Pope] will rather expect to hear. So I should like your Holiness to know that I was induced to think of a method of computing the motions of the spheres by nothing else than the knowledge that the Mathematicians are inconsistent in these investigations. . . .

Taking advantage of this I too began to think of the mobility of the Earth; and though the opinion seemed absurd, yet knowing now that others before me had been granted freedom to imagine such circles as they chose to explain the phenomena of the stars, I considered that I also might easily be allowed to try whether by assuming some motion of the earth, sounder explanations than theirs for the revolution of the celestial spheres might so be discovered.

Nicolaus Copernicus, *De Revolutionibus* (1543)

9.1 THE PERFECTION OF CIRCULAR MOTION

In the fourth century B.C., Greek philosophers turned to the sky and asked, How can we explain the cycles of change – the motions of the stars, sun, and planets? One such philosopher was Plato (427–347 B.C.), who believed that the senses are not to be trusted, that rather the universe can be understood only by pure thought. Understandably, he took little interest in science, but nevertheless he had a profound influence on it because everything he said and wrote was considered important. As a consequence, his work

Timaeus, which discussed a model of the universe, dominated theoretical astronomy for 20 centuries.

The stars, Plato said, represent eternal, divine, unchanging beings. They move at uniform speed around the earth in the most regular and perfect of all paths – an endless circle. Plato accepted motion in a circle with a constant speed – *uniform circular motion* – as so simple and natural an idea that it needed no explanation. All heavenly bodies, he proclaimed, have motions which if not a single perfect circle must be some combination of circles. According to historical tradition, Plato instructed the astronomers to ''save the phenomena'' of the heavenly bodies by using only combinations of uniform circular motions. In addition, Plato thought that it was impossible to know the real mechanism of the heavens. Consequently, the mathematical constructions created by the astronomers to save the phenomena were to be regarded as computational devices, not as pictures of celestial realities. As long as they predicted planetary positions accurately and used only combinations of uniform circular motion, they were satisfactory.

The first geometrical model of the universe of this kind was constructed by Eudoxus, a student of Plato. To explain the sometimes erratic motions of the planets as well as the simpler motions of stars, Eudoxus imagined a series of concentric spheres centered on an immobile Earth. The motion of a particular planet or star was due to the combinations of uniform circular motion of the rotating spheres. To account for all observed motions in the heavens, Eudoxus employed 26 spheres.

Eudoxus was the first in a line of technical astronomers whose task was to explain the apparently confused irregularity of the motions of the planets with the ideal mathematical system of uniform circular motion. Their responsibility was to predict eclipses, the positions of the planets, and the passing of comets. The need for predictions was practical; they were used in agriculture, navigation, and, most important in those days, casting horoscopes. For the fixed stars, simple models were enough. But the planetary motions were more complex: some planets are observed to stop and reverse their motion across the sky. It was difficult to see how such retrograde motion could be due to motion in a circle at a uniform rate about the earth. Figure 9.1 illustrates an example of this problem which confronted early astronomers – the motion of Mercury as observed from the earth. By measuring variations in brightness, the astronomers deduced that Mercury periodically approached and receded from the earth as the figure shows. But how could they explain it?

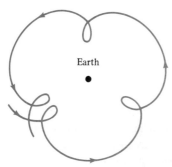

Figure 9.1 Motion of the planet Mercury as observed from Earth.

Late in the third century B.C., the Greek astronomer Apollonius developed epicycles in an attempt to describe the motion of the planets in terms of uniform circular motion. A planet moves at a uniform rate around a small circle, called the epicycle. The epicycle itself moves at a uniform rate around another circle, called the deferent, which is centered on the earth. By adjusting the relative size of the epicycle and deferent as well as the speeds and directions of the two motions, many different types of planetary motion are generated. Figure 9.2 shows an epicycle construction for the observed motion of Mercury (Fig. 9.1).

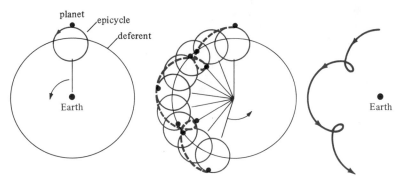

Figure 9.2 Epicycle construction for the orbit of Mercury.

The construction of epicycles was a mathematical technique for generating any curve out of combinations of circles, and early astronomers learned well how to do these calculations. Every time someone made a new and more precise observation, a new epicycle was required. Since these mathematical constructions were not intended to be pictures of reality, astronomers were satisfied, and nobody stopped to wonder why the machinery was so complicated. The peak of this approach to astronomy was reached in the second century A.D. with the publication of the *Almagest* by Claudius Ptolemaeus – Ptolemy – an astronomer from Alexandria, Egypt, which was then a province of the Roman Empire.

The *Almagest* was a compilation of all the astronomical knowledge of antiquity. It became the standard for 1400 years, up to the time of Copernicus. Within its pages, Ptolemy introduced two additional technical ideas to explain the motion of the planets. He placed the center of the deferent away from the earth so that it was eccentric. In addition, the epicycle was to move with uniform speed with respect to the equant, which was yet another point in space, as illustrated in Figs. 9.3 and 9.4. His system was complex, but nevertheless, it worked. It could accurately predict future positions of the heavenly bodies without straying too far from the Platonic dictum of perfectly uniform circular motion.

9.2 DERIVATIVES OF VECTOR FUNCTIONS

Before we discuss uniform circular motion and its application to the heavens, it is convenient to introduce the derivative of a vector function. If a particle moves along a

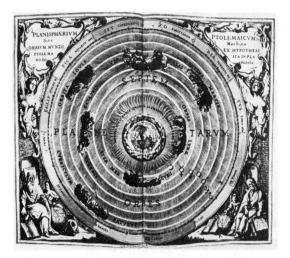

Figure 9.3 The Ptolemaic System. (Courtesy Royal Astronomical Society.)

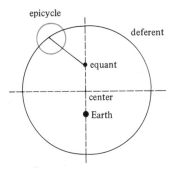

Figure 9.4 Ptolemaic eccentric and equant.

plane curve, its position coordinates (x, y) at time t can be specified by two scalar equations expressing each of x and y as functions of t:

$$x = x(t), \qquad y = y(t).$$

In vector notation, the position vector is

$$\mathbf{r}(t) = x(t)\hat{\mathbf{i}} + y(t)\hat{\mathbf{j}} \tag{9.1}$$

and is illustrated in Fig. 9.5. This is an example of a vector function of a real variable, which here is time.

As t varies through some interval, the position vector $\mathbf{r}(t)$ might change both its magnitude and direction. To study this change, we introduce the idea of the derivative of a vector function. As for ordinary scalar functions (which we encountered in Chapter 3),

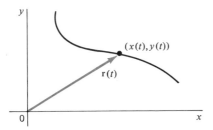

Figure 9.5 Position vector tracing out a plane curve.

we consider the position vector at time t, $\mathbf{r}(t)$, and at some later time $t + h$, $\mathbf{r}(t + h)$. The change in position, shown in Fig. 9.6a, is the vector $\mathbf{r}(t + h) - \mathbf{r}(t)$. The average velocity over that time interval is

$$\bar{\mathbf{v}} = \frac{\mathbf{r}(t + h) - \mathbf{r}(t)}{h},$$

obtained by multiplying the vector difference in position by the scalar $1/h$. The average velocity is parallel to $\mathbf{r}(t + h) - \mathbf{r}(t)$. Now if we allow h to shrink to zero, the two positions become closer together, and in the limit we obtain the derivative of $\mathbf{r}(t)$, which we call the velocity vector $\mathbf{v}(t)$:

$$\mathbf{v}(t) = \frac{d\mathbf{r}}{dt} = \lim_{h \to 0} \frac{\mathbf{r}(t + h) - \mathbf{r}(t)}{h}.$$

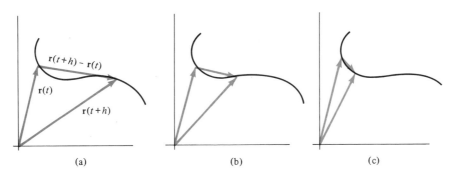

(a)   (b)   (c)

Figure 9.6 Shrinking of $\mathbf{r}(t + h) - \mathbf{r}(t)$.

As Figs. 9.6b and c illustrate, in the limit the velocity vector is tangent to the curve. Expressing the difference quotient in terms of components, we find

$$\frac{\mathbf{r}(t + h) - \mathbf{r}(t)}{h} = \frac{x(t + h) - x(t)}{h}\hat{\mathbf{i}} + \frac{y(t + h) - y(t)}{h}\hat{\mathbf{j}}.$$

As h tends to zero, the components on the right tend to dx/dt, dy/dt. By defining the limit of a vector sum to be the sum of the limits, we find

$$\frac{d\mathbf{r}}{dt} = \frac{dx}{dt}\hat{\mathbf{i}} + \frac{dy}{dt}\hat{\mathbf{j}}.$$ (9.2)

If we know the Cartesian components of the position vector, we can obtain the velocity vector by the prescription of Eq. (9.2), which contains the two scalar equations for its components:

$$v_x = \frac{dx}{dt}, \qquad v_y = \frac{dy}{dt}.$$

The *magnitude* of the velocity vector $\mathbf{v}(t)$ is called the speed $v(t)$. Thus

$$v(t) = |\mathbf{v}(t)| = \sqrt{v_x^2 + v_y^2} = \sqrt{\left(\frac{dx}{dt}\right)^2 + \left(\frac{dy}{dt}\right)^2}.$$

Example 1

The position of a particle varies according to

$$\mathbf{r}(t) = (3 \text{ m/s}^2)t^2\hat{\mathbf{i}} - (2 \text{ m/s})t\hat{\mathbf{j}}$$

where t is in seconds and r in meters. Find the velocity of the particle and its speed at any time t.

The components of the velocity are

$$v_x = \frac{dx}{dt} = \left(3 \frac{\text{m}}{\text{s}^2}\right)\frac{d}{dt}t^2 = \left(6 \frac{\text{m}}{\text{s}^2}\right)t,$$

$$v_y = \frac{dy}{dt} = -\left(2 \frac{\text{m}}{\text{s}}\right)\frac{d}{dt}t = -2 \frac{\text{m}}{\text{s}}.$$

In the above, we used our knowledge of derivatives. Writing this in vector notation, we have

$$v(t) = (6 \text{ m/s}^2)t\hat{\mathbf{i}} - (2 \text{ m/s})\hat{\mathbf{j}}.$$

The speed at time t is

$$v(t) = |\mathbf{v}(t)| = \sqrt{36t^2 + 4} \quad \text{m/s}.$$

The velocity $\mathbf{v}(t)$ is itself a vector function, and we often want to know its derivative, the acceleration $\mathbf{a}(t)$. Figure 9.7a shows the velocity vectors $\mathbf{v}(t)$ and $\mathbf{v}(t + h)$, a short time later. The change in velocity is $\Delta\mathbf{v} = \mathbf{v}(t + h) - \mathbf{v}(t)$, so the average acceleration is

$$\bar{\mathbf{a}} = \frac{\mathbf{v}(t + h) - \mathbf{v}(t)}{h}$$

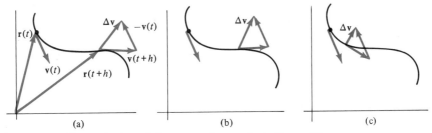

Figure 9.7 Change in the velocity vector $\Delta\mathbf{v} = \mathbf{v}(t + h) - \mathbf{v}(t)$ as the time interval shrinks to zero.

As the time h gets smaller, $\Delta\mathbf{v}$ changes its length and direction, as indicated by Figs. 9.7b and c. In the limit as h shrinks to 0 the average acceleration vector becomes the instantaneous acceleration vector, the derivative of velocity:

$$\mathbf{a}(t) = \lim_{h \to 0} \frac{\mathbf{v}(t + h) - \mathbf{v}(t)}{h} = \frac{d\mathbf{v}}{dt}.$$

In terms of components, we have

$$\mathbf{a}(t) = \frac{d\mathbf{v}}{dt} = \frac{dv_x}{dt}\hat{\mathbf{i}} + \frac{dv_y}{dt}\hat{\mathbf{j}}. \tag{9.3}$$

Unlike the velocity vector, the acceleration vector is not necessarily tangent to the curve $\mathbf{r}(t)$. [It is tangent to the curve described by $\mathbf{v}(t)$.]

Since the acceleration vector is obtained from the position vector by differentiating it twice, we often use the notation

$$\mathbf{a}(t) = \frac{d^2\mathbf{r}}{dt^2} = \frac{d^2x}{dt^2}\hat{\mathbf{i}} + \frac{d^2y}{dt^2}\hat{\mathbf{j}}$$

to indicate this relationship.

Example 2

Calculate the acceleration vector for the particle in Example 1.

Using Eq. (9.3) we form the derivatives and obtain

$$\mathbf{a}(t) = (6 \text{ m/s}^2)\,\hat{\mathbf{i}}.$$

This is a case of constant acceleration because both the magnitude and direction of the acceleration vector are constant.

The same rules of differentiation of scalar functions also hold for vector functions. The derivative of the sum of two vector functions $\mathbf{A}(t)$ and $\mathbf{B}(t)$ is the sum of the derivatives:

$$\frac{d}{dt}(\mathbf{A}(t) + \mathbf{B}(t)) = \frac{d\mathbf{A}}{dt} + \frac{d\mathbf{B}}{dt}.$$

The derivative of the dot product of two vector functions follows from the product rule for scalar differentiation:

$$\frac{d}{dt}(\mathbf{A} \cdot \mathbf{B}) = \mathbf{A} \cdot \frac{d\mathbf{B}}{dt} + \frac{d\mathbf{A}}{dt} \cdot \mathbf{B}.$$

Finally, if $\mathbf{A}(t)$ is a vector function of t and t is a scalar function of another variable, say $t = t(u)$, then for $\mathbf{A}[t(u)]$, the chain rule becomes

$$\frac{d}{du}\mathbf{A}[t(u)] = \frac{d\mathbf{A}}{dt}\frac{dt}{du}.$$

Example 3

What is the second derivative of $\mathbf{A}(t) = 5e^{-t^2}\,\hat{\mathbf{i}} + 3t\hat{\mathbf{j}}$?

We use the chain rule on the first component and find the first derivative to be

$$\frac{d\mathbf{A}}{dt} = 5e^{-t^2}(-2t)\hat{\mathbf{i}} + 3\hat{\mathbf{j}}.$$

Taking the derivative of this function using the product rule on the first component, we get

$$\frac{d^2\mathbf{A}}{dt^2} = [5e^{-t^2}(4t^2) - 2(5e^{-t^2})]\hat{\mathbf{i}} = 10e^{-t^2}(2t^2 - 1)\hat{\mathbf{i}}.$$

Taking the derivatives of vector functions is no more difficult than differentiating scalar functions.

Questions

1. In forming the vector derivative, why were changes in the unit vectors $\hat{\mathbf{i}}$ and $\hat{\mathbf{j}}$ not considered?

2. A particle moves according to $\mathbf{r}(t) = 2t^4\hat{\mathbf{i}} + 5t\hat{\mathbf{j}}$, where t is in seconds and r is in meters.

    (a) Find the velocity of the particle and its speed at $t = 2$ s.
    (b) Calculate the particle's acceleration at $t = 2$ s.

3. The path of an object is described by $\mathbf{r}(t) = 4t^3\hat{\mathbf{i}} - 2t^2\hat{\mathbf{j}}$, where t is in seconds and r in meters.

    (a) When is the object at rest?
    (b) Is the acceleration of the object constant?

4. If the object in Question 3 has a mass of 5 kg, what force is acting on it at any instant?

5. A particle is described by $\mathbf{r}(t) = -2t^3\hat{\mathbf{i}} + 5t\hat{\mathbf{j}}$.

  (a) Find the velocity of the particle at any instant.
  (b) Evaluate $\mathbf{r}(t) \cdot \mathbf{v}(t)$.

6. Compute the first and second derivatives of each of the following vector functions:

  (a) $\mathbf{r}(t) = 2e^t\hat{\mathbf{i}} + 3e^{-t}\hat{\mathbf{j}}$,
  (b) $\mathbf{r}(t) = \ln t^2\hat{\mathbf{i}} - t^{1/2}\hat{\mathbf{j}}$,
  (c) $\mathbf{r}(t) = (\cos \alpha t)\hat{\mathbf{i}} + (\sin \alpha t)\hat{\mathbf{j}}$.

7. A sprinter runs along the oval track shown below. His position at any time t is

$$\mathbf{r}(t) = 4(\cos 2t)\hat{\mathbf{i}} + 2(\sin 2t)\hat{\mathbf{j}}.$$

  (a) Draw the position vector at time $t_1 = \pi/8$ and at time $t_2 = \pi/4$.
  (b) Sketch the average velocity vector during the time interval from t_1 to t_2.
  (c) Draw the instantaneous velocity vectors at times t_1 and t_2.
  (d) Sketch the average acceleration vector during the time interval.
  (e) Show that the acceleration vector $\mathbf{a}(t)$ always has the opposite direction of $\mathbf{r}(t)$.

9.3   UNIFORM CIRCULAR MOTION

To Plato, motion in a circle was natural, perfect, and primary. Uniform circular motion, whether it be of a star, planet, or any object, is an important type of motion in physics, and to describe it we shall use the ideas of vector derivatives. Let's select a Cartesian coordinate system with its center at the center of the circular motion we wish to describe. Figure 9.8 shows our coordinate system.

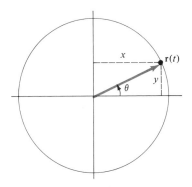

Figure 9.8   Coordinate system to describe uniform circular motion.

Any position in the plane is described by the *radius vector*

$$\mathbf{r} = x\hat{\mathbf{i}} + y\hat{\mathbf{j}}.$$

This vector can also be described by specifying its length $|\mathbf{r}| = r$ and the angle θ that the vector makes with the positive x axis. The two numbers r and θ are called the polar coordinates of the point, and they are related to the Cartesian coordinates by

$$x = r \cos \theta, \qquad y = r \sin \theta.$$

Expressed in terms of polar coordinates, the radius vector is

$$\mathbf{r} = (r \cos \theta)\,\hat{\mathbf{i}} + (r \sin \theta)\,\hat{\mathbf{j}}. \tag{9.4}$$

Up to this point, our description is completely general; it applies to any plane motion in which rectangular coordinates (x, y) are expressed in terms of polar coordinates.

Ancient cultures discovered that if you drive a stake into the ground, attach a rope to it, and draw a figure with the other end of the taut rope as you walk around the stake, then you have a circle. In our mathematical language, this means that the length of the vector $\mathbf{r}$ is the same everywhere on the circle. The vector $\mathbf{r}$ is *not* constant, since its direction changes, but its *magnitude* is constant:

$$|\mathbf{r}| = \text{const} = r. \tag{9.5}$$

So far we've described what we mean by circular motion, but the motion we want to describe is *uniform* circular motion. This means that the object moves around a circle in such a way that the angle θ increases at a constant rate; it is proportional to the time t. Calling the constant of proportionality ω (the lowercase Greek letter omega), we write

$$\theta = \omega t. \tag{9.6}$$

The derivative $d\theta/dt = \omega$ tells us that the rate of change of the angle θ is a constant, implying uniform circular motion. The quantity ω is known as the *angular speed* of the object and has units of radians per second. In addition, ω is related to the time it takes to complete one revolution. This time is called the period T. In one revolution an object moves through 2π radians in time T, so from (9.6) we find its angular speed to be

$$\omega = 2\pi/T. \tag{9.7}$$

Equations (9.5) and (9.6) describe uniform circular motion. By putting $\theta = \omega t$ in (9.4), we can summarize this motion as

$$\mathbf{r}(t) = (r \cos \omega t)\,\hat{\mathbf{i}} + (r \sin \omega t)\,\hat{\mathbf{j}}, \tag{9.8}$$

where r and ω are constants.

The velocity of the object at any instant is the ·derivative of the radius vector. We can easily compute this derivative using Eq. (9.2) and obtain

$$\mathbf{v}(t) = \frac{d\mathbf{r}}{dt} = (-r\omega \sin \omega t)\,\hat{\mathbf{i}} + (r\omega \cos \omega t)\,\hat{\mathbf{j}}. \tag{9.9}$$

In calculating the derivative we used our knowledge of the derivatives of sine and cosine, along with the fact that r and the unit vectors $\hat{\mathbf{i}}$, $\hat{\mathbf{j}}$ are constant.

Note that the dot product of the velocity and radius vectors is

$$\mathbf{v}(t) \cdot \mathbf{r}(t) = -r^2\omega \sin \omega t \cos \omega t + r^2\omega \cos \omega t \sin \omega t = 0.$$

This tells us that the velocity vector is always perpendicular to the radius vector $\mathbf{r}(t)$, as shown in Fig. 9.9. This was to be expected because the velocity vector is always tangent to the curve, and on a circle the tangent is perpendicular to the radius.

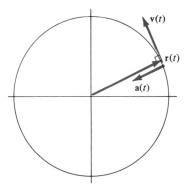

Figure 9.9 Directions of $\mathbf{r}(t)$, $\mathbf{v}(t)$, and $\mathbf{a}(t)$ for uniform circular motion.

Example 4

Prove that for any circular motion, uniform or not, the vectors $\mathbf{r}(t)$ and $\mathbf{v}(t)$ are perpendicular.

Consider the scalar product $\mathbf{r} \cdot \mathbf{r} = r^2$ and take the derivative of both sides of this equation. Using the product rule to take the derivative of the left side, we have

$$\frac{d}{dt}\mathbf{r} \cdot \mathbf{r} = \mathbf{r} \cdot \frac{d\mathbf{r}}{dt} + \mathbf{r} \cdot \frac{d\mathbf{r}}{dt} = 2\mathbf{r} \cdot \mathbf{v}.$$

Since r^2 is a constant, the derivative of the right-hand side is zero. Therefore, we have $2\mathbf{r} \cdot \mathbf{v} = 0$. Excluding the trivial cases when either $\mathbf{r}(t) = \mathbf{0}$ or $\mathbf{v}(t) = \mathbf{0}$, our result implies that $rv \cos \gamma = 0$, where γ is the angle between $\mathbf{r}$ and $\mathbf{v}$, so $\gamma = 90°$. In other words, $\mathbf{r}(t)$ and $\mathbf{v}(t)$ are always perpendicular.

Using Eq. (9.9) and the Pythagorean theorem, we can find the speed:

$$v = \sqrt{v_x^2 + v_y^2},$$

$$v = \sqrt{(-r\omega \sin \omega t)^2 + (r\omega \cos \omega t)^2},$$

$$v = \sqrt{r^2\omega^2(\sin^2 \omega t + \cos^2 \omega t)};$$

because $\sin^2 \theta + \cos^2 \theta = 1$, we have the simple result

$$v = r\omega. \tag{9.10}$$

This tells us that the *speed* is constant as the particle undergoes uniform circular motion. Although the velocity is changing in direction, its magnitude is constant.

We can easily calculate the acceleration by differentiating the velocity given in Eq. (9.9):

$$\mathbf{a}(t) = \frac{d\mathbf{v}}{dt} = \frac{d}{dt}\,[(-r\omega \sin \omega t)\hat{\mathbf{i}} + (r\omega \cos \omega t)\hat{\mathbf{j}}],$$

$$\mathbf{a}(t) = (-r\omega^2 \cos \omega t)\hat{\mathbf{i}} + (-r\omega^2 \sin \omega t)\hat{\mathbf{j}}.$$

Factoring out $-\omega^2$,

$$\mathbf{a}(t) = -\omega^2[(r \cos \omega t)\hat{\mathbf{i}} + (r \sin \omega t)\hat{\mathbf{j}}],$$

and recalling Eq. (9.8),

$$\mathbf{r}(t) = (r \cos \omega t)\hat{\mathbf{i}} + (r \sin \omega t)\hat{\mathbf{j}}, \tag{9.8}$$

we cast our result into a simpler form,

$$\mathbf{a}(t) = -\omega^2\mathbf{r}. \tag{9.11}$$

This tells us that **a** *rotates around with* **r**, *always pointing radially inward*, and is an example of a *centripetal acceleration*.

The magnitude of the acceleration is simply

$$a = \omega^2 r, \tag{9.12}$$

and using Eq. (9.10),

$$v = \omega r, \tag{9.10}$$

we obtain

$$a = v^2/r. \tag{9.13}$$

Any object moving uniformly on a circle has a centripetal acceleration which is directed radially inward and constant in magnitude, as shown in Fig. 9.9.

Example 5

A space shuttle orbits the earth (radius 6.4×10^6 m) at 240 km above the surface, making one revolution in 90 min. What is the acceleration of the shuttle?

From Eq. (9.13) we know that the acceleration has a magnitude

$$a = v^2/r. \tag{9.13}$$

Since the shuttle completes one orbit of radius $r = 6.4 \times 10^6$ m $+ 2.40 \times 10^5$ m $= 6.64 \times 10^6$ m in a time $T = 90$ min $= 5400$ s, its speed is the circumference of the circle $(2\pi r)$ divided by the time T:

$$v = 2\pi r/T.$$

Substituting this into Eq. (9.13), we obtain

$$a = 4\pi^2 r/T^2,$$

$$a = 4\pi^2(6.64 \times 10^6 \text{ m})/(5400 \text{ s})^2 = 8.99 \text{ m/s}^2.$$

Questions

8. A particle moves in a circle of radius 0.25 m in 20 s. Find

   (a) its speed,
   (b) the magnitude of its acceleration.

9. A jet can withstand an acceleration of $5g$, that is, an acceleration that is five times the acceleration due to gravity. What is the radius of the smallest circular path that the jet can safely follow at a speed of 200 m/s?

10. Using the data in Appendix D, calculate the acceleration of the moon.

11. A Ferris wheel that has a diameter of 12 m completes one revolution in 20 s. What is the vector acceleration of a passenger at

   (a) the top?
                                   (b) the bottom?

9.4 CIRCULAR ORBITS

Newton said that the moon stays in its orbit because it is always falling due to gravity. We can now describe the motion of the moon, or any other heavenly body that has a circular orbit, in a new way. According to Newton, the force of gravity on the moon is

$$\mathbf{F} = -G\frac{M_M M_E}{r^2}\,\hat{\mathbf{r}}, \tag{8.1}$$

According to the second law, this force causes the moon to have an acceleration

$$\mathbf{F} = M_M \mathbf{a}.$$

Substituting for the force from Eq. (8.1), we get

$$M_M \mathbf{a} = -G\frac{M_M M_E}{r^2}\,\hat{\mathbf{r}}.$$

The mass of the moon cancels out, leaving

$$\mathbf{a} = -G\frac{M_E}{r^2}\,\hat{\mathbf{r}}.$$

This tells us that the acceleration is radially inward, toward the center of the earth.

The orbit of the moon is very nearly a circle, and the acceleration we've found is a centripetal acceleration. From Eq. (9.13), the magnitude of this acceleration should be

$$a = v^2/r. \tag{9.13}$$

Here we can equate our two expressions and solve for the speed,

$$v = \sqrt{GM_E/r}. \tag{9.14}$$

At precisely this speed, the force of the earth's gravity makes the moon fall just the right amount to stay in its circular orbit. This principle is true not only of the moon but also for any satellite of any planet, including artificial ones. It is the basic mechanism of the solar system.

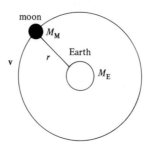

Figure 9.10 Quantities pertaining to the moon's orbital motion.

Example 6

What horizontal velocity would a golf ball need in order to orbit the earth?

Taking the radius of the orbit as the radius of the earth, and using Eq. (9.14), we have

$$v = \sqrt{GM_E/R_E}$$

$$= \sqrt{(6.67 \times 10^{-11} \text{ N m}^2/\text{kg}^2)(6.0 \times 10^{24} \text{ kg})/(6.4 \times 10^6 \text{ m})}$$

$$= 7900 \text{ m/s,}$$

which is over 20 times the speed of sound!

We can also obtain a relationship between the period of a circular orbit and the radius of the orbit. using Eqs. (9.7) and (9.12), we can express the centripetal acceleration as

$$a = \omega^2 r = 4\pi^2 r/T^2.$$

Equating this to the acceleration caused by gravity from a body of mass M,

$$a = GM/r^2,$$

we obtain

$$T^2 = \frac{4\pi^2 r^3}{GM}. \tag{9.15}$$

This relationship, a shining success of the Copernican revolution, is a special case of Kepler's third law, which we shall encounter in Chapter 25.

Questions

12. The planet Mars has a satellite, Phobos, which has an orbital radius of 9.4×10^6 m and a period of 7 h 39 min. Calculate the mass of Mars from these data.

13. Two satellites have circular orbits about the earth and initially one satellite is 100 km directly above the other. Will this satellite always remain directly above the other?

14. Two satellites, one having five times the mass of the other, orbit the earth with the same speed and radius. Which of the following statements must be true for this to be possible?

  (a) They are beyond the pull of the earth's gravity.
  (b) The net force on each satellite is zero.
  (c) There is a universal gravitational constant.
  (d) The gravitational force is directly proportional to the mass of each satellite.
  (e) The gravitational force is inversely proportional to the square of the distance from the center of the earth.

15. What is the radius of the orbit of a geosynchronous communications satellite which at all times is directly above a point on the equator? Can a spy satellite hover constantly over, say, Moscow? Explain.

16. Taking the radius of the orbit of Mars about the sun as 1.52 times that of the earth, determine the number of years for Mars to make one revolution about the sun.

17. Suppose a satellite could be put into orbit in an evacuated circular tunnel inside the earth. Would the speed of a satellite in such an orbit be greater or less than that of the golf ball in Example 6?

9.5 MOTION ALONG SPACE CURVES

The concepts of velocity and acceleration are easily extended to motion that is not always in the same plane. If a particle moves along a space curve its position coordinates (x, y, z) at time t can be specified by three scalar equations,

$$x = x(t), \qquad y = y(t), \qquad z = z(t),$$

or by one vector equation,

$$\mathbf{r}(t) = x(t)\hat{\mathbf{i}} + y(t)\hat{\mathbf{j}} + z(t)\hat{\mathbf{k}}.$$

The velocity vector in this case is given by

$$\mathbf{v}(t) = \frac{d\mathbf{r}}{dt} = \frac{dx}{dt}\hat{\mathbf{i}} + \frac{dy}{dt}\hat{\mathbf{j}} + \frac{dz}{dt}\hat{\mathbf{k}},$$

and the acceleration vector is

$$\mathbf{a}(t) = \frac{d\mathbf{v}}{dt} = \frac{d^2x}{dt^2}\hat{\mathbf{i}} + \frac{d^2y}{dt^2}\hat{\mathbf{j}} + \frac{d^2z}{dt^2}\hat{\mathbf{k}}.$$

Example 7

Uniform motion on a circular helix.

Suppose a particle moves in such a way that its x and y coordinates undergo uniform circular motion but the z coordinate increases at a constant rate. The position vector of such a motion is

$$\mathbf{r}(t) = (r\cos\omega t)\hat{\mathbf{i}} + (r\sin\omega t)\hat{\mathbf{j}} + bt\hat{\mathbf{k}},$$

where r, ω, and b are constants. The particle moves along a circular helix which winds around a circular cylinder of radius r, as shown in Fig. 9.11. Calculate the velocity and acceleration vectors.

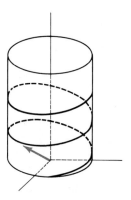

Figure 9.11 Motion on a helix.

Differentiating the position vector, we find the velocity to be

$$\mathbf{v}(t) = (-r\omega\sin\omega t)\hat{\mathbf{i}} + (r\omega\cos\omega t)\hat{\mathbf{j}} + b\hat{\mathbf{k}}.$$

Differentiating the velocity vector, we find

$$\mathbf{a}(t) = (-r\omega^2\cos\omega t)\hat{\mathbf{i}} - (r\omega^2\sin\omega t)\hat{\mathbf{j}}.$$

The velocity vector is tangent to the helix, but the acceleration vector has z component zero. It points radially and horizontally toward the central axis of the cylinder.

Questions

18. (a) Show that the speed and the magnitude of the acceleration of the helical motion in Example 7 are both constant.

   (b) Show that the velocity vector in Example 7 makes a constant angle with the z axis.

19. A particle moves in space with position vector

$$\mathbf{r}(t) = (\cos t)\,\hat{\mathbf{i}} + (\sin t)\hat{\mathbf{j}} + e^t\hat{\mathbf{k}}.$$

   (a) Calculate the velocity vector at any time t.

   (b) Calculate the speed at any time t.

   (c) Calculate the acceleration vector at any time t.

20. A particle moves in space with position vector

$$\mathbf{r}(t) = (t - \sin t)\hat{\mathbf{i}} + (1 - \cos t)\hat{\mathbf{j}} + (4 \sin \tfrac{1}{2}t)\,\hat{\mathbf{k}}.$$

   (a) Calculate the velocity vector at any time t.

   (b) Show that the speed is constant.

   (c) Calculate the acceleration vector at any time t.

9.6 A FINAL WORD

In the latter days of the Roman Empire, Ptolemy surveyed the work of his predecessors and then set out to provide a more complete and exact description of the motion of the ''wanderers,'' the planets. In preserving the centuries-old Platonic idea that heavenly bodies execute uniform motion about the earth, Ptolemy out of necessity had to introduce a new device, the equant, to describe retrograde motion. Despite the complexity of circle moving on circle, the Ptolemaic system fitted available observations with remarkable accuracy. It was the culmination of centuries of attempts to describe complex motions of the heavens. A technical triumph, the Ptolemaic system gave unparalleled certainty in predictions of planetary positions without straying too far from the ideal of uniform circular motion.

Copernicus, however, believed that the Ptolemaic system of the *Almagest* was not sufficiently Platonic. In trying to describe the motion of the sun around the earth, for example, Ptolemy was unable to describe the motion as uniform circular motion. The motion was circular, but it was not uniform – the speed of the sun was not constant unless it was viewed from the equant. Unsatisfied with Ptolemy's system, Copernicus thought that perhaps he could restore the uniform circular motion and perfection of the heavens by placing the sun at the center of the universe. Most significant, Copernicus believed that astronomy should not only predict positions of heavenly bodies but also describe the real structure of the heavens. The result, as we know, was a profound revolution in physics as well as astronomy.

CHAPTER 10

FORCES

I don't know what I may seem to the world, but, as to myself, I seem to
have been only like a boy playing on the sea shore, and diverting myself in
now and then finding a smoother pebble or a prettier shell than ordinary, whilst
the great ocean of truth lay all undiscovered before me.

Sir Isaac Newton

10.1 THE FUNDAMENTAL FORCES

Using crude water clocks to time balls rolling down inclined planes, Galileo searched
for and found a description of how bodies fall. His law of falling bodies, however, wasn't
a fundamental law of nature. Within half a century it was superseded by a deeper insight
into nature – Newton's universal law of gravity. Through the genius of Newton, the
force of gravity,

$$\mathbf{F} = -G\,\frac{M_1 M_2}{r^2}\,\hat{\mathbf{r}}, \tag{8.1}$$

was revealed as a fundamental force of nature.

Gravity acts on all matter, a fact reflected in its dependence on the masses of objects. Although its strength diminishes with distance, the effects of gravity are nevertheless felt across the far reaches of the universe. Gravity holds together planets and stars, organizes solar systems and galaxies; it orders the universe.

Inspired by Newton, scientists in the eighteenth century sought to identify, classify, and mathematically describe the numerous forces observed in nature. Knowledge of these forces provided physics with a certain predictive power, because according to Newton's second law, $\mathbf{F} = m\mathbf{a}$, forces shape the motion of all things. Through painstaking experiments, these scientists developed empirical descriptions of forces in the world about them: tensions, spring forces, viscosity, friction, electricity, magnetism, heat, light, chemical action. As the number of forces grew, so did the applications in an increasingly industrialized world. Yet there was a question confronting these physicists. Were all these forces fundamental, or could they be reduced to more basic forces?

Not until late in the eighteenth century did another force emerge as fundamental – the electrical force. The French engineer Charles Augustin Coulomb assumed that, analogous to the gravitational force between two masses, the electric force between two charges is proportional to the product of the charges. Experimentally he found that the electric force is similar to gravity in another way: the force between two charges decreases as the square of the distance between them. Summarized mathematically, the electric force $\mathbf{F}$ between two charges q_1 and q_2 that are separated by a distance r is known as Coulomb's law and written

$$\mathbf{F} = K_e\,\frac{q_1 q_2}{r^2}\,\hat{\mathbf{r}}. \tag{10.1}$$

Just as G is a universal constant for gravity, K_e is a universal constant for electricity.

Magnetism was also identified as a fundamental force of nature. The attraction or repulsion between two magnets could be described by a force similar to Coulomb's law. The progress of physics was a triumph of Newtonian mechanics: the forces of nature were successively reduced to attractions and repulsions between particles.

The first 40 years of the nineteenth century, however, saw a growing reaction against such a division of phenomena, in favor of some kind of correlation of forces. The turn inward to unification of forces was spearheaded by James Clerk Maxwell. By the second half of the nineteenth century Maxwell succeeded in unifying two hitherto disparate forces, electricity and magnetism, into one – electromagnetism. Maxwell's unification of electricity and magnetism was expressed by a set of equations that interrelate electric and magnetic phenomena. Soon tensions, spring forces, friction, viscosity, chemical actions, and even light were recognized as arising fundamentally from the electromagnetic force; this force dominates the everyday world about us. Based on Maxwell's success the search for a common mathematical description, or unification, of forces had begun.

With the twentieth century came the discovery of radioactivity, the probing of atoms, and the subsequent realization that more than just gravity and electromagnetism would be needed to explain this new world. Experiments probing invisible atoms revealed that inside an atom there is a compact center – the nucleus – composed of positively charged protons and neutral neutrons. Negatively charged electrons orbit the nucleus, held by the electric force from the protons. This naturally led scientists to ask what held the nucleus together, since the protons in it would be expected to repel one another, and the neutrons in it should neither repel nor attract the protons or one another. Physicists realized that neither gravity nor electromagnetism held the compact nucleus together, but that a new force was at work. Aptly named the *strong force*, it overcomes the electric repulsion between protons and holds the nucleus together. Unlike gravity and electricity, the strong force does not extend to distant corners of the universe; it has a limited range – the size of a nucleus, 10^{-13} cm. Outside this range, the strong force has no effect. If it did, the universe would be one very dense lump of subatomic particles.

Natural radioactivity could not be explained by any of the known forces – strong, electromagnetic, or gravitational. Another force was responsible for the decays of nuclei – the weak force. This force is 10^5 times weaker than the strong force, but like the strong force it has a limited range, which is the size of a nucleus. Nevertheless, the weak force causes some stars ultimately to explode.

Table 10.1 summarizes the four fundamental forces of nature, the strong, electromagnetic, weak, and gravitational; their relative strengths; and their respective ranges.

Table 10.1 Characteristics of the Four Fundamental Forces

Force	Relative strength	Range	Importance
Strong	1	10^{-13} cm	Holds nucleus together
Electromagnetic	10^{-2}	Infinite	Friction, tensions, etc.
Weak	10^{-5}	10^{-13} cm	Nuclear decay
Gravitational	10^{-39}	Infinite	Organizes universe

The behavior of each of the four forces is reasonably understood, but nobody knows why there should be four of them. Albert Einstein spent the last 20 years of his life unsuccessfully searching for a way to unify two of the forces, gravity and electromagnetism. In the twentieth century a search for unification of the fundamental forces has become an important part of physics. The water clocks and inclined planes of Galileo have been replaced by increasingly larger, more energetic particle accelerators. Emerging are unified and grand unified theories that present a coherent account of how these forces may have evolved from simpler laws in the infancy of the universe. The early universe ultimately may be the only experimental test for such theories. It may be the great ocean of truth that still lies undiscovered before us.

10.2 GRAVITATIONAL AND ELECTRIC FORCES

One of the great and deep mysteries of physics is that the laws describing gravitational force and electric force have the same mathematical form:

$$\mathbf{F} = -G\frac{M_1 M_2}{r^2}\,\hat{\mathbf{r}} \qquad (8.1)$$

and

$$\mathbf{F} = K_e \frac{q_1 q_2}{r^2} \hat{\mathbf{r}}.$$ (10.1)

It seems almost a minor point that each has an unknown universal constant in it. Yet for real-world applications, it is essential to know what those constants are.

For gravity, we already know that G is related to the acceleration of a falling body near the surface of the earth, g, and the mass and radius of the earth:

$$g = GM_E/R_E^2.$$

The radius of the earth has been known for a long time, and we also know g, so measuring G amounts to finding the mass of the earth, M_E. The determination of G was one of the classic experiments of physics.

Henry Cavendish, a British physicist, performed the historic experiment to measure G in 1798. Cavendish was deeply inspired by Newton and regarded the *Principia* as the model for exact sciences, and the search for the forces between particles guided his scientific explorations. But Cavendish had fitful habits of publication; he left unpublished whatever did not fully satisfy him. Luckily the determination of G was an experiment of which he was proud.

Figure 10.1, which is taken from Cavendish's 1798 article, shows the apparatus he used for his delicate experiment to determine the value of G. In that experiment, he measured forces equal to one-billionth of the weights of the bodies involved. The two small lead balls are attached to a rigid rod, forming a dumbbell that is suspended by a thin fiber that allows the dumbbell to rotate freely. When the two larger lead balls are placed near the ends of the dumbbell, the smaller masses are attracted to the larger ones by the gravitational force. This force, although extremely small, nevertheless rotates the dumbbell and twists the fiber, which opposes the twisting. Using a telescope, Cavendish sighted the balls against scales illuminated by candles and thereby measured the amount of twisting. From that, he determined the force between the two balls, and then through Eq. (8.1) found the value of G. The accepted value is

$$G = 6.67 \times 10^{-11} \quad \text{N m}^2/\text{kg}^2.$$

By weighing the world, Cavendish rendered the universal law of gravitation complete. The law was no longer a proportionality as Newton had stated it, but an exact law through which quantitative analyses could be made. It was the most important contribution to gravitation since Newton.

Questions

1. Why are lead masses used instead of, say, rubber masses?

2. Newton's universal law of gravity holds for particles, but the spheres used in the experiment are not particles but extended objects. Does this affect the result of the experiment?

3. Using the values of G, g, and the radius of the earth, calculate the mass of the earth. From your value for the mass, determine the density of the earth by treating

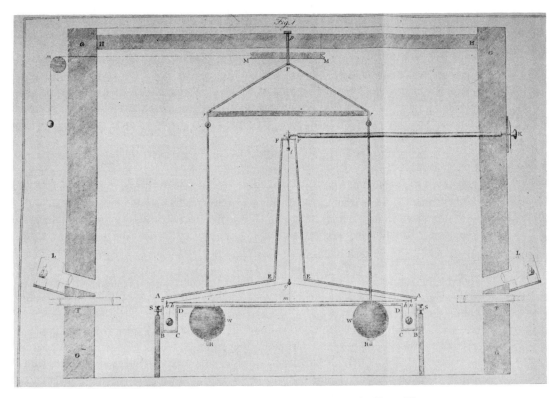

Figure 10.1 Cavendish's apparatus for measuring G. (From *The Scientific Papers of the Honourable Henry Cavendish*, Cambridge University Press.)

the earth as a solid sphere. The average density of rocks on the earth's surface is about 2.0 g/cm³. Comparing this value to your calculated value, what can you conclude about the interior of the earth?

Through a similar experiment in 1787, Coulomb showed that the electric force between two charges is similar to gravity: it decreases as the inverse square of the distance between the charges. In fact, Cavendish himself had already performed an even better experiment than Coulomb's, but, characteristically, he never published it. In any case, the value of the electric constant K_e could be measured; its value is 9.0×10^9 N m²/ C², where one coulomb (1 C) is the unit of charge. But what is charge?

The early Greeks had discovered that amber attracts bits of straw, and they identified that property of amber with charge. Charge is that which creates electric forces; even today that's all we can say about it. We don't know exactly what charge is any more than we know what mass is. Unlike mass, charge comes in two different varieties – positive and negative. There also is a smallest unit of charge – the charge of the proton (or electron). All charges come in multiples of this unit of electricity; the charge of a proton is 1.6×10^{-19} C. Furthermore, like charges repel, whereas opposite charges attract. Consequently, there can be attractive or repulsive forces between charges, and electricity can be neutralized. Because the electric force is comparatively strong, opposite charges attract and neutralize each other. Gravity, on the other hand, is always attractive.

Atoms consist of positively charged nuclei and negatively charged electrons. A proton is about 2000 times heavier than an electron. The simplest atom, hydrogen, consists of one proton and one electron in a region of 10^{-8} cm. This region is much bigger than the nucleus, so we imagine a low-density cloud of negative charge attached to the positive nucleus. The force that holds the electron cloud to the proton to make a hydrogen atom is the electric force.

To construct a model of heavier atoms, we first construct nuclei with more protons and neutrons in them. Since the overall charge of atoms is zero, there are as many electrons in clouds about the nucleus as there are protons in the nucleus. Even though the electrons repel one another, they are held to the nucleus by the electric force.

Atoms can in turn attract other atoms to make larger composites called molecules. The force that holds the atoms together to form molecules is again the electric force. Atoms and molecules can form larger agglomerations, which we see as liquids and solids. These too are held together by electric forces. Electricity is a fundamental force that governs the nature of the world around us. At distances small compared to the nucleus, the strong and weak forces dominate. Over distances large compared to the earth, gravity dominates. For the world of matter as we know it, electric forces are dominant.

The gravitational and electric forces have the property that they act on distant objects through seemingly empty space. The moon, for example, feels a gravitational force from the earth 240,000 mi (390,000 km) away. Newton was bothered by the action-at-a-distance character of gravity and thought that there should be a physical mechanism for transmitting this force. Yet when pressed for such a mechanism, he declared, "I make no hypotheses."

Question

4. Compare the electric force between the proton and electron in a hydrogen atom with the gravitational force between them. Does your answer explain why gravity is not responsible for binding atoms together? (Use $m_p = 1.67 \times 10^{-27}$ kg, $m_e = 9.11 \times 10^{-31}$ kg.)

10.3 CONTACT FORCES

The fundamental forces of nature act at a distance: their effects can be experienced when the particles are not in contact. A second category is *contact* forces. These are forces that two objects exert on each other when they are physically in contact with each other, as, for example, when a book rests on a table. Contact forces are not fundamental forces; instead, they arise fundamentally from electric forces acting in complicated ways.

The force a spring exerts on an object is an example of an electric force. Inside the spring are metal atoms that are bound together by electric forces. These electric forces keep the metal atoms a certain distance apart, called the equilibrium distance. When you stretch a spring, each atom is pulled a tiny bit out of the equilibrium distance. The electric forces try to pull the atoms back into the equilibrium position. The net result of all the electric forces acting on the atoms is what causes the end of the spring to pull on you, that is, to exert a macroscopic force.

Trying to describe how a spring works by examining the electric forces acting between the atoms is impossible, because of the sheer numbers of atoms to consider; there might be 10^{24} atoms in a mousetrap spring. Rather than attempting to describe all these complicated interactions in terms of a fundamental force, we describe them by *empirical rules*, which are experimental summaries of the net result of all the complications. Most of these empirical descriptions were deduced by eighteenth-century scientists.

The empirical law for a spring is simple: the force exerted by a spring (on an object) is proportional to the change in length of the spring. This is known as Hooke's law (after Robert Hooke, a contemporary of Newton) and may be expressed as

$$\text{Hooke's law:} \quad F = -kx, \tag{10.2}$$

where x is the change in length of the spring. The constant k is called the spring constant and is a measure of the stiffness of a spring; the stiffer the spring, the larger the value of k. The direction of the force is always opposite to the displacement of the end of the spring from its unstretched position. When $x > 0$, the spring is stretched and F is negative; when $x < 0$, the spring is compressed and F is positive. The spring force always acts to restore the spring to its unstretched length, as Fig. 10.2 illustrates.

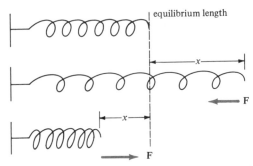

Figure 10.2 Force exerted by a spring described by Hooke's law.

Example 1

A 1.5-kg block on a frictionless table is attached to a spring with spring constant $k = 0.5$ N/m. If the spring is stretched 3.0 cm and released, what is the acceleration of the block at the instant it is released?

From Hooke's law, we know that the force exerted on the block by the spring is $F = -kx$, where $x = 0.03$ m. By Newton's second law, this force accelerates the block according to $F = ma$. Therefore the acceleration of the block at the instant it is released is

$$a = F/m = -kx/m,$$

$$a = -(0.5 \text{ N/m})(0.03 \text{ m})/(1.5 \text{ kg}) = -0.01 \text{ m/s}^2.$$

Since the force $F = -kx$ changes as the spring contracts, the acceleration is not constant.

The tension in a rope or string is another example of an electric force. As in a spring, the atoms in the rope have equilibrium positions at which electric forces tend to keep them. When you pull on one end each atom electrically tugs on its neighbors, but unlike a spring, the rope doesn't stretch because it can't uncoil. The pull is transmitted to the other end of the rope, usually undiminished in force, much like a chain link, as Fig. 10.3 illustrates.

Figure 10.3 Tension in a rope arises from electric interactions.

Whenever any object is pressing against another there is a contact force between the two objects, known as the normal force. (Here *normal* means perpendicular, not the opposite of *abnormal*.) This force is a result of repulsion between the atoms of the two objects. The magnitude of the normal force depends on how hard the two objects press against each other. The direction of the normal force acting on an object, however, is always perpendicular to the surface. Figure 10.4 indicates normal forces between different objects.

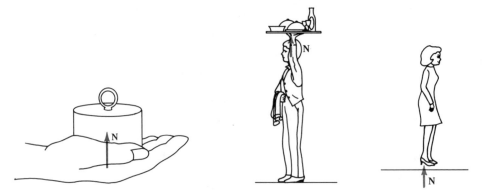

Figure 10.4 Illustration of the normal force between different objects.

Friction is an inescapable example of an electric force. At times we wish that we could do away with it, so as, for example, to improve engine performance, yet without it we couldn't walk. Even though a highly polished object may appear smooth, when

examined through a microscope it appears very rough, having many tiny surface irregularities, as shown in Fig. 10.5. When two objects are placed in contact, the many contact
points resulting from the (microscopic) rough edges actually become welded together by
electric forces. When one object moves across another, these tiny welds rupture and
continually reform. The net result is friction – a force parallel to the surface which
opposes the motion of the object. Since the number of welds is proportional to pressure
from the object on the surface, the force of friction is proportional to the normal force
on the object.

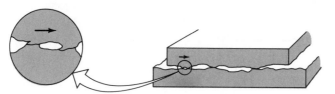

Figure 10.5 Microscopic examination of a highly polished surface
reveals irregularities.

Example 2

A mover pulls on a 20.0-kg crate resting on a floor with a force of 80.0 N at an angle
of 37°, but the crate does not move. What is the normal force of the floor on the crate?

Let's first list the forces that act *on the crate*:

(a)   tension $T = 80.0$ N from the rope,
(b)   weight $W = mg = 196$ N,
(c)   normal force N from the floor,
(d)   friction f from the floor.

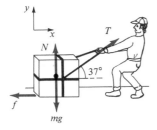

In the diagram we indicate these forces as well as a set of coordinate axes. Since the
block has zero acceleration (it doesn't move), Newton's second law implies $\Sigma \mathbf{F} = \mathbf{0}$,
which gives us two scalar equations,

$$\Sigma F_x = 0, \qquad \Sigma F_y = 0.$$

The normal force is in the y direction, so let's add up the components of all the forces
in that direction and, by the second law above, set them equal to zero. The result is

$$\Sigma F_y = N - mg + T \sin 37° = 0.$$

Solving for N, we get

$$N = mg - T \sin 37° = 196 - (80) \sin 37° = 148 \text{ N}.$$

Does this make sense? If the crate were simply sitting on the floor with no rope pulling on it, the floor would support the entire weight of the crate and then $N = mg$. But the rope has a vertical component, which helps support the weight of the crate and decreases the normal force from the floor, just as we found.

In the *Principia* Newton considered the motion of objects in resistive mediums – things like cannonballs speeding through air or marbles falling in water. The resistive force in these cases is due to *viscosity* and is extremely complicated to work out in detail. But when an object moves at a relatively low velocity through a fluid, such as a gas or liquid, the viscous force may be approximated by assuming that it is proportional to the velocity. Mathematically we write

$$\mathbf{F}_{\text{vis}} = -K\eta\mathbf{v}. \tag{10.3}$$

The minus sign indicates that this force is always opposite to the velocity of the object; K is a proportionality constant that depends on the shape of the object and η (the Greek letter eta) is the coefficient of viscosity, which depends on the internal friction between different layers of the fluid.

Questions

5. Is the normal force that acts on an object resting on a surface the reaction force to its weight? Explain.

6. Do tensions in rope always pull on objects to which they are connected? In other words, can ropes ever push?

7. A parachutist approaches a constant velocity, called the terminal velocity, when falling through air. What condition between forces is satisfied when this occurs?

10.4 APPLICATION OF NEWTON'S LAWS

The great sixteenth-century humanist Erasmus, when he was a student, wrote a letter to a friend saying how dull mechanics lectures were. Generations of students have learned Newton's laws by solving all sorts of dull as well as interesting problems in mechanics. Nobody will ever know how many minds, eager to learn the secrets of the universe, found themselves studying inclined planes and pulleys instead, and decided to become businessmen. Nonetheless, through solving problems you can come to understand physics.

The success of Newtonian mechanics was the identification of forces and the subsequent dynamical explanation of the motion of objects influenced by these forces. Our task is to apply Newton's laws and analyze the motion of objects. Let's list a few extremely useful steps, which once mastered will allow you to solve practically any problem in mechanics:

1. Draw a *free-body diagram* for every object whose motion is to be analyzed. This entails drawing each object separated from all others and clearly indicating the forces acting on each by arrows which either start or end on the object.

2. Label all forces acting on the objects. Use the same symbol for those forces that are the same through Newton's third law.

3. Indicate the direction of the acceleration of each object.

4. Choose a coordinate system for each object under consideration. It is often useful to place one axis along the direction of the acceleration.

5. Apply Newton's second law in component form:

$$\sum F_x = ma_x,$$

$$\sum F_y = ma_y,$$

$$\sum F_z = ma_z,$$

to *each* object until you obtain as many equations as unknown quantities. This requires resolving forces into components.

6. Additional geometric and other constraints may need to be considered in some cases in order to have enough equations.

7. Algebraically solve for the unknowns first, then substitute numbers to obtain quantitative answers. This allows you to check your work more easily and reduces errors in calculations.

The following examples illustrate the method used to apply Newton's laws. Each body is treated as a point mass, so the forces are assumed to act at one point. In addition, the masses of pulleys and strings are considered negligible. Although these assumptions may appear artificial (where can you buy a massless string?), it is understanding the method that is important now.

Example 3

A 60-kg passenger is riding in an elevator. Find the force exerted by the floor on the passenger when the elevator is

(a) accelerating upward at 3.0 m/s^2,
(b) accelerating downward at 3.0 m/s^2,
(c) moving downward with a constant speed of 4.0 m/s.

In drawing a free-body diagram of the passenger (unkindly represented by a block), we have only two forces to consider, gravity mg and the normal force N.

Choosing the positive z axis to be pointed vertically upward, the second law $\Sigma F_z = ma$ implies

$$N - mg = ma,$$

which tells us that the normal force is $N = m(g + a)$. For case (a), we use $a = +3.0$ m/s^2 and get $N_a = 770$ N. For case (b), the acceleration is downward and therefore negative, $a = -3.0$ m/s^2. Substituting this into our expression for N, we get $N_b = 410$ N. For case (c), the acceleration is zero, and the normal force is simply equal to the weight $N_c = 590$ N.

Example 4
A parcel slides down a chute so smooth that friction is negligible. Calculate the acceleration of the parcel as well as the normal force from the chute.

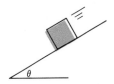

      The free-body diagram for the parcel is shown below. Note that the normal force N is perpendicular to the inclined chute.

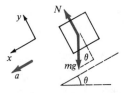

Choosing the x axis to point down the plane, in the direction of the acceleration, we have the following:

$$\sum F_x = ma \quad \text{implies} \quad mg \sin \theta = ma,$$

$$\sum F_y = 0 \quad \text{implies} \quad N - mg \cos \theta = 0.$$

The first equation tells us that the acceleration of the parcel is $a = g \sin \theta$, and depends only on the angle of the incline. As θ increases to $90°$, the acceleration becomes g (Galileo knew this). The second equation tells us that the normal force is

$$N = mg \cos \theta.$$

Example 5

Two blocks, one of mass $m_1 = 1.0$ kg and the other with $m_2 = 2.0$ kg, are pushed along a frictionless surface by a force of 2.0 N. Find the acceleration of the blocks and the force of block 1 on block 2.

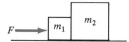

Following the steps outlined in the text, we first draw a free-body diagram for each block. The forces acting on block 1 are gravity m_1g; normal force N_1; outside push F; and, as 1 pushes on 2, 2 pushes back on 1 (by the third law) with a force we'll call P. A similar set of forces act on block 2, as shown in the free-body diagram. Note that F does not act directly on block 2, so it is not shown acting on it. The effect of F is the force P from block 1 being in contact.

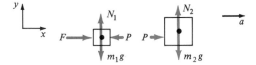

The blocks accelerate to the right, so we'll choose that to be the direction of the x axis. Applying Newton's second law $\Sigma F = ma$ to block 1, we obtain

$$F - P = m_1 a.$$

In this equation there are two unknowns, P and a, and mathematicians tell us that we need another equation. Applying the second law to block 2, we get

$$P = m_2 a.$$

Substituting this value of P into our first equation, we have

$$F - m_2 a = m_1 a,$$

which allows us to solve for a:

$$a = F/(m_1 + m_2).$$

(We could also have obtained this result by thinking of the two blocks as one block of mass $m_1 + m_2$ acted on by only the force F in the x direction.)

Substituting numbers, we find $a = 0.67$ m/s^2.

To find the force which block 1 exerts on 2, that is, P, we simply substitute the value of a into the equation for P:

$$P = m_2 a = (2.0 \text{ kg})(0.67 \text{ m/s}^2) = 1.3 \text{ N}.$$

The direction of this force is as shown in the free-body diagram.

Example 6

A 3.0-kg monkey holds on to a light rope, which passes over a frictionless pulley and is attached to a 4.0-kg bunch of bananas. What is the acceleration of the system?

Since we have two objects, we draw a free-body diagram for each. Because tensions in strings always pull, the tension T acting on the monkey is vertically upward. The same tension T pulls vertically upward on the bananas as well. Since the bananas (mass M) are heavier than the monkey (mass m), they accelerate downward and the monkey accelerates upward with the same magnitude of acceleration a. Let's choose the positive x axis to be vertically upward for the monkey, and for the bananas we'll choose the positive x axis to be vertically downward.

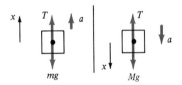

Applying $\sum F_x = ma$ to the monkey, we have $T - mg = ma$. Applying the second law to the bananas, we get $Mg - T = Ma$. We have two equations and two unknowns. Solving for T from one equation and substituting into the other, then solving for the acceleration, we obtain

$$a = (M - m)g/(M + m).$$

Substituting values, we find that the acceleration $a = 1.4$ m/s^2. The monkey accelerates upward with an acceleration of 1.4 m/s^2, while the bananas accelerate downward at that rate.

Example 7

A puck of mass m on a frictionless table is attached to a mass M by a light string which passes through a hole in the table. Of the many types of motion possible in this situation, we are interested in the case in which the puck, having been given an initial push, is found moving in a circle of radius r, while the mass M remains at rest.

We want to know the speed of the puck in this case. We know that a centripetal force must be acting on the puck to keep it in a circular path. Here the tension in the string (created by the weight Mg) provides that force. The free-body diagrams are shown below:

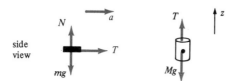

Since the puck is moving in a circle, it has a centripetal acceleration $a = v^2/r$. That's how the speed of the puck enters the problem.

For M, let's choose the positive z axis to be vertically upward, and for the puck, we choose the radial direction inward to be the positive r direction (the direction of the centripetal acceleration).

Applying the second law $\Sigma F_z = 0$ to the mass M (which is not accelerating), we find

$$T - Mg = 0.$$

For the puck, $\Sigma F_r = ma$ implies

$$T = mv^2/r.$$

Substituting for $T = Mg$ and solving for the speed, we find

$$v = \sqrt{Mgr/m}.$$

Questions

8. As the ball rolls down the hill shown, which of the following holds?

(a) Its speed decreases and its acceleration increases.
(b) Its speed increases and its acceleration decreases.

(c) Both speed and acceleration decrease.
(d) Both speed and acceleration increase.

9. Two blocks of ice of different mass are released on a smooth (frictionless) inclined plane initially separated by a distance d.

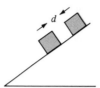

As the blocks slide down the plane, the distance d

(a) increases,

(c) remains the same,

(b) decreases,

(d) first increases then decreases.

10. Rework Example 5 for the case of F applied to the other block. Explain why the force of one block on the other is different in this case.

11. Three identical boxes each of mass $m = 2.0$ kg are pulled along a horizontal frictionless table by a force $F = 18$ N.

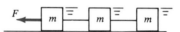

Find

(a) the acceleration of the blocks,

(b) the tension of each string.

12. A strongman suspends a 10-kg brick by a light rope (negligible mass) so that the string makes an angle of 30° as shown.

(a) Calculate the tension in each string.

(b) Could the strongman support the brick by pulling the rope horizontally? Explain.

13. A lamp hangs vertically from a cord in a posh elevator that is accelerating upward at 2.5 m/s². If the tension in the cord is 35 N, what is the mass of the lamp?

14. In Example 6, suppose the monkey and bananas have equal mass. What then would be the acceleration of the system? If in this case the monkey doesn't simply hold on to the rope but climbs up it with a constant speed of 1.0 m/s, what will happen to the bananas?

15. Inside a revolving space station, a container appears to rest on the outside wall as shown. If the mass of the container is 5.0 kg and the radius of the outer wall is 12.0 m, with what angular speed must the station be revolving so that the normal force on the container is 49 N?

16. Blocks A and B, of masses 3.0 kg and 5.0 kg, respectively, are attached by a light cord which passes over a frictionless pulley as shown. Block A is on a frictionless table.

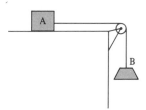

Find

   (a) the acceleration of the system,
   (b) the tension in the cord.

17. A 0.5-kg stone is twirled about in a vertical circle of radius 0.5 m. If the speed of the stone is 3.0 m/s, determine the tension in the string at the lowest point.

18. For the stone in Question 17, is the string more likely to break when the stone is at the top or at the bottom of the circle? Why?

19. For the stone in Question 17, determine the minimum speed it must have so that the string won't become slack at the highest point.

10.5 FRICTION

In 1699 the French scientist Guillaume Amontons investigated the losses caused by friction in machines. From his studies, he found the empirical relationship that frictional forces from a surface do not exceed an amount proportional to the normal force exerted by the surface on the object,

$$f \leq \mu N,$$

where μ (the Greek letter mu) is the coefficient of friction. Later Coulomb noted that μ depends on the two materials that are in contact. As we noted, friction is the result of a very large number of electric interactions between molecules of the two bodies in contact.

The frictional forces acting between surfaces at rest with respect to each other are called forces of *static friction*. Suppose that you have a block at rest on a horizontal surface. By Newton's second law, the force of friction is zero, as Fig. 10.6a illustrates. Now suppose you apply a small measurable force **F** to it as in Fig. 10.6b, and observe that the block doesn't move. By Newton's second law, the force of static friction is equal in magnitude to F $(= -F)$. Now suppose that you increase **F** and note that the block still doesn't move. The force of static friction increases as well, being equal to $-F$ always, as Figs. 10.6c and d show. If **F** is increased, there will be a definite value of **F** for which the block slips, as Fig. 10.6e illustrates. The smallest force necessary to start

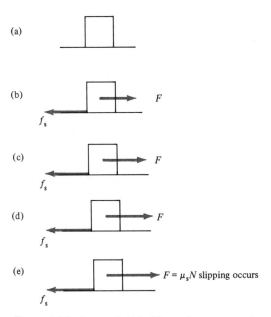

Figure 10.6 Force of static friction increases up to a maximum value equal to $\mu_s N$.

motion is the maximum force of static friction $\mu_s N$. These observations can be summarized by the relation for the magnitude of the force of static friction f_s:

$$\boxed{\text{Static Friction:} \quad f_s \leq \mu_s N,} \tag{10.4}$$

where μ_s is the coefficient of static friction, which depends on the two surfaces in contact, and N is the normal force. Static friction always opposes the intended motion of an object in its rest frame. Table 10.2 lists values of μ_s for various materials.

Once the block begins to move, *kinetic friction* acts on the block. This frictional force is usually less than the static friction. The magnitude of the kinetic friction f_k obeys the empirical relationship

$$\boxed{\text{Kinetic Friction:} \quad f_k = \mu_k N,} \tag{10.5}$$

where N is the normal force and μ_k is the coefficient of the kinetic friction, which depends on the two surfaces in contact. Table 10.2 lists a few values of μ_k. The force of kinetic friction is always opposite to the velocity of the object.

Example 8

A block rests on an inclined plane that has a variable angle θ. The angle θ is increased from zero, and at 40° the block slips. What is the coefficient of static friction?

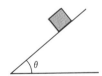

It seems that we don't have much to go on, but if we patiently follow our steps and apply Newton's laws, we should be able to find μ_s. In a free-body diagram for the block, static friction is directed up the plane (because the block tends to slip down the plane).

Table 10.2 Coefficients of Static and Kinetic Friction

Material	μ_s	μ_k
Steel on steel	0.78	0.42
Nickel on nickel	1.10	0.53
Teflon on Teflon	0.04	0.04
Wood on wood	0.35	0.15
Ice on ice	0.05	0.04

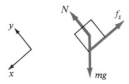

Choosing the positive x axis along the plane as shown and the positive y axis perpendicular to the plane, and remembering that just before the block slips the acceleration is of course still zero, we have the following:

$$\sum F_x = 0 \quad \text{implies} \quad mg \sin \theta - f_s = 0,$$

$$\sum F_y = 0 \quad \text{implies} \quad N - mg \cos \theta = 0.$$

When $\theta = 40°$, the force of static friction is maximum and we can substitute $f_s = \mu_s N$. But from our second equation, the normal force $N = mg \cos 40°$. Substituting all this into our first equation, we get

$$mg \sin 40° - \mu_s mg \cos 40° = 0,$$

which tells us that

$$\mu_s = \tan 40° = 0.84.$$

Note that only when *maximum* static friction is acting on an object can you use $f_s = \mu_s N$.

Example 9
A bug walks radially outward on a record that is rotating at an angular speed $\omega = 45$ rpm (revolutions per minute). If the coefficient of static friction between the bug and the record is 0.08, how far can it walk before it slips?

At each point of its path, a centripetal force must keep the bug moving in a circle. In this case that force is static friction, as shown in the free-body diagram.

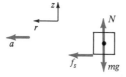

From Chapter 9, the centripetal acceleration of the bug is $a = \omega^2 r$, where r is the distance from the center of the record to the bug. Applying the second law to the radial direction, we have

$$\sum F_r = ma \quad \text{implies} \quad f_s = m\omega^2 r.$$

In the z direction, we simply have

$$\sum F_z = 0 \quad \text{implies} \quad N = mg.$$

When the bug is about to slip, the force of static friction is maximum and we can use $f_s = \mu_s N$, which here becomes $f_s = \mu_s mg$. Substituting this into our first equation and solving for r, we find

$$r = \mu_s g/\omega^2.$$

Before we can insert numbers, we need to express ω in radians per second:

$$\omega = [45 \text{ rpm}] \times [2\pi \text{ rad/rev}] \times [1 \text{ min/60 s}] = 4.7 \text{ rad/s}.$$

Therefore we get

$$r = 0.08(9.8 \text{ m/s}^2)/(4.7 \text{ rad/s})^2 = 0.04 \text{ m}.$$

Example 10
A 10-kg crate is pushed with a constant speed up a rough incline by a horizontal force **F**. If the coefficient of friction between the plane and crate is 0.3, find the magnitude of **F**.

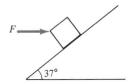

As shown in the free-body diagram, kinetic friction is directed down the incline (opposite to the velocity of the block).

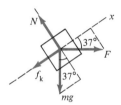

Choosing the x axis along the incline, we see that **F** has components in both the x and y directions. Applying Newton's second law, we have the following:

$$\sum F_x = 0 \quad \text{implies} \quad F \cos 37° - f_k - mg \sin 37° = 0,$$

$$\sum F_y = 0 \quad \text{implies} \quad N - F \sin 37° - mg \cos 37° = 0.$$

Solving our second equation for N, using $f_k = \mu_k N$, and substituting into the first equation, we get

$$F \cos 37° - \mu_k (F \sin 37° + mg \cos 37°) - mg \sin 37° = 0.$$

Solving for F, we obtain

$$F = \frac{mg \sin 37° + \mu_k mg \cos 37°}{\cos 37° - \mu_k \sin 37°}.$$

Inserting numbers, we find $F = 133$ N.

Questions

20. If a horse is seen walking to the right, what is the direction of static friction on its hooves?

21. A 15-N cart rests in the aisle of a jet airplane which is cruising horizontally with constant speed. The coefficient of static friction between the cart and the floor is 0.4; the kinetic coefficient is 0.2. The frictional force acting on the cart is

    (a) 0 N,   **(b)** 6.0 N,   **(c)** 3.0 N   **(d)** 15 N.

22. A horizontal force of 20 N is applied to a 4.0-kg block resting against a wall. The block is on the verge of slipping. Determine the coefficient of static friction between the block and the wall.

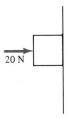

23. A delivery truck is loaded with crates having a coefficient of static friction with the floor of 0.3. If the truck is moving at 50 km/h, what is the shortest distance in which the truck can stop without the crates sliding?

24. A hockey puck slides 20.0 m across a frozen pond in 8.0 s before stopping. Find the coefficient of kinetic friction between the ice and puck.

25. A 2.0-kg block of ice slides down a chute that is inclined at 53°. If the coefficient of friction between the ice and chute is 0.1, calculate the acceleration of the block.

26. A person rides in a rotor ride at an amusement park. The rotor consists of a hollow cylinder that rotates about a vertical axis with an angular speed ω. With the person against the spinning wall of the rotor, the floor drops out, but the person remains "pinned" to the wall.

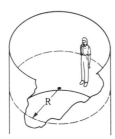

   (a) Draw a free-body diagram for the person pinned to the wall.
   (b) Determine the minimum coefficient of static friction between the person and the wall.

27. Two blocks are connected over a massless pulley as shown. The mass of A is 8.0 kg and the coefficient of kinetic friction is 0.20. If block A slides down the plane with constant speed, what is the mass of B?

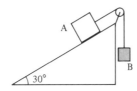

28. A car rounds an unbanked curve with a constant speed of 12 m/s. If the radius of the curve is 30 m, what must be the minimum value of the coefficient of friction between the tires and road so that the car doesn't skid? (*Hint:* Static friction provides the centripetal force to keep the car moving in a circle.)

10.6   A FINAL WORD

By investigating the forces of nature, we've discovered that they shape our universe, whether it be the structure of the nucleus or the motion of a car. But we've also learned something subtle and far-reaching about the nature of physics itself. Before Newton, the science of physics, as we understand it today, did not exist. The problem was not just that no one had hit upon the right answers. The real problem was that nobody had figured out what questions to ask.

Of course it is true that a great genius like Galileo could make progress, essentially by making up the rules as he went along. And there were others who thought they had the situation under control. Descartes thought it wouldn't take more than a few years to figure everything out, just by following rational principles, which he was gracious enough to prescribe for us. But he was wrong. The fact is, our knowledge of the universe was in a state of chaos, and there was not even any clear idea of what questions had to be answered in order to get anywhere. Then along came Isaac Newton, who said "*F* equals *ma*," and suddenly the mists started to clear. Why? Because we could now calculate accelerations? Obviously, that's not the point. The real point lies at the heart of the philosophical enigma we flirted with earlier. If "*F* equals *ma*" is to be the keystone of our understanding of nature, then we must understand what is meant by *F* and what is meant by *m*. Those are clear questions that must be answered. The questions are, What are the fundamental forces of nature? What does matter ultimately consist of? Those are the questions that Newton's law poses. They remain to this day the central questions in all of physics.

CHAPTER 11

GRAVITY, ELECTRICITY, AND MAGNETISM

In order to obtain physical ideas without adopting a physical theory we must make ourselves familiar with the existence of physical analogies. By a physical analogy I mean that partial similarity between the laws of one science and those of another which makes each of them illustrate the other. Thus all the mathematical sciences are founded on relations between physical laws and laws of numbers, so that the aim of exact science is to reduce the problems of nature to the determination of quantities by operations with numbers. Passing from the most universal analogies to a very partial one, we find the same resemblance in mathematical form between two different phenomena giving rise to a physical theory of light.

James Clerk Maxwell, "On Faraday's Lines of Force" (1855)

11.1 FINDING THE CONNECTION BETWEEN ELECTRICITY AND MAGNETISM

The flood of forces identified and classified in the eighteenth century was reduced to a trickle after it was realized that electric forces were responsible for many phenomena. Nature, it seemed, acknowledged only a handful of forces. Each force had its own "universal constant." For electric forces, it was K_e; for gravity, G; for magnetism, there was an additional constant K_m; and for light, there was the speed of light, known since 1630 to be 3×10^8 m/s. Surely many physicists wondered whether these constants and the forces they represent are somehow related. Similarities in mathematical forms of electric and magnetic forces, an 1820 lecture demonstration, and an experimental virtuoso led the way in the search for such relationships.

In the days of the ancient Greeks, it was known that a certain substance found in the ground has the power of attracting pieces of iron. The name *magnet* was derived from the district in Greece, Magnesia, where the material was found in great quantities. As time went on many fabulous tales of the rare and magic properties of magnets arose. Certain varieties were recommended as love potions; in the presence of diamonds or of garlic, a magnet's property was believed to be lost, but fortunately the attractive power could be restored by the timely use of goat's blood.

By the eleventh century magnets appeared as compass needles and became an indispensable tool in navigation. However, credit for the first scientific studies of magnets usually goes to William Gilbert, court physician to Queen Elizabeth I of England. Gilbert rejected the idle tales of magnetic virtues and determined the properties of magnets through experiments. He suggested that the earth itself is a large magnet, and he pointed out that every magnet has definite points in it, which he called the north pole and south pole, each of the same strength. By floating two magnets he showed that unlike poles attract and like poles repel each other. In addition, Gilbert found that if a magnet is cut in half, it acquires poles where it had been neutral. Although there is no doubting the importance of Gilbert's contribution, much of what he wrote had already appeared in a thirteenth-century manuscript, written during a leisurely siege of a town in Southern Italy by a mysterious scholar named Pierre of Merricourt.

In 1750 a young Cambridge theology student, John Michell, aided by a torsion balance like the one Cavendish had independently developed, measured the magnetic force between two magnet poles at different distances. From his experiments he concluded that the force follows an inverse square law similar to the force between two charges. If p_1 and p_2 are the pole strengths, we can describe the force between the two poles as

$$\mathbf{F} = K_m \frac{p_1 p_2}{r^2} \,\hat{\mathbf{r}}. \tag{11.1}$$

This equation is the same as for electric or gravitational forces, except that now the force is between two poles and there is a new constant K_m.

Coulomb, like Michell, verified the inverse square law for magnetic poles, and observed that a magnet when broken in pieces, however small, still has poles at the broken ends. Magnetism, unlike electricity, apparently always comes in pairs of poles. Physicists today are still searching for single magnetic poles, called magnetic monopoles, because some unified theories of the forces of nature predict that magnetic monopoles were created in the early stages of the universe. If they exist, however, they are certainly rare, so we shall assume that every pole comes attached to an equal and opposite partner.

By the end of the eighteenth century certain similarities between electric and magnetic phenomena were realized, yet there was no clearly established relationship between the two forces. Electricity and magnetism were regarded as forces obeying similar laws but as being fundamentally different in nature. However, in 1820 a Danish scientist, Hans Christian Oersted, not a consummate experimentalist, but rather nearsighted and bumbling, nevertheless found a connection that opened a new epoch in physics. Oersted was an ardent popularizer of science whose public lectures attracted curious citizens of Copenhagen. Folklore about his discovery places him in one such lecture on electricity, electric currents (known then as galvanism), and magnetism, during which he placed a wire conducting a current over a compass and at right angles to the needle. No effect was observed and obviously the audience was unimpressed. After the lecture he tried the experiment again with the wire parallel to the compass needle as shown in Fig. 11.1. This time the compass needle moved.

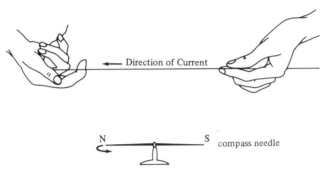

Figure 11.1 Oersted's experiment relating electricity and magnetism.

What Oersted discovered was that electric charges in motion – electric current – can exert forces on a compass needle. The news of Oersted's discovery spread throughout Europe like wildfire, and dozens of more dexterous experimentalists explored the effect and derived the laws describing it. Although Oersted did not quantitatively investigate the phenomenon he had discovered, he did speculate about its ultimate cause. Somehow, he believed, the medium surrounding the conducting wire plays a role in the transmission of force from a wire to a magnet.

Oersted's discovery showed that electricity and magnetism are not separate, independent forces, and it fueled the search for unification. It showed that a force exists between magnetic poles and an electric current. But the direction of the force depends in an unexpected way on the direction of the magnet and the direction of their current. Electromagnetism seemed very complicated.

11.2 FARADAY'S FIELDS

No one more vigorously explored the connection between electricity and magnetism than Michael Faraday, a nineteenth-century English chemist. Faraday, whose passion for science started from reading an article on electricity in the *Encyclopaedia Britannica*, had little formal education, never attended a university, and never learned mathematics. Yet he consistently made the most important discoveries of his time, because he had an intuition that grasped the essence of things far better than the formulas of the most powerful mathematician.

It had been long known that if iron filings are sprinkled on a sheet of paper and a magnet is held underneath, the filings will arrange themselves in definite curves, as shown in Fig. 11.2a. Faraday, of course, had noticed this effect and imagined the curves of iron

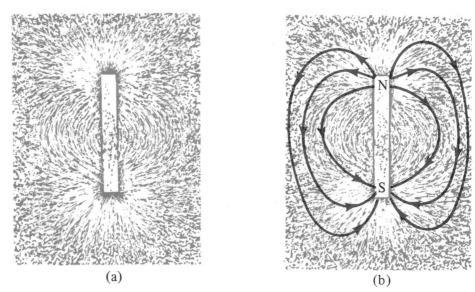

(a)   (b)

Figure 11.2 (a) Iron filings form curves around a magnet.
(b) Magnetic field lines around a magnet.

filings as following lines of force that spread out from the north pole to the south pole.
In his mind's eye, he saw all space as filled with such lines, never crossing or tangling,
so that at every point there was a potential force ready to act on a magnet pole if one
were placed there. From these speculations the idea of the *magnetic field*, a new vector
quantity, was born. The direction of the field at any point is the direction of the line,
and the strength of the field depends on how densely packed the lines are. The field is
strong where the lines are closely packed, as near the poles of a magnet, and weaker
where they are spread apart, as in the center, as Fig. 11.2b illustrates.

The force a magnetic field **B** exerts on a magnetic pole with strength p_0 can be
described by

$$\mathbf{F} = p_0\mathbf{B}. \tag{11.2}$$

Since magnetic poles come in pairs, the magnetic field pushes a north pole one way and
a south pole the opposite way, as shown in Fig. 11.3a. The result of the forces is to
rotate the magnet so that it lines up with the field, as Fig. 11.3b illustrates.

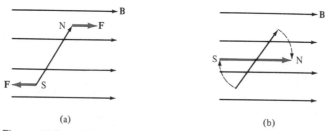

(a)   (b)

Figure 11.3 (a) Forces on poles of a magnet from a magnetic field **B**.
(b) Forces acting on a magnet align it with the field.

To explain Oersted's discovery, Faraday conjectured that a current-carrying wire creates a magnetic field around itself. This field, in turn, causes a magnet to become aligned. Using a small compass needle, Faraday mapped out the magnetic field around a current-carrying wire. As Fig. 11.4 indicates, the field characteristically forms circles about the wire. The direction of the field can be found by pointing the thumb of your right hand in the direction of the current; your fingers then curl in the direction of **B**.

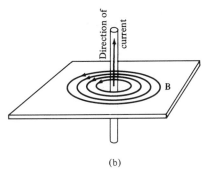

(b)

Figure 11.4 Magnetic field due to a current-carrying wire.

Not only did Faraday imagine that magnets and currents create fields, but also that electric charge creates an electric field. Electric field lines reach out from negative charges and end on positive charges, as shown in Fig. 11.5. The direction of the force on a small charge at any point is the same as the way in which the field line is directed. We write the force on a charge q_0 in an electric field **E** as

$$\mathbf{F} = q_0\mathbf{E}. \tag{11.3}$$

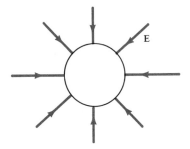

Figure 11.5 Electric field around a negatively charged sphere.

In the same way, we can imagine a gravitational field **G** created by mass at each point in space. If a mass m_0 is placed in this field, the force due to the gravitational field **G** is

$$\mathbf{F} = m_0\mathbf{G}. \tag{11.4}$$

Knowing that the gravitational force between two masses is

$$\mathbf{F} = -G\frac{m_0 M}{r^2}\hat{\mathbf{r}}, \tag{8.1}$$

we can mathematically describe the gravitational field around a mass M by

$$\mathbf{G} = -\frac{GM}{r^2}\hat{\mathbf{r}}.$$

Figure 11.6a illustrates the gravitational field of the earth, where the lines of force are radially inward. If we look very close to the surface, all the lines go straight down, as shown in Fig. 11.6b. A mass placed in this field feels a force which causes it to fall vertically downward. By Eq. (11.4) the magnitude of the force is proportional to the mass of the object, so the larger the mass of the object, the greater the force on it. Consequently, all objects fall with the same acceleration.

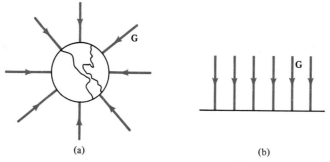

(a)   (b)

Figure 11.6 (a) Gravitational field $\mathbf{G}$ of the earth. (b) Field lines near the surface of the earth.

What we've just found out is not new; we discovered why all bodies fall with the same acceleration in Chapter 8. What is new is the idea of a field, which was Faraday's answer to action-at-a-distance forces, for which Newton made no hypothesis. In Faraday's vivid imagination, the space surrounding charges, masses, or magnets is filled with a field which transmits forces to other charges, masses, or magnets. The fertile idea of fields led Faraday to unravel the connection between electricity and magnetism.

We still have one puzzle to unravel: How is an electric current related to a magnetic pole? Since we know how a pole reacts to a magnetic field, let's ask how an electric current reacts to a magnetic field. Suppose we have a charge q, moving with velocity $\mathbf{v}$ (that constitutes an electric current), in a magnetic field $\mathbf{B}$. What is the force on the charge?

Experimentally the magnitude of the force on a moving charge is simply found to be

$$F = qvB\sin\theta,$$

where θ is the angle between $\mathbf{v}$ and $\mathbf{B}$. The direction of this force, however, is not so simple. The direction of $\mathbf{F}$ depends upon the direction of $\mathbf{v}$ as well as the direction of $\mathbf{B}$.

Furthermore, the force is in neither of these directions; it is perpendicular to both. Therefore the cross product (introduced in Chapter 5) provides a natural mathematical tool to help us describe the situation.

In terms of the cross product, the force **F** on a charge q moving with a velocity **v** through a magnetic field **B** is

$$\mathbf{F} = q\mathbf{v} \times \mathbf{B}. \qquad\qquad (11.5)$$

Using this force and the idea of fields, we can predict what will happen when two current-carrying wires are placed near each other. As shown in Fig. 11.7, we can think of one wire creating a magnetic field **B** in circles around it. The current in the second wire consists of charges moving in the magnetic field of the first wire. From Eq. (11.5), the direction of the force on any charge in the second wire is directly toward the first wire. Consequently, the wire is attracted. The same thing happens to the first wire: the second wire creates a magnetic field, which exerts a force on the charges moving in the first wire; the result is that the wires are attracted. Experimentally this is precisely what happens: wires carrying currents in the same direction are attracted.

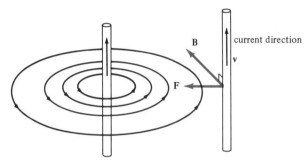

Figure 11.7 Two parallel current-carrying wires attract each other.

Questions

1. Knowing that the magnetic field around a current-carrying wire forms circles about the wire, explain why Oersted's first experiment did not produce a deflection of the compass needle.

2. Using the idea of a gravitational field, show why all bodies fall with the same acceleration.

3. According to Faraday's field idea, the strength of a field is indicated by the number of field lines in a region. Using Fig. 11.6, discuss how the strength of the gravitational field is indicated.

4. What is the net force on a magnet placed in a uniform (constant everywhere in space) magnetic field? Does the magnet accelerate?

5. Suppose you are given two identical metal bars, of which one is a magnet. How can you tell which one is the magnet, using only the two bars?

6. The diagram below represents the tracks of three particles that passed through a region where there is a uniform magnetic field perpendicular to the plane of the paper and indicated by the x's. What is the sign of the charge of each particle?

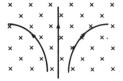

11.3 A PREDICTION FROM ELECTROMAGNETISM

The researches of Faraday into electromagnetic effects excited nineteenth-century physicists. One person stimulated by this work was James Clerk Maxwell, a Scottish student at Cambridge. Like Faraday, Maxwell possessed a keen power of visualizing physical ideas, but in addition he had mathematical prowess. His first contributions to electromagnetic theory amounted to a translation of Faraday's ideas into mathematical language. Like Faraday, he abandoned the notion of action at a distance and sought to interpret electric and magnetic phenomena in terms of fields.

Maxwell succeeded in expressing Faraday's ideas in a compact set of equations. Maxwell's equations unite electric and magnetic phenomena in one common description. But the success was not simply mathematical cunning. Based upon his theory, Maxwell found a connection between electricity and magnetism and something else – light.

If electricity and magnetism are related, Maxwell thought, then there should exist a relationship between the constant K_e, which pertains to electric phenomena, and K_m, the magnetic constant. The constant K_e had been determined from Coulomb's experiment by measuring the force between two charges,

$$\mathbf{F} = \frac{K_e q_1 q_2}{r^2}\ \hat{\mathbf{r}}.$$

(10.1)

The result of measurements is

$$K_e = 9 \times 10^9 \text{ N m}^2/\text{C}^2.$$

Can we repeat the same measuring process for the force between magnetic poles,

$$\mathbf{F} = K_m \frac{p_1 p_2}{r^2} \hat{\mathbf{r}}?$$

(11.1)

The problem here is that somehow magnetic poles are created by moving electric charges. We simply can't adopt a standard pole, lock it away in a vault in Paris, and compare all other poles to it. That was the whole point of Oersted's experiment.

There is, however, a way of defining a unit pole. For magnets, we found that the force exerted on a pole by a magnetic field is

$$\mathbf{F} = p_0 \mathbf{B}.$$

(11.2)

For a moving charge, the force exerted by a magnetic field is

$$\mathbf{F} = q\mathbf{v} \times \mathbf{B}.$$

(11.5)

The situation is complicated by the strange vector character of the relation between charges and magnets. Nevertheless, one thing is clear: a pole has the units of a charge multiplied by a velocity. If $\mathbf{v}$ is perpendicular to $\mathbf{B}$, the force is qvB, while the force on a pole is p_0B. Therefore we can define one pole as charge times speed:

1 pole = 1 C m/s.

If we measure the force between two unit poles, then we can determine the constant K_m. In practice this is not done. Instead we measure the force between two wires carrying current which we discussed earlier. The idea, however, is the same: we measure a force to determine a constant, and the result is

$$K_m = 1.00 \times 10^{-7} \text{ N m}^2/(\text{pole})^2 = 1.00 \times 10^{-7} \text{ N s}^2/\text{C}^2.$$

Looking at the units of the two constants K_e and K_m, we see

$$K_e \sim \text{N m}^2/\text{C}^2,$$

$$K_m \sim \text{N s}^2/\text{C}^2,$$

and so the ratio has units of

$$\frac{K_e}{K_m} \sim \frac{\text{m}^2}{\text{s}^2}.$$

In other words, the ratio is the square of a speed. Inserting the values of the constants we find that

$$\text{speed} = \sqrt{\frac{K_e}{K_m}} = \sqrt{\frac{9 \times 10^9}{1 \times 10^{-7}}} \frac{\text{m}}{\text{s}} = 3 \times 10^8 \frac{\text{m}}{\text{s}}.$$

That is the speed of light!

When Maxwell discovered that the speed of light was buried in the forces between charges and magnets, he knew that he had in his hands a discovery of vast importance, or as he put it in 1865, "hold to be great guns." By purely theoretical investigations he deduced that an electromagnetic disturbance – electric disturbances which create magnetic disturbances and vice versa – travels with the speed of light. By the time he finished his investigation, he realized that visible light was only a tiny portion of the kinds of disturbances that could occur in the electromagnetic field, and that all would travel at the same speed, 3×10^8 m/s. Maxwell's theory is beautiful, perfect, and complete, and stands unchallenged to this day.

Maxwell's theory was not widely accepted by the more conservative physicists during his own lifetime. But ten years after Maxwell's early death from stomach cancer at the age of 48, electromagnetic waves were actually produced and detected, their speed was measured, and the result was found to agree with the prediction of his theory. Almost immediately thereafter, the radio was invented; radio waves are only one of the many examples of Maxwell's electromagnetic radiation. Radio, television, the total world of communications can be traced to Maxwell's stunning discovery that the speed of light is buried in the forces between charges and magnets.

Questions

7. A wire carries a current as shown and is immersed in a uniform magnetic field **B**. What is the direction of the force on the wire?

8. If the direction of current in a wire is reversed, does the magnetic field due to that current also reverse direction?

9. If two parallel wires carry currents in opposite directions, will they be attracted to each other or repelled?

11.4 A FINAL WORD

Maxwell's research into the fundamental nature of electricity and magnetism led directly to the invention of radio, television, and the whole revolution in long-range communication that has changed all our lives so much. But suppose Maxwell had started out with the specific goal of improving long-distance communication. Then he wouldn't have fiddled around with the fundamental nature of electricity and magnetism. Instead he would have conducted experiments with giant megaphones and relay stations and the like. Or would he?

History does hold some surprises, because the fact is Maxwell had been interested in long-distance communications since he was a young man. At that time, the 1850s, the world was undertaking a great technological adventure, comparable to our own voyages to the moon. The first means of instantaneous long-distance communication was the telegraph, which was invented in 1838. By the 1850s an attempt was being made to lay a cable under the Atlantic Ocean. When completed, it would reduce the time for communication between New York and London from a period of weeks to a fraction of a second.

This first attempt to lay a cable failed, however. The cable snapped somewhere in the middle of the Atlantic. The young James Clerk Maxwell marked that sad occasion by writing a poem about it, in a letter to a friend. The poem is called "Song of the Atlantic Telegraph Company:"

> Under the sea, under the sea
> Something has surely gone wrong.
> What is the cause of it does not transpire
> But something has broken the telegraph wire
> Or else they've been pulling too strong.
> Under the sea, under the sea
> So many hundred miles long
> How could they spin out such durable stuff,
> Line all wiry, elastic and tough
> In the salt water so strong.

Under the sea, under the sea
There'll be lots of cables 'ere long
For they'll spin a new cable and try it again
And settle our bargains of cotton and grain
With a line that will never go wrong.

Figure 11.8   James Clerk Maxwell. (Courtesy of the Archives, California Institute of Technology.)

CHAPTER

THE MILLIKAN OIL-DROP EXPERIMENT

A prominent literary writer, I think it was Chesterton, once spoke of the electron as "only the latest hypothesis which will in its turn give way to the abra-ca-da-bra of tomorrow." This sort of ignorance will disappear in time, just as will Kipling's "village that voted the earth was flat, flat as my hat, flatter than that." In any case, the most direct and unambiguous proof of the existence of the electron will probably be generally admitted to be found in the oil-drop experiment here under discussion.

Robert A. Millikan, *Autobiography* (1950)

12.1 THE DISCOVERY OF THE ELECTRON

In the 1890s physicists were puzzled by cathode rays, by glowing gases, and by apple-green fluorescence that appeared in their investigations of light. To study the light emitted by various elements, researchers had developed a specially designed glass apparatus called the cathode-ray tube. To two pieces of metal, called the cathode and anode, which pass through the tube, would be connected a powerful electric voltage as illustrated in Fig. 12.1. When the gas was at atmospheric pressure, nothing happened, but when the pressure was reduced to one-hundredth of atmospheric pressure, the gas began to glow. The end of the tube near the anode glowed, or fluoresced, with an eerie green color, except where a shadow of the anode appeared. Evidently, the cathode emitted mysterious rays – cathode rays – that streamed from the cathode to the anode. Some of these rays hit the glass, creating the green light, but others were blocked by the anode, creating a shadow.

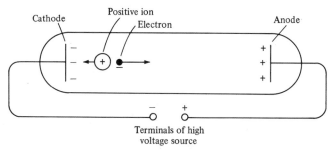

Figure 12.1 Diagram of a cathode-ray tube.

The British physicist Sir Joseph John Thomson was one of the many who set out to investigate the mysterious rays. Thomson designed a special cathode tube, shown in Fig. 12.2, in which the rays passed through a narrow slit and hit a fluorescent screen at the other end. Because the beam produced a tiny bright spot where it hit the screen, Thomson could observe the behavior of the beam.

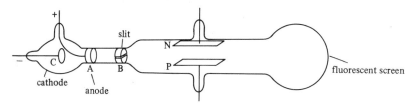

Figure 12.2 J. J. Thomson's cathode ray tube.

By passing a cathode-ray beam between two metal plates charged by a battery that created an electric field between the plates, he succeeded in deflecting the beam, as Fig. 12.3a indicates. The implication was that the cathode rays have electric charge, which causes them to be deflected by the electric field according to

$$\mathbf{F} = q\mathbf{E}. \tag{11.3}$$

When the beam passed through a region containing a magnetic field, it was also deflected by the magnetic force

$$\mathbf{F} = q\mathbf{v} \times \mathbf{B}, \tag{11.5}$$

as Fig. 12.3b shows. Both observations were consistent with cathode rays being negatively charged particles.

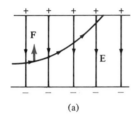 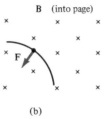

Figure 12.3 (a) Deflection of cathode rays in an electric field **E**.
(b) Deflection of cathode rays in a magnetic field **B**.

Thus the cathode-ray apparatus, which eventually evolved into the modern television tube, has also served a useful purpose in the history of science. Thomson, as discoverer of the new particle, should have had the honor of naming it, but his choice, the *corpuscle*, did not stick, because the existence of the electron was already widely suspected and the name had long since been coined.

When Thomson "discovered the electron" in 1896, the cathode-ray tube had already been around for nearly 40 years. It was suspected that cathode rays were charged particles, especially because they were easily deflected by a magnet, but they did not seem to respond to electric fields as they should have. The reason was that the vacuum in the tube was not good enough. Ions (charged atoms) in the residual gas would interfere with the action of the electric field, confusing the result of the experiment.

Thomson realized that the poor vacuum was due to molecules adhering to the glass walls of the tube when it was sealed by the glass blower. These molecules would escape from the walls afterward; they could be expelled from the tube by heating the walls. Thomson had his apparatus pumped out and sealed while it was being baked in an oven. This technique worked, and permitted him to show the rays had charge and to measure their mass-to-charge ratio, which he found to be about 10^{-11} kg/C. Repeating his experiments with various metals for the cathode (the source of the rays) he found the same mass-to-charge ratio. Replacing the air inside the tube with other gases, he still found the same ratio.

All atoms, he conjectured, contain electrons. No longer could the atom be viewed as the smallest, indivisible object in the universe. Thomson had split the atom and discovered the electron by baking his tubes.

Thomson himself eventually came to prefer the name electron to corpuscle. That choice started a tradition in physics of giving quantized units names ending in -on: the quantum of light is the photon; of sound, the phonon; there is the proton, the neutron, and so on. It has recently been discovered that there is even a quantized unit of human population, called the person.

Example 1
By observing the bending of an electron beam in a magnetic field **B**, what quantities, if measured, would reveal the mass-to-charge ratio?

An electron beam is observed to move in a circle when in a uniform magnetic field, as shown in the diagram, where the x's represent the magnetic field lines going into the page.

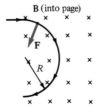

B (into page)

We can understand why the electrons move in a circle by applying Newton's second law. The force $\mathbf{F} = q\mathbf{v} \times \mathbf{B}$, being perpendicular to both $\mathbf{v}$ and $\mathbf{B}$, causes each electron to be deflected when it enters the field. The force changes the direction of the velocity, but not the magnitude because it is perpendicular to $\mathbf{v}$. Consequently, the direction of $\mathbf{F}$ changes as the particle moves. Like a string pulling on a whirling rock, the magnetic field causes the electrons to move in a circle. The centripetal force provided by the magnetic field is $F = qvB$. As we saw in Chapter 8, anything moving in a circle with a constant speed has a centripetal acceleration $a = v^2/R$, where R is the radius of the circle. Therefore, by Newton's law $F = ma$,

$$qvB = mv^2/R.$$

Solving for m/q, the mass-to-charge ratio, we find

$$m/q = BR/v.$$

Knowing the speed of the electrons, the magnetic field, and measuring the radius of their circular motion, we can find the mass-to-charge ratio.

Example 2

Thomson did not know the speed of his electrons. How could the electric field be used to solve that problem?

As we saw in Fig. 12.3a, electrons are deflected in an electric field as well as in a magnetic field. The magnetic force causing a deflection depends on the velocity,

$$\mathbf{F} = q\mathbf{v} \times \mathbf{B},$$

whereas the electric force producing a deflection,

$$\mathbf{F} = q\mathbf{E},$$

does not. Now if the magnetic and electric fields are arranged so that the total force on the electron is zero, there will be no deflection. Below is one combination of crossed electric and magnetic fields which produces no net force on a moving charged particle.

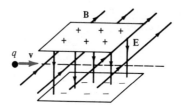

Using the forces given above, with **v** perpendicular to **B**, the condition for no net force is

$$qvB - qE = 0,$$

from which we can solve for the speed,

$$v = E/B.$$

Therefore, by using crossed electric and magnetic fields and adjusting them so that the electron beam is not deflected, Thomson could determine the speed of the electrons.

Questions

1. If the entire cathode-ray beam bent the same amount in an electric field, what could Thomson conclude about the charges on individual electrons, assuming that all electrons had the same speed? (In actual experiments, electrons do not all have the same speed but rather a distribution of speeds.)

2. Use the right-hand rule to verify the direction of the force on the electrons in Fig. 12.3b.

3. Suppose that a beam of particles in a magnetic field follows a circle that has half the radius of a beam of electrons of the same speed. Can you conclude that these particles have twice the charge of electrons? Why or why not?

12.2 MOTION IN A RESISTIVE MEDIUM

Thomson's experiments, proving that a charged particle a thousandth of the mass of the hydrogen atom both exists inside atoms and comes out of all kinds of atoms, quickly gained universal acceptance by physicists. But Thomson's experiments revealed no information about a quantity essential for understanding the relationship of electrons to gross matter, the charge of the electron. Do all particles of cathode rays have the same charge? If so, what is this value? Or does a cathode-ray beam have some overall, statistical average of charge, being composed of electrons with different charges?

After the discovery of the electron, several scientists, including Thomson himself, attempted to measure the charge of the electron. One method, devised by one of Thomson's students, involved falling clouds of charged water droplets. But the falling droplets were not subject to electric forces and gravity alone; air resistance must be taken into account. Before we can find out how such experiments lead to the charge of the electron, we need to understand the physics of motion in a resistive medium.

A water droplet falling through air experiences the force of viscosity, which, as we discussed in Chapter 10, is a result of friction between different layers of the fluid. The force can be empirically described by

$$\mathbf{F} = -K\eta\mathbf{v}, \tag{10.3}$$

where η is the coefficient of viscosity and K a constant which depends on the shape of the object. Finding K for an object requires a laborious calculation, but in the last century George Stokes found the result for a sphere of radius R (known as Stokes's law):

$$K = 6\pi R.$$

The viscous force acting on a sphere takes the simple form

$$\mathbf{F} = -6\pi R \eta \mathbf{v}. \tag{12.1}$$

From Stokes's law we see that K has units of meters, so by Eq. (12.1) the coefficient of viscosity η has units of newton-seconds per square meter. At a given temperature and pressure, any fluid has an intrinsic, measurable coefficient of viscosity. Table 12.1 lists the coefficients of viscosity for several fluids.

Table 12.1 Coefficients of Viscosity at 20°C (Unless Noted) and at Standard Atmospheric Pressure

Liquid	$\eta \ (10^{-3} \ \text{N s/m}^2)$	Gas	$\eta \ (10^{-5} \ \text{N s/m}^2)$
Water (0°C)	1.792	Air (0°C)	1.71
Water	1.005	Air	1.81
Castor oil	9.86	Ammonia	0.97
Glycerin	833	Hydrogen	0.93

The important feature of the viscous drag force is that it is proportional to the velocity and in the opposite direction. A marble dropped into a beaker of a very viscous liquid, like glycerin, appears to fall with a constant speed. The forces acting on the marble are gravity, $m\mathbf{g}$ downward, and viscous force, $-6\pi R \eta \mathbf{v}$ upward. If the marble is falling with a constant speed v_L, its acceleration is zero, and so by Newton's second law

$$mg - 6\pi R \eta v_L = ma = 0.$$

In other words, the viscous force is equal to the weight. Solving for the speed v_L, called the *limiting or terminal velocity*, we find

$$v_L = \frac{mg}{6\pi R \eta}. \tag{12.2}$$

Example 3

Find the terminal velocity of a raindrop, assuming it to have a radius of 0.2 mm and using the density of water to be 1.0×10^3 kg/m³. Compare the value with a speed of a raindrop falling in a vacuum from 1 mi ($= 1.6$ km).

According to Eq. (12.2), we need to know the mass of the raindrop; the radius $R = 2.0 \times 10^{-4}$ m is given, and from Table 12.1 we have $\eta = 1.8 \times 10^{-5}$ N s/m² for air. The mass is simply the density of water ρ times the volume, which is assumed to be spherical:

$$m = \rho\tfrac{4}{3}\pi R^3.$$

Substituting, we have

$$v_{\mathrm{L}} = \rho\tfrac{4}{3}\pi R^3 g/6\pi R\eta = \tfrac{2}{9}\rho g R^2/\eta.$$

Inserting values, we find $v_{\mathrm{L}} = 4.8$ m/s, which is about 10 mph. If there were no air resistance, in time t the raindrop would fall a distance $x = \tfrac{1}{2}gt^2$ and its speed would be $v = gt$. Solving for t in the distance equation and substituting into the speed equation we have

$$v = (2gx)^{1/2}.$$

Using the values given ($x = 1.6$ km), we find $v = 180$ m/s or 400 mph! If raindrops kept on accelerating as they fell and didn't approach a terminal velocity, a rainstorm would be a deadly place to be.

Equation (12.2) tells us that terminal velocity is proportional to the weight of an object. In other words, that heavier bodies fall faster! Can this be true? Is the world really Aristotelian? The answer is that we are now *including* air resistance – precisely what Aristotle thought must always be present and Galileo preferred to ignore. To understand what is happening, we turn to Isaac Newton.

For any spherical object falling in a viscous medium, as illustrated in Fig. 12.4, Newton's second law implies

$$m\frac{dv}{dt} = mg - 6\pi R\eta v. \qquad (12.3)$$

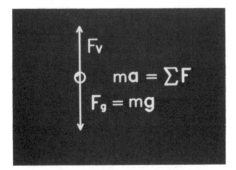

Figure 12.4 Forces acting on a sphere falling in a resistive medium.

This is a differential equation describing the motion of any spherical object moving in a viscous medium. Presently we will solve it and see how the velocity depends on time. But first, let's see what we can learn from the differential equation without solving it.

Let's look at what Eq. (12.3) describes. At the instant we drop the object, it is at rest, which means $v = 0$; therefore, the viscous force $-6\pi R\eta v$ is also *momentarily* zero. Consequently, we have $m\, dv/dt = mg$ at that instant. In other words, the object starts with acceleration $dv/dt = g$, just as Galileo said it should. Because the ball accelerates, the speed and viscous force increase. As a result the right-hand side of Eq. (12.3), $mg - 6\pi R\eta v$, becomes smaller than mg, and the acceleration decreases. As the velocity increases (of course, the velocity increases ever more slowly as time goes on because the acceleration is getting smaller) the right-hand side approaches zero. If it were zero, gravity and the viscous force would balance each other, so the object would have zero acceleration and would fall with the terminal velocity.

The object starts out being Galilean, and ends up being Aristotelian. A key question is, how long does this take? If it takes hours to reach terminal velocity, we can forget about Aristotle; the effects of viscosity can be ignored. But if it takes only a fraction of a second, then the object spends most of its time falling at terminal velocity.

We can figure out whether the velocity v approaches the terminal velocity slowly or quickly by a dimensional analysis of the differential equation. This means that we take a look at the units of each term. If we divide Eq. (12.3) by the mass m we get

$$\frac{dv}{dt} = g - \frac{6\pi R\eta}{m} v.$$

Since each term must have the same units, distance/time^2, the factor multiplying v on the right must have units 1/time, so its reciprocal has units of time. Let's denote this reciprocal by t_0. Thus, by definition,

$$t_0 = \frac{m}{6\pi R\eta}, \tag{12.4}$$

and t_0 has units of time. The differential equation now becomes

$$\frac{dv}{dt} = g - \frac{v}{t_0}.$$

What "time" does t_0 represent physically? To find out we let t approach infinity in the differential equation. The velocity approaches the terminal velocity v_L, and dv/dt approaches 0, so in the limit we get

$$0 = g - \frac{v_L}{t_0} \quad \text{or} \quad v_L = gt_0.$$

In other words, t_0 is the time it would have taken to reach the terminal velocity v_L if the acceleration were always equal to g. The number t_0 is called the *characteristic time*. We shall now solve the differential equation and see how the value of t_0 will tell us whether the velocity v approaches terminal velocity v_L quickly or slowly, depending on whether t_0 is small or large.

The differential equation may appear more complicated than any we have seen so far, but actually it is a familiar one in disguise. Since $g = v_L/t_0$, we can rewrite the differential equation as follows:

$$\frac{dv}{dt} = -\frac{v - v_L}{t_0},$$

Now let $w = v - v_L$, the difference between the actual velocity v and the terminal velocity v_L. Then $dw/dt = dv/dt$ since v_L is constant, so w satisfies the simpler differential equation

$$\frac{dw}{dt} = -\frac{1}{t_0}w.$$

This is the differential equation for the exponential function which we've seen earlier in Chapter 3. Its solution is

$$w(t) = w(0)e^{-t/t_0}$$

where $w(0) = v(0) - v_L$. But $v(0) = 0$, so $w(0) = -v_L$. Replacing $w(t)$ by $v(t) - v_L$ we see that

$$v(t) - v_L = -v_L e^{-t/t_0}$$

so the solution is

$$v(t) = v_L(1 - e^{-t/t_0}).$$

The characteristic time t_0 appears in the denominator of the exponential term, so it governs the rate at which this exponential tends to zero. If t_0 is small, the exponential decays very rapidly and v quickly approaches the terminal velocity v_L. If t_0 is large the exponential decreases more slowly and it takes a longer time for v to approach v_L.

We can now unravel many mysteries. A heavy ball (large m) falling in air (very small η), according to Eq. (12.4), takes a very long time to reach terminal velocity; it tends to behave in the way Galileo described, unless it falls very far. On the other hand, a marble (let's say with $m = 0.01$ kg, $R = 0.01$ m) in glycerin ($\eta = 0.8$) takes a time $t_0 = 0.1$ s to reach terminal velocity; in other words it is always terminal.

Example 4

Verify by differentiation that $v(t) = gt_0(1 - e^{-t/t_0})$, where t_0 is specified by Eq. (12.4), is a solution to Eq. (12.3),

$$m\frac{dv}{dt} = mg - 6\pi R\eta v. \tag{12.3}$$

Differentiating the proposed solution, we find

$$\frac{dv}{dt} = -gt_0\left(-\frac{1}{t_0}\right)e^{-t/t_0} = ge^{-t/t_0}.$$

Substituting for $v(t)$ in the right-hand side of Eq. (12.3), and after dividing through by m, we get

$$g - (1/t_0)(gt_0)(1 - e^{-t/t_0}) = ge^{-t/t_0}.$$

But this is exactly dv/dt, so the solution satisfies Eq. (12.3). From the solution $v(t)$ we see that $v(0) = 0$. As t gets infinitely large, e^{-t/t_0} goes to zero, and $v(\infty) = gt_0$, which is the terminal velocity v_L. The graph below illustrates $v(t)$.

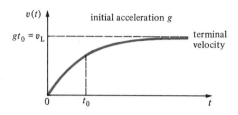

Example 5

From the solution $v(t) = v_L(1 - e^{-t/t_0})$, what is $v(t_0)$?

Substituting $t = t_0$, we find

$$v(t_0) = v_L(1 - e^{-1}).$$

Evaluating the exponential, we find $v(t_0) = 0.63 v_L$. We have a quantitative interpretation of the time t_0: it is the time required for the object to reach 63% of its terminal velocity.

Example 6

How far is the velocity from v_L at time $10t_0$?

What we want to know is $v_L - v(10t_0)$. We already know that

$$v(t) = v_L(1 - e^{-t/t_0}),$$

so

$$v_L - v(10t_0) = v_L - v_L(1 - e^{-10}),$$

or

$$v_L - v(10t_0) = e^{-10}v_L,$$

which is $4.5 \times 10^{-5} v_L$. In other words, at time $10t_0$ the velocity is within 0.004% of the terminal velocity.

Questions

4. If a stone is thrown vertically upward in air, does it take more time to go up to its highest point or to come down? Explain.

5. Considering the viscosity of air, would the arc of a cannonball be higher or lower than Galileo's parabola? Would the range be greater or smaller?

6. Referring to Eq. (12.3), why doesn't the viscous force ever become greater than *mg* and cause the particle to accelerate upward?

7. How many times greater is the terminal velocity of a sphere falling in castor oil than if it were falling in glycerin?

8. From the solution $v(t)$ given in Example 4, determine the acceleration as a function of time. Sketch a graph of your result.

9. Using the solution $v(t)$ given in Example 4, find $x(t)$, the distance fallen at time t. What is $x(t_0)$?

10. Suppose another constant force, say F, were acting on a falling sphere. What then would be the terminal velocity?

11. In an amusing essay, "On Being the Right Size," the biologist J. B. S. Haldane wrote: "You can drop a mouse down a thousand-yard mine shaft; and, on arriving at the bottom, it gets a slight shock and walks away. A rat would probably be killed, though it can fall safely from the eleventh story of a building; a man is killed; a horse splashes." Why do you think there are qualitative differences?

12.3 THE OIL-DROP EXPERIMENT

In 1906 Robert A. Millikan, then a lowly assistant professor at the University of Chicago, devised an ingenious experiment that for the first time made it possible to measure the charge on an individual droplet rather than on a cloud. Through Millikan's experiment it became possible to determine whether or not electricity in gases and chemical solutions is built out of electrons and whether each electron has the same amount of charge.

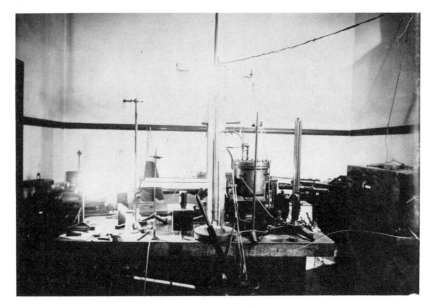

Figure 12.5 Robert A Millikan's original apparatus to measure the electron charge. (Courtesy of the Archives, California Institute of Technology.)

Millikan's original apparatus is shown in Fig. 12.5. Millikan used oil droplets for the very same reason mankind spent 300 years improving clock oils: oil droplets scarcely evaporate. Unlike water droplets, the mass of an oil droplet does not change with time. Sprayed from an atomizer, the droplets would acquire a charge due to friction as they passed through the nozzle. The charged droplets fell through a hole in one of two metal plates, which Millikan connected to a room full of electric batteries. While between the plates the droplets experience an electric force in addition to gravity, as shown in Fig. 12.6. By adjusting the voltage on the plates (and hence the electric field) certain droplets could be suspended when the upward electric force equaled the weight of the droplet:

$$qE = mg.$$

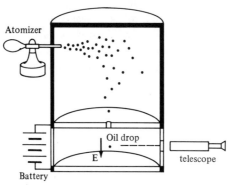

Figure 12.6 Schematic of Millikan's oil-drop apparatus.

The electric field is downward but the force qE is upward on those droplets having negative charge.

Through an optical device placed near the chamber, Millikan could watch individual droplets. Because of their extremely small size, the droplets appeared as stars on a black background. By increasing the voltage he could make droplets rise; those droplets with greater charge rose more quickly. By reversing the voltage, he could make them fall faster. In addition, he could change the charge on a droplet by sending a stream of ions into the chamber. Millikan's fascination with the acrobatic motion of droplets may have lightened the long, solitary hours in the lab he spent squinting through the device and recording hundreds of measurements.

What measurements did Millikan make that revealed the charge of the electron? Using Newton's second law, the motion of a droplet drifting upward between the plates of Fig. 12.6 can be described by

$$m\frac{dv}{dt} = qE - mg - 6\pi R\eta v. \tag{12.5}$$

The electric force pushes the negatively charged droplets upward but the viscous force is downward (opposite v) just like gravity.

By setting $dv/dt = 0$ in Eq. (12.5), we find the terminal velocity to be

$$v_1 = \frac{qE - mg}{6\pi R\eta}. \tag{12.6}$$

The characteristic time to reach this terminal velocity turns out to be the same as when there is no electric field and is given by

$$t_0 = \frac{m}{6\pi R\eta}. \tag{12.4}$$

A typical droplet approaches terminal velocity rapidly (in about 10^{-5} s).

Using a stopwatch to time a droplet moving between marks etched in the optical device, Millikan could measure the terminal velocity. As Eq. (12.6) indicates, the larger the charge on a droplet, the greater its terminal velocity. By observing the motion of the hundreds of droplets with different charges on them, Millikan uncovered the pattern he expected: the charges were multiples of the smallest charge he measured.

Measuring v_1 and knowing E, η, and g, you might think that Eq. (12.6) could be solved for the charge q on any droplet. But there are two other unknown quantities in that equation: the droplet's mass m and its radius R. These quantities, however, are not independent, but are related by the density of oil ρ. The oil drop has volume $\frac{4}{3}\pi R^3$, so its mass is

$$m = \tfrac{4}{3}\pi R^3 \rho.$$

To be very precise, Millikan actually used $\rho - \sigma$ in place of ρ, where σ is the density of air. The reason for this is that the air provides an additional upward buoyant force on a droplet which is equal to the weight of the air displaced by the droplet (this is known as Archimedes's law). The weight of air displaced is just the density of air times the volume of the droplet. Accounting for this force is equivalent to saying that the weight of the droplet is reduced to $mg - m_a g = (m - m_a)g$, where m_a is the mass of air displaced: $m_a = \frac{4}{3}\pi R^3 \sigma$. Since the density of air is about one-thousandth that of oil, the correction is barely necessary. With this correction taken into consideration, Eq. (12.6) becomes

$$v_1 = \frac{qE - \tfrac{4}{3}\pi R^3 (\rho - \sigma)g}{6\pi R\eta}. \tag{12.7}$$

Millikan could not measure the radius R of a droplet directly because the drops are too small to be seen clearly. But he had a clever way to find R indirectly. What he did was first measure the terminal velocity of a droplet drifting upward in the electric field, and then measure the terminal velocity of the same droplet falling without the electric field on. With the field on, the terminal velocity is given by Eq. (12.7). When the droplet is simply falling under the force of gravity, the terminal velocity is given by

$$v_2 = \frac{mg}{6\pi R\eta}, \tag{12.3}$$

which when corrected for the buoyant force of air (just as we did earlier) becomes

$$v_2 = \frac{2(\rho - \sigma)R^2 g}{9\eta}. \tag{12.8}$$

In effect, this second measurement is used to find the size of the drop, R.

From (12.7) and (12.8) we find

$$v_1 + v_2 = \frac{qE}{6\pi R\eta},$$

which, when solved for q, gives us

$$q = \frac{6\pi R\eta}{E}(v_1 + v_2).$$

But we can express R in terms of v_2 from Eq. (12.8) and we find, after some algebra, that

$$q = \frac{18\pi\eta^{3/2}}{E\sqrt{2g(\rho - \sigma)}}\, v_2^{1/2}(v_1 + v_2). \tag{12.9}$$

By measuring v_1 and v_2 Millikan determined the charge of a droplet.

Shown in Fig. 12.7 is a page from Millikan's research notebook, dated Friday, 15 March 1912. The column labels G and F refer to the times for a droplet to move between calibration marks in gravity alone and in the electric field, respectively. Through hundreds of such delicate measurements, Millikan, the patient experimentalist, discovered that the charge on a droplet always comes out an integral multiple (i.e., 1, 2, 3, etc.) of the smallest charge he found. Here was the first evidence that charges come in integral multiples of a fundamental charge – the charge of the electron.

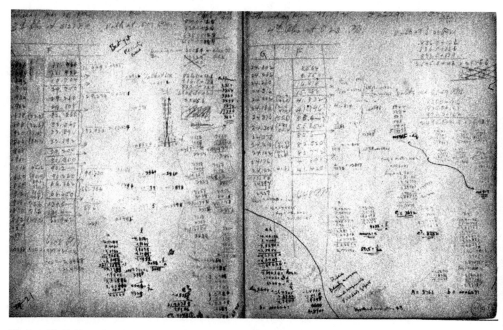

Figure 12.7 Page from Robert A. Millikan's research notebook.
(Courtesy of the Archives, California Institute of Technology.)

By reevaluating the coefficient of viscosity for air and reducing errors caused by temperature variations and air currents, Millikan succeeded in determining the charge e of the electron with an error of 0.1%. The value he published in 1913 was $e = -(4.774 \pm 0.005)$ electrostatic units, equivalent to $e = -(1.603 \pm 0.002) \times 10^{-19}$ C, which served physics for a generation and is within the experimental-error bounds he gave of the most recent value. Millikan had observed the electron itself, and for his momentous efforts he received the Nobel prize in 1923.

Today physicists are searching for fractionally charged particles called quarks. Based on a symmetry classification for elementary particles, quarks are thought to be the building blocks of particles that exist inside nuclei and carry charges of $+\frac{2}{3}e$ and $-\frac{1}{3}e$. Modifications of Millikan's historic experiment are used by some of these quark hunters.

Questions

12. Fill in the steps leading to Eq. (12.9).

13. Suppose that in Millikan's oil-drop experiment it were possible to replace all the excess electrons on a given droplet with particles having twice the charge of an electron and twice the mass. Which of the following statements is true?

   (a) In order for the particles to be suspended between the capacitor plates, the voltage would have to be doubled.

   (b) With no electric field on, the drops would fall with a terminal velocity twice that of droplets having electrons.

   (c) The characteristic time to reach terminal velocity would be doubled.

14. Using Eq. (12.9) what percentage of error would be introduced if the correction for the buoyant force of air were not included? To find the answer, compare the charge obtained for a given v_1, v_2, etc., with and without σ considered.

15. Suppose that the electric field was such as to create a force downward rather than upward on a droplet. What then would replace Eq. (12.5)? What would be the terminal velocity in this case?

16. Show that $v(t) = t_0(qE/m - g)(1 - e^{-t/t_0})$ is a solution of Eq. (12.5), where t_0 is given in Eq. (12.4). Prove that it is consistent with the terminal velocity given in Eq. (12.6).

17. Why do you expect the characteristic time for a droplet to reach terminal velocity to be the same with or without the electric field on?

18. Estimate t_0 and v_2 in Millikan's experiment. A typical drop had a radius $R \approx 10^{-6}$ m.

12.4 A FINAL WORD

When Millikan made his measurements, alone in his laboratory, he had to have a notebook like any scientist to record what he had done. Afterward, he would gather his results together, write a scientific paper, and publish it for all the world to see. But his notebooks, the raw data of his experiments, were for his own eyes only. Figure 12.7 shows a page from Millikan's notebook. Before we criticize what we see, let's remember what Millikan was doing. He was measuring, for the first time ever, one of the fundamental constants of nature. His task was to make his measurements in the most careful, dispassionate way possible, then publish all of his results so that other scientists could judge whether he'd done it properly. The page in Fig. 12.7 is dated 15 March 1912. Here he writes down the temperature and barometric pressure, then he starts recording data, the times for a droplet: F means in the field, and G means in gravity. Then he calculates the velocities, uses logarithms to multiply them together (he didn't have a hand calculator), and finally he gets his result.

On one page he writes: "One of the best ever . . . almost exactly *right.*" – What's going on here? How can it be right if he's supposed to be measuring something he doesn't *know*? On another page he writes: "Beauty. Publish!" One might expect him to publish everything! On another page, the usual stuff, then: "4% too low – something wrong." Not 4% too low but publish anyway, like a good scientist. Then something very revealing: ". . . distance wrong." He's found an excuse for not publishing it. More pages: "Beauty, one of the best," and so on for pages and pages.

Now, you shouldn't conclude that Robert Millikan was a bad scientist. He wasn't – he was a great scientist, one of the best. What we see instead is something about how real science is done in the real world. What Millikan was doing was not cheating. He was applying scientific judgment. He had a pretty clear idea of what the result ought to be – scientists almost always think they do when they set out to measure something. So, when he got a result he didn't like, he wouldn't just ignore it – *that* would be cheating. Instead, he would examine the experiments to see what went wrong. Now that seems reasonable, but it's actually a powerful bias to get the result he wants, because you can be sure that when he got a result he liked, he didn't search as hard to see what went right. But experiments must be done in that way. Without that kind of judgment, the journals would be full of mistakes, and we'd never get anywhere. So, then, what protects us from being misled by somebody whose "judgment" leads to a wrong result? Mainly, it's the fact that someone else with a different prejudice can make another measurement. Every scientist believes there *is* a real answer; it's part of nature. That's the belief that keeps scientists rigorously honest, causing them to temper and guard against their own prejudices. Dispassionate, unbiased observation is supposed to be the hallmark of the scientific method. Don't believe everything you read. Science is a difficult and subtle business, and there is no method that assures success.

13

THE LAW OF CONSERVATION OF ENERGY

You see, therefore, that living force [kinetic energy] may be converted into heat, and that heat may be converted into living force, or its equivalent attraction through space. All three, therefore – namely, heat, living force, and attraction through space (to which I might also add light, were it consistent with the scope of the present lecture) – are mutually convertible into one another. In these conversions nothing is ever lost. The same quantity of heat will always be converted into the same quantity of living force. We can therefore express the equivalency in definite language applicable at all times and under all circumstances.

James Prescott Joule, "On Matter, Living Force, and Heat" (1847)

And in each of these decades [the 1950s and 1960s] more oil was consumed than in all of man's previous history combined.

President Jimmy Carter (18 April 1977)

13.1 TOWARD AN IDEA OF ENERGY

The law of conservation of energy is a fundamental law of physics. No matter what you do, energy is always conserved. The total amount of energy in the universe is, has been, and always will be the same as it is right now. So why do people tell us to conserve energy? Evidently the phrase *conserve energy* has one meaning to a scientist and quite a different meaning to other people, for example, to the president of a utility company or to a politician. What then, exactly, is energy?

The notion of energy is one of the few elements of mechanics not handed down to us from Isaac Newton. The idea was not clearly grasped until the middle of the nineteenth century. Nevertheless, we can find its germ even earlier than Newton. The essence of the idea of the conservation of energy can be seen in the incredibly fertile experiments that Galileo performed with balls rolling down inclined planes.

It is astonishing how many results Galileo squeezed out of his simple experiments. Bodies fall much too fast to be timed by the crude water clocks of the seventeenth century, but by slowing down the falling motion with his inclined planes, Galileo showed that uniformly accelerated motion was a part of nature. That alone was an achievement to crown him as a genius.

Galileo, of course, did much more besides his inclined-plane experiments, but he also did much more *with* his experiments. He arranged that once a ball had finished rolling down one inclined plane it would proceed to roll back up another, which could be more or less inclined than the first. Here he discovered a suggestive fact: no matter what the incline of the first plane and no matter what the incline of the second, the ball would finally come to rest on the second plane at the same vertical height above the table as that at which it had started on the first.

He concluded that if the second plane were horizontal the ball would continue rolling with the same speed forever. In other words, he discovered the law of inertia, which we discussed in Chapter 4.

Once we have grasped the idea of inertia, we can easily see why the ball starts up the second inclined plane after rolling down the first. But that does not tell us why it always reaches the same vertical height at which it started. The ball almost seems to remember its origin. We prefer to say that something is conserved, rather than remembered. The name we give to the conserved quantity is *energy*.

When Galileo lifted a ball from the table to the height of its starting point, as illustrated in Fig. 13.1, he endowed it with a form of energy called *potential energy*. *Energy that a body has by virtue of its location is called potential energy*. Then Galileo released the ball and it started rolling, picking up speed. By the time it reached the bottom of the incline, it was at the level of the table. If it previously had potential energy because of its height above the table, that energy was now gone. In place of the potential energy, the ball was in motion. The potential energy had not been lost, but rather had been transformed into another form. *Energy of motion* is called *kinetic energy*.

As the ball continued, ascending the second inclined plane, it slowed down. It was losing kinetic energy, but in return it was regaining energy of position, potential energy, as its height increased. When the ball finally came to rest, all its kinetic energy had been

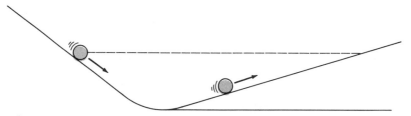

Figure 13.1 Galileo's experiment with inclined planes and rolling balls.

transformed back into potential energy. That happened when the ball's height above the table was precisely what it was at the beginning of the experiment.

Earlier we would have described the same experiment in different terms. We would have said: To lift the ball, Galileo applied a force opposing gravity. When he released the ball, the force of gravity made it roll down the plane. At the bottom of the plane, its inertia made the ball roll back up the second plane. We have a valid description without the idea of energy, so why talk about energy?

We need the idea of energy because it expresses one fact our old description didn't prepare us for: the ball ends up at the same height it started at, never any higher, and, ignoring air resistance and friction, never any lower. Something is the same at the end of the ball's motion as it was in the beginning. That something is energy.

The idea of energy was invented precisely because it is conserved. Then why do politicians and gas company executives tell us to conserve energy? Before answering this burning question, we need precisely and quantitatively to define energy.

13.2 WORK AND POTENTIAL ENERGY

The word *work* is used to describe the transfer of energy from one thing to another. We say that when Galileo lifted a ball onto the inclined plane, he was doing work on it. The clue to the definition of work is that we have two descriptions of Galileo's inclined-plane experiment, both of which must be correct: Galileo applied a force opposing gravity, lifting a ball to a certain height; in terms of energy, he did work on the ball, thereby giving it a certain potential energy equal to the work done. By comparing these two views, we see that work is related to both force and distance.

Our intuition suggests that it takes twice as much work to lift a ball twice the weight to the same height, and that it takes three times as much work to lift a given ball three times as high. That indicates how work is related to both force and distance. The simplest relation is to define the work W done by a constant force F to be the product of the force times the distance h through which it acts. Thus, for a constant force,

$$W = Fh. \tag{13.1}$$

For example, if you lift an object of mass m in such a way that it doesn't accelerate, then the force is equal to the weight mg of the object. Since mg is constant, the work done to raise the object up to height h is

$$W = mgh.$$

What if the force is not constant? Suppose a force that changes with position acts on a particle moving in a straight line. Say the force is $F(z)$ (in the direction of the line) when the particle is at z (as shown in Fig. 13.2). How much work is done by this variable force in moving the particle from $z = a$ to any other position $z = x$?

The work will depend on x and we denote it by $W(x)$. We haven't defined $W(x)$ yet but we can give an argument which suggests that a reasonable definition is

$$W(x) = \int_a^x F(z)\, dz. \tag{13.2}$$

This is a *definition*, so it cannot be proved, but it can be motivated as follows:

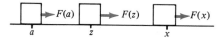

Figure 13.2 Work done by force $F(z)$ moving an object from $z = a$ to $z = x$ is

$$W(x) = \int_a^x F(z) \, dz.$$

First, we want $W(a) = 0$ since no work is done if we don't move the particle at all. Now suppose we move the particle from the position $z = x$ to a nearby position $z = x + h$, where h is a small positive number. The work done by the force from x to $x + h$ is the difference

$$W(x + h) - W(x).$$

We can express this same work in another way. Over the interval from x to $x + h$, the force $F(z)$ will change, but it has some average value, which we can call $F(\bar{z})$. If we treat the force as though it had the constant value $F(\bar{z})$ over the interval from x to $x + h$, then the work it does is $F(\bar{z})h$, the product of force and distance. Therefore

$$W(x + h) - W(x) = F(\bar{z})h$$

or, dividing by h,

$$\frac{W(x + h) - W(x)}{h} = F(\bar{z}).$$

Now we let h shrink to zero on both sides of the equation. The left-hand side becomes dW/dx and the right-hand side becomes $F(x)$, so in the limit we find

$$\frac{dW}{dx} = F(x).$$

In other words, the derivative of the work is $F(x)$. Therefore, by the second fundamental theorem of calculus, we have

$$W(x) - W(a) = \int_a^x F(z) \, dz.$$

Since $W(a) = 0$, this gives us Eq. (13.2).

Note that if the force is constant, the integral is the constant force times the length of the interval, which agrees with (13.1).

The dimensions of work are those of force times distance. The SI unit of work is one newton-meter, called one joule (after a British physicist whom we shall encounter later) and abbreviated 1 J. In the cgs system the unit of work is one dyne-centimeter, or 1 erg. In the British system, one foot-pound is the basic unit of work and has no other name. Using conversions between fundamental units, you can verify that

$$1 \text{ J} = 10^7 \text{ erg} = 0.738 \text{ ft lb}.$$

An important remark is appropriate at this point: The integral

$$W(x) = \int_a^x F(z)\, dz$$

is the work done *by the force* $F(z)$ acting on a particle in moving it from $z = a$ to $z = x$. The negative of this quantity,

$$-W(x) = -\int_a^x F(z)\, dz,$$

is called the work done *against the force* $F(z)$. For example, suppose a car is rolling slowly along a road and a man pushes backward on it with a constant force of 50 lb, bringing it to rest within a distance of 10 ft. The work done *by the car* on the man is 500 ft lb, a positive quantity. But the work done *by the man* on the car is -500 ft lb, a negative quantity because the force and the displacement are in the opposite directions.

Now that we have a definition of work, we need to understand its connection with energy and why energy is conserved. To do so, let's think about Galileo's experiment. To get the ball rolling, Galileo first lifted it up to a certain height and in doing so performed work on it, endowing it with a certain amount of potential energy. The work he did and the potential energy of the ball are each equal to *mgh*. Once he released it, the ball rolled down the plane. Its potential energy, equal to *mg* times its height above the table, was converted into kinetic energy. As it rolled up the second incline, the kinetic energy was converted back into potential energy. This continued until its original potential energy *mgh* was restored. Thus it rolled to a height equal to that at which it started, and where it again had the same potential energy as it did initially.

The ball would continue to roll up and down the inclined planes without Galileo's attention. So once Galileo lifted the ball, once he did work on it, he was out of the picture. We no longer need to consider Galileo, but rather we can focus entirely on the energy and motion of the ball.

Let us reiterate: *work* is the name we use to describe a process in which energy is transferred from one part of the universe to another. Here the first part of the universe is Galileo, and the second system is the ball. Energy entered the system in the form of potential energy from the universe outside of it by means of work done on it. Work is merely a bookkeeping device to keep track of transfer of energy from one thing to another.

We identify the work done to lift an object without accelerating it as the transfer of potential energy U to the object:

$$U = W. \tag{13.3}$$

It follows that in Galileo's experiment, the ball receives an amount of potential energy

$$U = mgh.$$

Aside from depending on the mass of the ball, the potential energy depends only on where it is, which here is its height h above the table. This connection is plausible because when the ball rolls up the second plane it reaches the same height as it started from, h, where it has exactly the same potential energy. Thus energy defined in this way is indeed a conserved quantity.

So far our definition of work as a force times displacement in the same direction appears feasible. But what happens if the force and resulting displacement are not in the same direction? For example, when gravity works on a ball on Galileo's inclined plane, the force of gravity is always vertically downward, but the ball is displaced not straight downward, but along the inclined plane. How do we handle this?

Since gravity should do exactly as much work on the ball moving it down the plane as Galileo did to put it up there, we can find the answer. As shown in Fig. 13.3, mg is the force and r is the displacement along the inclined plane. The work done by gravity to bring the ball to the bottom of the inclined plane must be equal to mgh. From the figure we see that $h = r \cos \theta$, so we can write the work done as

$$W = mgh = mgr \cos \theta.$$

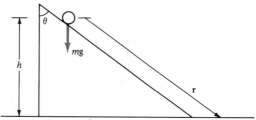

Figure 13.3 Work done by gravity is $m\mathbf{g} \cdot \mathbf{r} = mgh$.

Taking another look at Fig. 13.3, we realize that $mg \cos \theta$ is the component of the weight in the direction of the displacement $\mathbf{r}$. We already have a notation prepared exactly for this purpose, the dot product. As you recall from Chapter 5, the dot product is equal to the product of the component of one vector along a second vector times the magnitude of the second vector. In summary, we describe the work done here as

$$W = \mathbf{F} \cdot \mathbf{r} = m\mathbf{g} \cdot \mathbf{r} = mgr \cos \theta.$$

In general, the work done by a constant force $\mathbf{F}$ acting through a displacement $\mathbf{r}$ is

$$W = \mathbf{F} \cdot \mathbf{r}. \tag{13.4}$$

Example 1

Show that the work done against gravity depends only on the vertical distance an object is moved.

Suppose we wish to move an object from point A to point B shown in the figure.

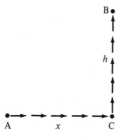

We can think of moving the object from A to B by first moving it horizontally, a distance x to point C, then vertically a distance h to B. The work done from A to B is simply the sum of the work from A to C and from C to B:

$$W_{AB} = W_{AC} + W_{CB}.$$

Because the force we apply to the object is vertically upward and equal to mg, Eq. (13.1) tells us that

$$W_{AC} = mgx \cos 90° = 0,$$

$$W_{CB} = mgh \cos 0° = mgh.$$

Therefore the total work done is simply

$$W_{AB} = mgh.$$

In fact, the work done in moving from A to B will always be mgh, no matter what path is taken. If the path is replaced by vertical and horizontal steps, the work done along any horizontal path is zero. The work done against gravity depends only on the difference in vertical height between the initial and final points.

Example 2

What amount of potential energy is stored in a spring that is stretched from equilibrium to a distance d?

We'll use Eq. (13.2) to calculate the work, which in turn is equal to the potential energy stored in the spring. When we stretch a spring a distance x, the spring exerts a force

$$F = -kx. \tag{10.2}$$

Therefore, to stretch it, we need to apply a force $F = +kx$, which is, of course, in the same direction as the displacement of the spring. Thus the work we do is

$$W = \int_0^d kx \, dx = \tfrac{1}{2}kd^2,$$

where we know the value of the integral from Chapter 7. Since the potential energy stored in the spring is equal to the work done on it, we have

$$U = \tfrac{1}{2}kd^2.$$

The potential energy is the same whether the spring is stretched or compressed.

Questions

1. If you were to hold a book with an outstretched arm, you would soon tire and claim that it is hard work to keep the book there. According to Eq. (13.1), are you doing work on the book?

2. Suppose that two forces, $\mathbf{F}_1$ and $\mathbf{F}_2$, act on an object, with $F_1 = 10$ N, $F_2 = 16$ N, such that the two forces are acting in opposite directions. What is the total work done on the object when it moves a distance of 2 m under the action of these forces?

3. Two springs A and B are identical except that A is stiffer than B (larger spring constant). On which spring is more work done if (a) they are stretched by the same force, (b) they are stretched by the same amount?

4. If work represents the energy transferred to a system from the outside, where does the energy come from that enables you, for example, to lift a ball onto a table?

5. A gardener pushes a lawn mower with a force of 20.0 N at an angle of 37° down from the horizontal for a distance of 15.0 m. Calculate the amount of work done by the gardener.

6. A force $\mathbf{F} = (3.0 \text{ N})\hat{\mathbf{i}} - (7.0 \text{ N})\hat{\mathbf{j}}$ moves an object through a displacement $\mathbf{r} = (4.0 \text{ m})\hat{\mathbf{i}} + (3.0 \text{ m})\hat{\mathbf{j}} + (2.0 \text{ m})\hat{\mathbf{k}}$. Find the amount of work done on the object by this force.

7. The force exerted on an object is described by $F = F_0(x/b - 1)$, where F_0 and b are constants. Find the work done as the object moves from $x = 0$ to $x = 3b$ by

   (a) plotting $F(x)$ and graphically determining the area under the curve,
   (b) evaluating the integral analytically.

8. The work done to stretch a certain spring a distance of 1.0 cm from equilibrium is 0.2 J. If 0.2 J additional work were done on the stretched spring, how far from equilibrium would the spring be stretched?

13.3 THE LAW OF CONSERVATION OF ENERGY

When the ball reaches the bottom of Galileo's incline, where $h = 0$, it will have lost all of its potential energy U, converting it to kinetic energy K. We know exactly how large K will be: it will be equal to U. But K is the energy of motion; it should depend not on where the ball was, as U does, but on how fast it is going. What exactly is the connection between kinetic energy and speed?

Let's imagine an experiment slightly different from Galileo's. Suppose we don't start with an inclined plane. Instead we have a smooth block on a smooth table, at height $h = 0$. If we push the block horizontally, we set it in motion; the force merely overcomes the block's inertia. While we are pushing on the block, we keep it on the table, so h and therefore the potential energy remain equal to zero. However, suppose we cleverly arrange that by the time the block reaches the second incline, as shown in Fig. 13.4, it has the same speed v_0 it would have acquired in sliding down an incline starting at height h.

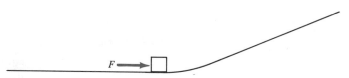

Figure 13.4 Sliding block on inclined planes.

Since the kinetic energy should depend on the block's speed, and we are arranging things so that the speed is the same as it would have been had it slid down the plane, K must be equal to U, which is equal to the work W. In other words, the work we have done pushing the block horizontally must be the same as W. But how can we express that in terms of speed? We don't even know how much force in the horizontal direction we need to get the block up to the right speed. We do know, however, that

$$F = \frac{dW}{dx}. \tag{13.5}$$

As we push the block, we do work over some period of time, so it is reasonable to think of the work done per unit time (known as power) rather than how much work we do at a given place. Thinking of the work W as a function of time t we use the chain rule to get a relation between F and dW/dt:

$$\frac{dW}{dt} = \frac{dW}{dx}\frac{dx}{dt} = F\frac{dx}{dt}.$$

But dx/dt is the block's speed v. Therefore we have

$$\frac{dW}{dt} = Fv.$$

Using Newton's second law, $F = m\,dv/dt$ to substitute for F, we obtain

$$\frac{dW}{dt} = mv\frac{dv}{dt}.$$

Because $d(v^2)/dt = 2v\,dv/dt$ we can write this last result as

$$\frac{dW}{dt} = \frac{d}{dt}(\tfrac{1}{2}mv^2).$$

This means that the time rate of change of work is the same as the rate of change of the quantity $\frac{1}{2}mv^2$. In other words, the total work we do by the time we get the block to the base of the second incline is $\frac{1}{2}mv^2$. Since the block started at rest, $v = 0$, the change is simply $\frac{1}{2}mv_0^2$, where v_0 is the speed it finally acquires; hence

$$W = \tfrac{1}{2}mv_0^2. \tag{13.6}$$

We want this to be equal to the work it takes to lift the block,

$$W = U = mgh,$$

which implies that

$$mgh = \tfrac{1}{2}mv_0^2.$$

We've succeeded in finding a quantitative expression for the conservation of energy applied to our own Galileo-like experiment. We have used a sliding block instead of a rolling ball so that we could ignore the fact that some energy is necessary just to get a ball rotating. When we originally lifted the block a distance h above the table, our muscles did work on it and gave the block a potential energy equal to mgh. That initial amount of energy remains in the system even though it can change into a mixture of both potential and kinetic energy. When the block has reached the bottom of the incline, its potential energy has been completely transformed into kinetic energy, which we can now quantiatively express as $\frac{1}{2}mv^2$:

$$K = \tfrac{1}{2}mv^2. \tag{13.7}$$

This formula for kinetic energy was derived for a sliding block having the speed v it would have acquired by sliding down an incline under the constant force of gravity. Using this example as motivation, we now *define* the kinetic energy K of any body of mass m moving with speed v by the equation

$$K = \tfrac{1}{2}mv^2.$$

The foregoing analysis which showed that

$$\frac{d}{dt}\left(\frac{mv^2}{2}\right) = \frac{dW}{dt} \tag{13.8}$$

is valid not only for a sliding block but for the work done by any variable force F acting in a fixed direction. The work done *against* this force is $-W$, the potential energy U imparted to the body. Since $K = \frac{1}{2}mv^2$ and $W = -U$, Eq. (13.8) states that

$$\frac{dK}{dt} = -\frac{dU}{dt},$$

or

$$\frac{d}{dt}(U + K) = 0. \tag{13.9}$$

Therefore the quantity $U + K$ remains constant in time. This quantity,

$$E = U + K,$$

is called the *total energy* of the particle. Both the kinetic energy K and the potential energy U can change as the particle moves, but their sum – the total energy – remains constant. This is the law of conservation of energy. It is described mathematically by Eq. (13.9).

Let's see what the law tells us about the sliding block. Let h denote the maximum or initial height above the table, and let z be an arbitrary height, as indicated in Fig. 13.5. Then mgz is the potential energy when the block is at height z, and $\frac{1}{2}mv^2$ is the kinetic energy. The conservation law says that

$$mgz + \tfrac{1}{2}mv^2 = \text{const.}$$

By saying that the left-hand side is constant, we mean that its value does not change in time. Since this quantity is the same anywhere along the block's motion, we can determine

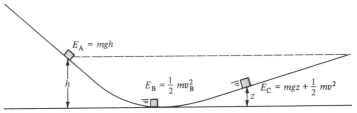

Figure 13.5 Conservation of energy applied to Galileo's experiment.

the constant by evaluating the left-hand side at any point where we know all the quantities. For example, if we take the initial point where $z = h$, we know that $v = 0$, so the value of the left-hand side is simply mgh, which is the constant. Thus we can write

$$mgz + \tfrac{1}{2}mv^2 = mgh. \tag{13.10}$$

From this statement of conservation of energy, we can find the speed of the block at any height z.

None of this depends in any way on how steeply the planes are inclined. In fact, the planes could be vertical; then we would have a freely falling body. Regardless of whether a body is slipping down an inclined plane or freely falling, if it starts at height h, its speed v can be found from (13.10), which when solved for v gives

$$v = \sqrt{2g(h - z)}\,.$$

The speed depends only on the vertical distance fallen. Does it agree with the law of falling bodies?

Recall that for a body in free fall,

$$s = \tfrac{1}{2}gt^2$$

and

$$v = gt.$$

If we eliminate t, we get

$$v = \sqrt{2gs}\,,$$

where s is the distance fallen and is just $h - z$. The law of conservation of energy predicts exactly the same speed as the law of falling bodies. This comparison helps to confirm that the way we defined work by Eq. (13.1) is a sensible choice; it leads to a quantitative expression of total energy, $E = U + K$, which is conserved.

The law of conservation of energy is useful for analyzing motion not only on inclined planes and for falling bodies but also for many other phenomena such as pendulums and springs. The following is a generalization of the procedure we used to apply the law of conservation of energy:

(a) Define the system.
(b) Write down the total energy of the system at the point, say A, where you want to determine some unknown quantity (like speed or height): $E_A = U_A + K_A$.

(c) Find another point, say point B, where you know everything about the object's motion and write down the total energy at that point: $E_B = U_B + K_B$.

(d) Conservation of energy implies that $E_A = E_B$; equate the two energies and solve for the unknown quantity.

So far, we know two types of potential energy which U can represent. One is the potential energy a mass m has due to its height z above the earth's surface:

$$U_{gravity} = mgz.$$

The other is the potential energy stored in a spring either stretched or compressed by an amount x:

$$U_{spring} = \tfrac{1}{2}kx^2.$$

The following examples apply the law of conservation of energy to various problems.

Example 3

A 2.0-kg block on a horizontal, frictionless surface is pressed against a spring of spring constant 1.5×10^3 N/m. The spring is compressed a distance of 8.0 cm. When released, what speed will the block acquire?

From Example 2, we know that the potential energy stored in a spring compressed by a distance d is

$$U = \tfrac{1}{2}kd^2$$

where k is the spring constant. The law of conservation of energy tells us that the total potential and kinetic energy of the system, which here consists of the block and spring, remains constant:

$$E = K + U = \tfrac{1}{2}kx^2 + \tfrac{1}{2}mv^2.$$

Let's apply this to the following two points: point A, where the spring is compressed an amount d and the block has $v = 0$; and point B, where the spring has no potential energy and the block has speed v; $E_A = E_B$ implies that

$$\tfrac{1}{2}kd^2 + 0 = 0 + \tfrac{1}{2}mv^2,$$

from which we can solve for the speed v:

$$v = (k/m)^{1/2}d = [(1.5 \times 10^3 \text{ N/m})/(2.0 \text{ kg})]^{1/2}(0.08 \text{ m}) = 2.2 \text{ m/s}.$$

Example 4

A pendulum bob is pulled aside from the vertical through an angle θ and released. Find the speed of the bob and the tension in the string at the lowest point of the swing, assuming $L = 0.3$ m, $\theta = 30°$, and $m = 0.5$ kg.

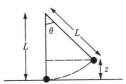

Since we anticipate using the law of conservation of energy, let's measure all vertical heights (and therefore potential energy) from the lowest point of the swing, point B. Now the total energy of the bob at point A, just before being released, is

$$E_A = mgz,$$

where z is the vertical height above B. Using geometry, we see that $z = L - L \cos \theta = L(1 - \cos \theta)$, and so the initial energy is

$$E_A = mgL(1 - \cos \theta).$$

The total energy at B is purely kinetic energy because we have chosen $z = 0$ there, so $E_B = \frac{1}{2}mv_B^2$. Therefore the law of conservation of energy, $E_A = E_B$, implies

$$mgL(1 - \cos \theta) = \frac{1}{2}mv_B^2.$$

Solving for the speed v_B we find

$$v_B = \sqrt{2gL(1 - \cos \theta)}.$$

Substituting the numbers, this turns out to be 0.9 m/s.

To find the tension in the string – a force – we need to use Newton's second law. Applying the second law with the free-body diagram for the bob at its lowest point, we see that a combination of weight and tension causes the bob to move in a circle of radius L.

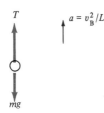

Therefore, we have

$$T - mg = mv_B^2/L,$$

which implies

$$T = mg + mv_B^2/L.$$

Substituting for v_B from our law of conservation of energy, we get

$$T = mg + 2mg(1 - \cos \theta) = mg(3 - 2 \cos \theta).$$

Numerically the tension turns out to be 6.2 N when $\theta = 0°$.

A combination of the law of conservation of energy and Newton's laws provides a powerful method for attacking a variety of problems in classical physics.

Example 5

A toy dart gun consists of a spring that when compressed 0.05 m can project a 20-g rubber dart vertically upward to a height of 3.0 m. What is the spring constant?

Here we have a dart and a spring, each of which can have energy. Initially the system (dart and spring) has potential energy stored in the spring. At its highest point, the dart possesses only gravitational potential energy and the spring, no longer compressed, has no potential energy stored in it. The law of conservation of energy implies that at these two points the energy must be the same. Let's formulate mathematically what we've just stated verbally. Initially the energy of the system is $E_A = U_s + U_d + K_d$, being the sum of the potential of the spring ($U_s = \frac{1}{2}kx^2$), the gravitational potential energy of the dart (in general $U = mgz$, but initially $z = 0$), and the kinetic energy of the dart (initially $K_d = 0$). At the highest point, we have $E_B = U_s + U_d + K_d = 0 + mgh + 0$. $E_A = E_B$ implies

$$\tfrac{1}{2}kx^2 = mgh.$$

Solving for k, we find

$$k = 2mgh/x^2 = 2(0.02 \text{ kg})(9.8 \text{ m/s}^2)(3.0 \text{ m})/(0.05 \text{ m})^2$$

$$= 4.7 \times 10^2 \text{ N/m}.$$

Questions

9. Suppose you were redesigning a roller coaster ride so that for added thrill, the roller coaster will be moving twice as fast at the bottom of the hill. How many times higher would you need to make that hill to achieve this?

10. A truck at rest on the top of a hill is allowed to roll down, and at the bottom it has a speed of 5 km/h. If it is allowed to roll down the hill again, but this time starting with an initial speed of 3 km/h what will be its speed at the bottom of the hill?

11. Three identical blocks slide down the frictionless surfaces shown below after being released from the same height A:

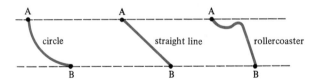

(a) How do the speeds of the blocks at point B compare?
(b) Which block do you think reaches point B first? Why?

12. A mass attached to a vertical spring is gently lowered to its equilibrium position, at which the spring is stretched a distance d. If the same object is attached to the same vertical spring but allowed to fall instead, what maximum distance does this stretch the spring?

13. A block of ice slides down the frictionless incline shown below and compresses the spring. Taking the mass of the block to be 1.5 kg, the spring constant to be 3.0×10^2 N/m, and the distance the block slides down the incline before striking the spring as 1.2 m, find the maximum distance the spring is compressed.

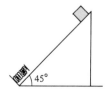

14. A spring with spring constant 3.0×10^3 N/m propels a small (0.5-kg) block up a frictionless plane inclined at 37°. If the spring were initially compressed 0.6 cm, how far up the plane from the point of release would the block travel before coming momentarily to rest? When the block returns and hits the spring, how much will it compress the spring?

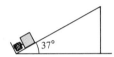

15. A pendulum bob of mass m hanging at the end of a string of length L is struck so that it has speed v. What must v be in order for the string to go slack (tension becomes zero) at the top of the swing (180° from where the bob began)?

13.4   HEAT AND ENERGY

Early attempts to formulate the idea of energy conservation ran into a rigid prejudice: Newton's laws embodied all truth about nature; if a law could not be based on Newton's mechanics, it was not physics but mysticism. Nevertheless, a dozen scientists in the first half of the nineteenth century proposed in some form that energy is conserved. Two of those were James Prescott Joule, a practical-minded brewery owner working in England, and Herman von Helmholtz, a German romantic. Through their work, the law of conservation of energy became scientifically respectable. But to achieve this, they had to confront the apparent nonconservation of energy in the world.

There are numerous phenomena for which energy – at least in the form of kinetic and potential energy – appears not to be conserved. A box sliding across the floor eventually comes to rest; a rubber ball dropped from some height bounces less and less until it too eventually comes to rest. However, if we could look at a sliding box or bouncing ball closely enough to see atoms and molecules, we would better understand what is happening. For example, as a box slides along the floor, atoms and molecules of the box and floor interact and are pulled from their equilibrium positions. However, the atoms aren't free to move very far; the electric forces pull them back to equilibrium position. When they spring back, they hit the adjacent atoms and set them moving, and so on. This motion goes on inside the box and inside the table. Energy is not really lost; it's converted into kinetic and potential energy of atoms and molecules.

Thermal energy is the name we give to energy in the form of hidden motion of atoms and molecules. Technically, heat is transferred between a system and its environment as a result of temperature differences. But the term *heat* is often used generically to encompass thermal energy as well. There are many processes in nature that turn kinetic energy, the organized energy of the motion of an entire large body, into thermal energy, the unorganized motion of atoms. Friction is one example; viscosity is another. In an automobile engine or a steam engine, the opposite process occurs: heat is turned into work.

If it weren't for the fact that energy can turn into heat, it would have been easier to discover the law of conservation of energy. But in the nineteenth century, our idea of atoms and molecules in constant motion had not yet developed. Instead, before the discovery of the law of conservation of energy, there was a different theory. Heat was thought to be a kind of fluid, called caloric. In this theory, heat itself was a conserved quantity.

The caloric theory was not idle speculation but a detailed mathematical theory. A caloric theorist would describe how an iron rod in a fire becomes hot by saying that caloric is flowing from the fire into the iron rod; he would know exactly how much heat, or caloric, was required to bring the rod to a given temperature. If the hot rod were immersed in water, a caloric theorist would say that some of the caloric leaks out of the metal into the water, thereby warming up the water; by applying the law of conservation of caloric, the precise rise in temperature of the water could be predicted. The missing element, the barrier to discovering the law of conservation of energy, was the fact that it is possible to change work or kinetic energy into heat and *vice versa*.

The credit for the law of conservation of energy goes not to the first of the many people who discovered it, but to the last because that person pinned it down so well that it didn't need to be rediscovered. James Prescott Joule, the son of a wealthy English brewer, is said to have become interested in heat by a desire to develop more efficient engines for the family brewery. In experiments conducted between 1837 and 1847 in the brewery and at his own expense, Joule established the law well enough that it never had to be discovered again.

Joule made careful measurements of exactly how much work turned into heat. The invention of the steam engine had led the way to measuring energy changes. Almost from their beginning, steam engines were rated according to their *duty*, which referred to how heavy a load an engine could lift by using a given supply of fuel. Joule used such a practical engineering approach to determine first if electric motors could be made economically competitive with steam engines (and improve the brewery) and later to quantify the relation between work and heat.

Joule's famous experiments involved an apparatus, as shown in Fig. 13.6, in which slowly descending weights turned paddle wheels in a container of water. As the weights fall and paddles turn, the water temperature rises. Through precise comparisons of the work done by the weights and the rise in temperature of the water, Joule uncovered the relationship between heat and work. The unit of heat is the calorie, which is the amount of heat required to raise the temperature of 1 g of water 1°C. Joule found that this unit of heat is related to units of energy according to

1 calorie = 4.18 J.

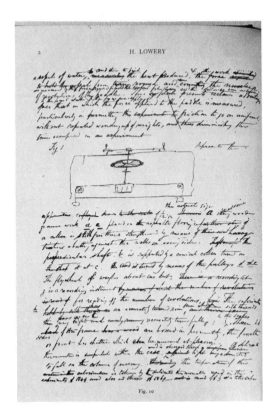

Figure 13.6 Joule's apparatus for measuring the conversion of work into heat.

(Of course, he didn't call them joules – that came later. A food calorie is equal to 1000 calories.)

Begin a practical man and possessing a limited mathematical background, Joule was content with experimenting in his lab. Despite his irrefutable experiments, many physicists of the time wanted a rigorous, mathematical theory of the law of conservation of energy. Such respectability was bestowed by the German physiologist and physicist, Herman von Helmholtz. In a paper published in 1847, Helmholtz showed, much as we did in Section 13.3, that the law of conservation of energy follows from Newton's laws. With his proof in hand, Helmholtz went on to apply the law to various cases, such as gravity, electricity, and friction.

We began our discussion with the observation that in many cases it appears that energy is not conserved; such cases involve friction, viscous force, or other means of dissipating energy into heat. Joule and Helmholtz showed us that energy is *always* conserved and that it can be *transformed* from one form into another.

Example 6

A ball of mass 0.5 kg is thrown vertically upward with an initial velocity of 20 m/s and reaches a height of 15 m. How much energy is dissipated by air resistance?

The ball begins with kinetic energy $\frac{1}{2}mv^2$ and winds up with potential energy mgh. The difference between these is the energy dissipated,

$$\tfrac{1}{2}mv^2 - mgh = \tfrac{1}{2}(0.5 \text{ kg})(20 \text{ m/s})^2 - (0.5 \text{ kg})(9.8 \text{ m/s}^2)(15 \text{ m}) = 26 \text{ J}.$$

Example 7

A spring with spring constant 5.0×10^2 N/m, initially compressed 0.07 m, fires a 0.4-kg block across a rough horizontal surface. If the coefficient of friction between the block and surface is 0.5, how far from its starting point will the block travel before coming to rest?

We can think of the block as doing work against the force of friction. This work is then dissipated into heat. The work done against friction as the block moves through a displacement x is, by definition [Eq. (13.2)],

$$W_{\text{AB}} = \mathbf{f} \cdot \mathbf{x} = \mu_k mgx.$$

Initially the block has only potential energy $\frac{1}{2}kd^2$ from the spring. Substituting into our energy equation, we have

$$\tfrac{1}{2}kd^2 = \mu_k mgx,$$

so we can solve for x and obtain

$x = kd^2/(2\mu_k mg),$

which, when values are inserted, gives 0.6 m.

Questions

16. How much work does a 60-kg athlete do in one pushup, which raises then lowers the body by 0.4 m? How many calories does this represent?

17. Suppose that in Joule's experiment a 1-kg mass falls through a distance of 0.5 m. By how many degrees would this work raise the temperature of 1 g of water?

18. Starting at a height of 25 m, a sled of mass 20 kg slides down a hill. If the sled starts from rest and has a speed of 15 m/s at the bottom of the hill, calculate the energy dissipated by friction along its path.

19. A half-ton pickup truck moving at 24 m/s runs into a giant haystack and penetrates 3 m into it. A similar half-ton truck with a half-ton load of grain runs into the same haystack. How far will this truck travel into the haystack?

20. A 25-g bullet fired at 400 m/s penetrates 10 cm into a block of wood, which remains stationary.

   (a)   What is the energy dissipated by friction between the block and bullet?
   (b)   What is the average force exerted on the block by the bullet?

21. A driver traveling at 30 km/h slams on the brakes and skids 10 m before stopping. If the car were moving at 90 km/h, how far would it skid?

22. A 0.4-kg block is given a speed of 1.5 m/s as it is projected up a plane inclined at 30°. If the coefficient of friction between the block and plane is 0.6, how far up the plane will the block slide?

13.5   A FINAL WORD

Once you understand something, it's almost impossible to put yourself in the position of not understanding it and trying to figure out how people thought about the problem before. Up to the time when Joule did his experiments, presumably heat itself was the conserved quantity and couldn't be created out of work. But how could people for thousands of years before Joule not realize that by rubbing their hands together they could warm them? Even more to the point, by the 1830s, long before Joule's experiments, railroads were strung across Europe. The burning of coal in locomotives eventually set the train in motion – converting heat into work. How could people living at this time, riding on railroads, not have believed that it was possible to turn heat into work?

It was not through lack of trying to understand, as we can see from the thoughts of a young French military engineer, Nicolas Léonard Sadi Carnot. Although Carnot died at the age of 32 from scarlet fever, he is one of the important figures of nineteenth-century science. Carnot discovered what we now call the second law of thermodynamics, which is perhaps the most profound law in physics. He succeeded in doing this without knowing the law of conservation of energy, which is the first law of thermodynamics. Yet it may have been easier to discover the second law of thermodynamics without knowing the first law.

Carnot reasoned by analogy. His idea of how heat worked was by analogy to a water wheel. As water runs down and over a water wheel, it makes the wheel turn, yet the water is conserved. That is, to get work out of the water doesn't require using up water. Instead water falling from a large height to a lower height turns the water wheel. Carnot thought that heat worked in exactly the same way. He thought that caloric, starting out at high temperature, was capable of doing work on the way down to low temperature, analogously to water flowing over the water wheel. His analogy also suggested the idea that came to be the second law of thermodynamics: once heat runs downhill from high to low temperature, it will not run uphill again, back up to the high temperature. That seemed obvious from the water wheel analogy.

A piece of coal has potential energy stored in it in the form of chemical bonds. The process of combustion, say, in a steam locomotive, turns that energy directly into heat at high temperature. Some of that heat is able to drive a piston, producing work and setting the locomotive in motion. Eventually all that heat, including the part turned into work, winds up as heat once again, after the locomotive goes through its frictional processes and so on. But the heat is no longer at the high temperature of the burning coal, but rather at the low temperature of the air outside. The net resut of all this, aside from getting you from one place to another, has been to turn potential energy into high-temperature heat and finally into low-temperature heat. And once that's done, it is no longer possible to get that heat back into the form of the original potential energy so it can be used again. That's what the second law of thermodynamics says: sometimes things happen that can never be undone, no matter what.

Although Carnot could not know that, unlike water, some of the heat actually is transformed into work before changing back into heat again, that did not prevent him from discovering the second law of thermodynamics.

Energy from the sun, at very high temperature, is stored temporarily in coal, oil, and other fossil fuels. We can release it as heat at high temperature (but lower than the temperature of the sun) and, whether we use it for useful work or not, it always winds up as low temperature heat (heat at ambient temperature). All of that energy is perfectly conserved, not a bit is ever lost. But the value of the energy has been lost – it has become useless. The world has run down a little bit, and it never will be wound up again.

This is the real nature of our energy crisis. We are not using up energy, we are just transforming it into useless forms.

Figure 13.7 James Prescott Joule. (Courtesy of the Manchester Literary and Philosophical Society.)

CHAPTER

ENERGY AND STABILITY

I do not consider these principles to be certain mysterious qualities feigned as arising from characteristic forms of things, but as universal laws of Nature, by the influence of which these very things have been created. For the phenomena of Nature show that these principles do indeed exist, although their nature has not yet been elucidated. To assert that each and every species is endowed with a mysterious property characteristic to it, due to which it has a definite mode in action, is really equivalent to saying nothing at all. On the other hand, to derive from the phenomena of Nature two or three general principles, and then to explain how the properties and actions of all corporate things follow from those principles, this would indeed be a mighty advance in philosophy, even if the causes of those principles had not at the time been discovered.

Roger Boscovich, *A Theory of Natural Philosophy* (1763)

14.1 FORMS OF ENERGY

The concept of energy, as we saw in Chapter 13, is subtle, elegant, and rich. It describes a dynamic property of the universe which is strictly and absolutely conserved; energy can neither be created nor destroyed. Not even in the presence of friction is energy ever lost; it is simply transformed into other forms. Nevertheless, the universe is winding down. Energy tends to be transformed from well-organized forms into more disorganized forms, until it becomes completely useless.

The organized forms of energy we discussed are potential and kinetic energy. In order for one object, or part of the universe, to receive such energy, another part must lose an equal amount of energy. To keep track of exchanges of energy between different parts of the universe, we have the bookkeeping device known as work.

Let's reexamine the ideas of work and energy, in part to tidy up a few loose ends, but mainly to generate deeper insights. Recall that work is the displacement of an object multiplied by the component of the force in the same direction.

If the force F varies in magnitude but always has the same direction, then the work done by $F(x)$ in moving a particle from $x = a$ to $x = b$ along that direction is the integral we introduced in Chapter 13:

$$W = \int_a^b F(x)\, dx. \tag{13.2}$$

Consider now a force $\mathbf{F}$, which may vary both in magnitude and direction, and suppose it acts on a particle moving along a curve from A to B, as illustrated in Fig. 14.1. The problem of determining the work done by this force is now more complicated. In fact, the real problem here is to decide on a reasonable *definition* of work. Again, we shall use an integral to define work. This integral must take into account both the force $\mathbf{F}$ and the curve along which the particle is moved.

Figure 14.1 A variable force $\mathbf{F}$ acting along a curve.

Here's an intuitive procedure for arriving at such an integral. Describe the curve by its position vector, say

$$\mathbf{r} = x\hat{\mathbf{i}} + y\hat{\mathbf{j}} + z\hat{\mathbf{k}}.$$

It seems reasonable to say that for a force $\mathbf{F}$ that produces a small displacement $d\mathbf{r}$ the amount of work done is the dot product

$$dW = \mathbf{F} \cdot d\mathbf{r}.$$

The total work done in moving the particle from point A to point B is obtained by adding all these small amounts of work. We indicate the summation process by the integral symbol and write

$$W = \int_A^B \mathbf{F} \cdot d\mathbf{r}. \tag{14.1}$$

This notation resembles that in (13.2), but the symbol in (14.1) is a new kind of integral. It is called a *line integral* or a *contour integral* because the integration takes place along a curve joining A and B. Such an integral can be defined in terms of ordinary integrals of the type we are familiar with. To do this, we consider the position vector as a function of time t and write

$$\mathbf{r} = \mathbf{r}(t).$$

The initial point A of the curve corresponds to some value of t, say $t = a$, so A $= \mathbf{r}(a)$. Similarly, B $= \mathbf{r}(b)$ for some time $t = b$.

We can interpret the dot product $\mathbf{F} \cdot d\mathbf{r}$ to mean $(\mathbf{F} \cdot d\mathbf{r}/dt)\, dt$. This suggests that we simply *define* the line integral $\int_A^B \mathbf{F} \cdot d\mathbf{r}$ by the equation

$$\int_A^B \mathbf{F} \cdot d\mathbf{r} = \int_a^b \mathbf{F} \cdot \frac{d\mathbf{r}}{dt}\, dt. \tag{14.2}$$

The integral on the right-hand side is our common garden variety of integral over an interval from $t = a$ to $t = b$. It incorporates the curve as well as the force because in the integrand on the right-hand side the derivative $d\mathbf{r}/dt$ is to be evaluated at time t and the force $\mathbf{F}$ is to be evaluated at the position $\mathbf{r}(t)$. Thus the integrand is the following function of t:

$$\mathbf{F}[\mathbf{r}(t)] \cdot \frac{d\mathbf{r}}{dt}.$$

Equation (14.2) is taken to be the definition of the line integral $\int_A^B \mathbf{F} \cdot d\mathbf{r}$, and then Eq. (14.1) is used to define the work done by $\mathbf{F}$.

If the force causes the acceleration $d\mathbf{v}/dt$ we can use this integral to prove that the work done is equal to the change in kinetic energy, the same result we obtained in Chapter 13 for linear motion. Using Newton's second law we replace $\mathbf{F}$ by $m\, d\mathbf{v}/dt$ in the integrand in (14.2) and $d\mathbf{r}/dt$ by $\mathbf{v}$ to obtain

$$\int_A^B \mathbf{F} \cdot d\mathbf{r} = \int_a^b m \frac{d\mathbf{v}}{dt} \cdot \mathbf{v}\, dt.$$

The dot product $d\mathbf{v}/dt \cdot \mathbf{v}$ is related to the speed v. Recalling that $v^2 = \mathbf{v} \cdot \mathbf{v}$ we know that

$$\frac{d}{dt}(v^2) = \frac{d}{dt}(\mathbf{v} \cdot \mathbf{v}) = \mathbf{v} \cdot \frac{d\mathbf{v}}{dt} + \frac{d\mathbf{v}}{dt} \cdot \mathbf{v} = 2\mathbf{v} \cdot \frac{d\mathbf{v}}{dt},$$

and hence

$$\frac{d\mathbf{v}}{dt} \cdot \mathbf{v} = \frac{1}{2}\frac{d}{dt}(v^2).$$

Substituting this into the last integral and denoting the work done by W_{AB}, we get

$$W_{AB} = \int_a^b \frac{d}{dt}(\tfrac{1}{2}mv^2)\, dt.$$

Since the integrand is the derivative of $\frac{1}{2}mv^2(t)$ we can evaluate the integral by the second fundamental theorem of calculus to obtain

$$W_{AB} = \tfrac{1}{2}mv^2(b) - \tfrac{1}{2}mv^2(a).$$

The scalar quantity

$$K = \tfrac{1}{2}mv^2(t)$$

is called the kinetic energy of the particle at time t, and we have just shown that the total work done by $\mathbf{F}$ is the change in kinetic energy,

$$W_{AB} = \int_A^B \mathbf{F} \cdot d\mathbf{r} = K_B - K_A. \tag{14.3}$$

Another way of performing work on the same particle is to apply a force that is equal and opposite to another force (gravity, for example) acting on the particle. The object is displaced without accelerating since there is no net force. The work done on it goes entirely into potential energy, the energy that results from the position of the object. So, if a force $\mathbf{F}$ acts on a particle and if we apply an equal and opposite force $-\mathbf{F}$ to it, the work we do goes into a change of the object's potential energy:

$$U_B - U_A = \int_A^B -\mathbf{F} \cdot d\mathbf{r}. \tag{14.4}$$

We now have two expressions relating the work done on an object, one to kinetic energy, the other to potential energy. They show how external forces can increase or decrease the energy of a system. To give a body potential energy, we apply a force to balance one that already exists (like gravity). If we move the particle from point B to point A, we give it potential energy $U_A - U_B$. Suppose we then remove our applied force (e.g., drop the body we were holding), leaving only the preexisting force. It gives the body kinetic energy $K_B - K_A$ as it moves the body from A to B. From (14.3) and (14.4) we see that

$$U_A - U_B = K_B - K_A.$$

The potential energy we had given the body is turned into kinetic energy. Rearranging terms, we find the conservation law

$$U_A + K_A = U_B + K_B.$$

In other words, once we stop doing work that transfers energy into our system, the total energy of the system, potential plus kinetic, is conserved.

The integral in (14.4) describes the *change* in potential energy, not the potential energy itself. The same type of integral can sometimes be used to define the potential energy itself (as will be done later), but there is a small technical difficulty that must be faced. Suppose we see a stone of mass m on a table and ask for its potential energy. One may reply that it is mgh. But from what level is h measured? From the floor? From the street level outside? Or from the center of the earth? In practice, potential energy is always calculated with respect to some arbitrary level of reference, usually the lowest level a body attains during a given discussion. The actual choice of the original level is often irrelevant because we are interested in *differences* or *changes* in the potential energy between two locations and the difference is independent of the choice of origin.

Example 1

A small block of mass m slides down a frictionless incline, starting from rest at a height h, and around a circular track of radius a. Find the speed of the block at point C.

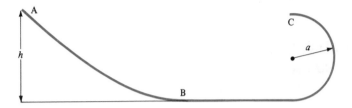

To find the speed v of the block at point C, we shall equate the total energy at points A and C. At A, the total energy is stored as potential energy,

$$E_A = mgh.$$

At point C, the block has both kinetic and potential energy:

$$E_C = mg(2a) + \tfrac{1}{2} mv^2.$$

Setting $E_A = E_C$ and solving for v, we get

$$v = \sqrt{2g(h - 2a)}.$$

Even in the absence of work transferring energy from or to another part of the universe, the total potential and kinetic energy of a system may not be conserved. There are forms of energy that are hidden from ordinary view. One example is heat, the disorganized motions of atoms and molecules. Heat is a kind of hidden kinetic energy.

There can also be hidden potential energy. For example, the potential energy of a carbon atom in a piece of coal is far greater than that in a molecule of carbon dioxide, produced when the coal burns. This is chemical energy, our name for the electrical potential energy built into the structure of molecules, crystals, and so on. When coal burns, of course, the released potential energy turns directly into heat.

Heat itself is not necessarily useless. For example, the intense heat in a jet engine is used to propel an airplane. The dilute heat dispersed by a ball bouncing on the floor, on the other hand, is not easy to recover. Such is the future of the energy of the universe. Eventually the world will run down, not run out of energy, as organized energy is transformed into useless thermal energy.

Example 2

A 0.03-kg marble is dropped into a cylinder containing glycerin. Starting from a height of 0.15 m, the marble reaches the bottom of the cylinder with a speed of 1.4 m/s. What fraction of the marble's initial energy is dissipated by the viscous force of the glycerin?

The marble starts out with potential energy mgh ($h = 0.15$ m) and ends up with kinetic energy $\frac{1}{2}mv^2$, where $v = 1.4$ m/s. The difference between these is the energy dissipated:

$$\text{energy dissipated} = mgh - \tfrac{1}{2}mv^2 = 1.5 \times 10^{-2} \text{ J}.$$

To find the fraction of the initial energy this amount represents, we divide the energy dissipated by mgh. The result is 0.33.

Not only can we follow the transformation of energy into the future, but we also can also trace its evolution backward in time. For example, the energy you use to lift a book was, at an earlier stage, chemical energy stored in muscles. And it was whatever brand of hamburger you had for lunch that provided that chemical energy. For jet engines, the energy to fly you to a vacation resort comes from potential energy stored in the jet fuel. The fossil fuel, in turn, came from creatures that inhabited primeval forests that covered the earth millions of years ago. Their energy ultimately came from sunlight. Every last joule of energy on earth came originally from the sun, with the sole exception of that generated by nuclear reactions. But the sun itself is a cosmic nuclear reactor.

Most energy starts out as nuclear energy, and all the energy that exists in the universe is a legacy to us from the instant in which the universe originated – the Big Bang. Further back than that, we can't trace it. But the amount of primordial energy from the Big Bang is still the same as it was at that first instant, 20 billion years ago, but that energy has been converted into various forms.

Questions

1. A spring is compressed and dropped into a vat of nitric acid which dissolves it. What happens to the potential energy of the spring?

2. Trace the following forms of energy back to their sources insofar as you possibly can:

   (a) the calories in an apple,
   (b) a lightbulb,
   (c) a wind-up wristwatch.

3. A pendulum bob is pulled aside and released. A peg is located as shown in the sketch. How high above the peg will the bob rise?

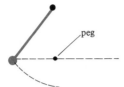

4. A dockworker is loading 100-kg drums onto a truck by rolling them up a ramp. The bed of the truck is 1 m above the ground and the ramp is 2 m long. What is the smallest force that must be exerted on a drum to get it up the ramp?

5. The chemical energy stored in a certain amount of gasoline is converted into kinetic energy as a car speeds up from 0 to 50 km/h. How does the energy to accelerate from 50 to 100 km/h compare to that used to go from 0 to 50 km/h? The answer is not "the same amount."

6. A child sits on a hemispherical mound of ice. Assuming the mound to be frictionless, if the child is given a slight nudge and slides off, find the angle from the horizontal at which the child leaves the mound. (*Hint:* The child leaves the mound when the normal force is zero.)

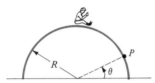

7. A penny given an initial speed of 1.5 m/s slides a distance of 25 cm across a horizontal table before stopping. Find the coefficient of friction between the penny and tabletop.

8. For the track in Example 1, determine the minimum height h can be (in terms of the radius a) so that the block barely makes it to point C. (*Hint:* Use the condition that if the block barely makes it to point C, the normal force exerted on the block by the track at that point will be zero.)

9. A toy car of mass 0.02 kg travels along the frictionless track shown below. The spring ($k = 3.0 \times 10^2$ N/m) is initially compressed 0.05 m.

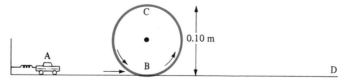

(a) How much work is done by the normal force as the car moves from B to C?
(b) What is the speed of the car at C?
(c) If at D a small parachute is released, bringing the car to rest, how much energy is dissipated by air resistance?

10. Suppose you are driving along a fog-shrouded road at a speed v, and suddenly a brick wall appears a distance R in front of you. Would it be better to slam on the

brakes (and hope that you stop before reaching the wall) or swerve in a circular arc of radius R to avoid the wall? To help you decide, compare the force necessary to change all the car's kinetic energy entirely into work done against friction with that necessary to turn it in a circle.

14.2 GRAVITATIONAL POTENTIAL ENERGY

A swinging pendulum, a plunging roller coaster, and a vibrating guitar string are all examples of potential energy changing into kinetic energy, and back into potential energy, and so on in an energy volley. In each of these examples, the system is endowed with potential energy by changing its position – pulling aside a pendulum, raising a roller coaster, plucking a guitar string. And once the system is released, the volley begins as potential energy is converted back and forth into kinetic energy. In all cases, all the energy is eventually transformed into heat, but it hasn't been totally useless; it may have told someone the time, provided a thrill, or pleased an ear. It is also possible to get a system started by giving it kinetic energy. A baseball thrown by a major league outfielder, an arrow fired from a hunter's bow, and a planetary probe launched by NASA are examples of giving a system kinetic energy and allowing that energy to change into gravitational potential energy.

For a rocket, however, we need to think more about its potential energy; it isn't mgh. Why? In Chapter 13 we assumed that the force of gravity on an object of mass m is simply mg, but we know (from Chapter 8) that this is only true near the surface of the earth. Newton's universal law of gravity expresses the general case:

$$\mathbf{F} = -G\frac{mM_e}{r^2}\,\hat{\mathbf{r}}, \tag{8.1}$$

where all distances are measured from the center of the earth. So we need to consider the change in the force of gravity as the rocket moves away from the earth.

Let's calculate the work done in sending a rocket from the surface of the earth, where $r = R_e$, to some distance R_f far away. First, to lift the rocket we must apply a force equal and opposite to the force of gravity given by Eq. (8.1). For simplicity, let's assume that we send the rocket straight out, in a radial direction. The force $-\mathbf{F}$ and displacement will be in the same direction as shown in Fig. 14.2. Since this is linear motion the work we do is

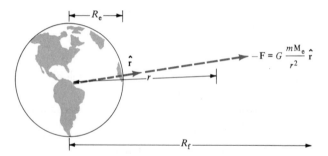

Figure 14.2 Work done to move a mass m from R_e to R_f.

$$W = \int_{r_e}^{r_f} -\mathbf{F} \cdot d\mathbf{r} = \int_{R_e}^{R_f} -F \, dr. \tag{13.3}$$

Substituting for F, which is a function of r, we have

$$W = \int_{R_e}^{R_f} G \frac{mM_e}{r^2} \, dr = GmM_e \int_{R_e}^{R_f} \frac{dr}{r^2}.$$

Knowing that an antiderivative of $1/r^2$ is $-1/r$, we can evaluate the integral and obtain

$$W = GmM_e \left(\frac{1}{R_e} - \frac{1}{R_f} \right).$$

Since $1/R_e$ is bigger than $1/R_f$, the work done is a positive quantity, which it had to be because we needed to supply energy to get the rocket to R_f.

This last result holds if the motion is along any path from the surface of the earth to any point in space a distance R_f away. We shall prove this in Example 3, but to get an idea of why this is so, consider a special path from distance R_e to distance R_f which consists of steps which alternate along radial lines and circular arcs (perpendicular to the radial lines) as shown in Fig. 14.3.

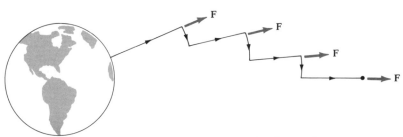

Figure 14.3 A path consisting of radial lines and circular arcs from the surface of the earth to any point in space.

Since the force $\mathbf{F}$ is always radial, no work is done along the circular arcs because $\mathbf{F}$ is perpendicular to the displacement there. So the total work done is obtained by adding work done in the radial directions alone, and this is the same result we got before,

$$W = GmM_e \left(\frac{1}{R_e} - \frac{1}{R_f} \right).$$

According to our bookkeeping, the work we do on the rocket increases its potential energy. The potential energy is higher at R_f than it would be on the surface of the earth by an amount $\Delta U = W$, or

$$\Delta U = U_f - U_e = G \frac{mM_e}{R_e} - G \frac{mM_e}{R_f}. \tag{14.5}$$

By conservation of energy, if the rocket stopped at R_f and fell back to the earth, it would arrive with a kinetic energy $\frac{1}{2}mv^2$ equal to ΔU. We've ignored one thing, though: in actual rocket propulsion, the mass of a rocket changes as it burns up fuel.

Example 3

Prove that the work done on the rocket is independent of the path from the surface of the earth to any point in space at distance R_f.

Let's say the rocket moves along a path with position vector $\mathbf{r}(t)$ at time t. At time $t = 0$ it is at the surface of the earth, so $R_e = r(0) = |\mathbf{r}(0)|$. At some later time t_1 it is at a distance R_f, so $R_f = r(t_1) = |\mathbf{r}(t_1)|$. The work we do to send the rocket from $\mathbf{r}(0)$ to $\mathbf{r}(t_1)$ is the line integral

$$W = \int_{\mathbf{r}(0)}^{\mathbf{r}(t_1)} -\mathbf{F} \cdot d\mathbf{r} = \int_0^{t_1} -\mathbf{F} \cdot \frac{d\mathbf{r}}{dt}\, dt,$$

where $\mathbf{F}$ is given by (8.1). The position vector satisfies $\mathbf{r} = r\hat{\mathbf{r}}$, so its derivative is

$$\frac{d\mathbf{r}}{dt} = r\frac{d\hat{\mathbf{r}}}{dt} + \frac{dr}{dt}\hat{\mathbf{r}}.$$

Taking the dot product of this with $-\mathbf{F}$ we get

$$-\mathbf{F} \cdot \frac{d\mathbf{r}}{dt} = -r\mathbf{F} \cdot \frac{d\hat{\mathbf{r}}}{dt} - \frac{dr}{dt}\mathbf{F} \cdot \hat{\mathbf{r}}.$$

But $\hat{\mathbf{r}}$, being a unit vector, has constant length so it is always perpendicular to its derivative $d\hat{\mathbf{r}}/dt$. Since $-\mathbf{F}$ has the same direction as $\hat{\mathbf{r}}$, it, too, is perpendicular to $d\hat{\mathbf{r}}/dt$, so $\mathbf{F} \cdot d\hat{\mathbf{r}}/dt = 0$ and the foregoing equation becomes

$$-\mathbf{F} \cdot \frac{d\mathbf{r}}{dt} = -\frac{dr}{dt}\mathbf{F} \cdot \hat{\mathbf{r}} = \frac{GmM_e}{r^2}\frac{dr}{dt},$$

where we have used (8.1) for $\mathbf{F}$. Therefore the integral for work becomes

$$W = \int_0^{t_1} \frac{GmM_e}{r^2}\frac{dr}{dt}\, dt = \int_0^{t_1} \frac{d}{dt}\left(\frac{-GmM_e}{r}\right) dt.$$

The integrand is now a derivative so we can evaluate the integral by the second fundamental theorem to obtain

$$W = GmM_e\left(\frac{1}{r(0)} - \frac{1}{r(t_1)}\right) = GmM_e\left(\frac{1}{R_e} - \frac{1}{R_f}\right),$$

the same formula we got for linear motion.

Example 4

A ballistic missile is fired vertically from the earth with a speed of 9.0 km/s. Neglecting atmospheric friction, how far above the earth's surface will it rise? (Use $R_e = 6400$ km, $M_e = 6.0 \times 10^{24}$ kg.)

As the projectile travels upward, its kinetic energy is converted into potential energy. Using Eq. (14.5) for the potential energy it gains, energy conservation implies

$$\tfrac{1}{2}mv^2 = GmM_e \left(\frac{1}{R_e} - \frac{1}{R_f} \right).$$

Solving for R_f, we get

$$R_f = \left(\frac{1}{R_e} - \frac{v^2}{2GM_e} \right)^{-1} = 1.8 \times 10^4 \text{ km.}$$

The distance above the earth's surface, $R_f - R_e$, is 1.2×10^4 km, almost two Earth radii.

In early science-fiction stories, like those of Jules Verne and H. G. Wells, human beings ventured into space, not in sleek rocket ships, but in shells fired out of colossal cannons. You might wonder, as these writers certainly did, is it really possible to release humanity from the shackles of the earth's gravity in such a way? How fast would a shell need to be fired to escape from the earth and never fall back?

We can figure out the speed necessary to escape from the earth by considering energy. If the shell starts out with a speed v, it has kinetic energy $\tfrac{1}{2}mv^2$. As it flies away from the earth, that energy gradually turns into potential energy. By the time the speed drops to zero, the shell is at a distance, say R_f, where the potential energy it gained is equal to its initial kinetic energy:

$$\tfrac{1}{2}mv^2 = GmM_e \left(\frac{1}{R_e} - \frac{1}{R_f} \right).$$

If the shell is to escape completely from the earth's gravitational field, it should be far, far away: R_f must be infinite. Then the term $1/R_f$ in the last equation is zero. Therefore, in order to escape, the shell must start with kinetic energy specified by

$$\tfrac{1}{2}mv^2 = GmM_e/R_e.$$

Solving for v, we find the *escape speed* (usually called the *escape velocity*),

$$v = \sqrt{2GM_e/R_e}. \tag{14.6}$$

Note that this value doesn't depend on the mass of the shell, or on its trajectory, so whether it is a molecule of air escaping from the atmosphere, or an interplanetary probe on its way to Neptune, all objects need the same initial speed to escape from the earth.

Example 5

Estimate the escape velocity from the earth.

We know from Chapter 8 that

$$g = GM_e/R_e^2,$$

so we can write the escape velocity from the earth in the convenient form

$$v = \sqrt{2gR_e}.$$

Knowing that g is about 10 m/s^2 and the radius of the earth is approximately 6.4 $\times$ 10^3 km, we easily find that the escape velocity is about 11 km/s, or 7 mi/s ($=$ 25,000 mph), a large but not impossible speed to attain.

Questions

11. A ballistic missile is fired from the earth with a speed v_0. At what distance from the center of the earth has its speed decreased to $0.5v_0$?

12. Back in 1958 many people were surprised when the first artificial satellite, Sputnik I, increased in speed as it fell back to Earth. As with all satellites, it fell to the earth because friction with the outer atmosphere caused it to lose energy, but as it spiraled closer and closer to the earth, its speed surprisingly increased. Through energy considerations, explain why its speeding up is no surprise.

13. Calculate the ratio of the escape velocity to the speed necessary for an object to orbit the earth at its surface.

14. Using the data in Appendix D, determine the escape velocity from

   (a) Mercury,   **(b)** Mars.

15. Find the speed needed for a satellite orbiting the earth at a speed of 4.5 km/s to escape from its orbit.

14.3 POTENTIAL ENERGY AND STABILITY

Now that we've mastered the ideas of work and potential energy, we shall use them to generate insights into the stability of things. Suppose that you have an object of mass m that has a particular force acting on it whenever it moves away from a certain point in space. To make it simple (which is the very idea of this sort of hypothetical problem), we'll assume that the object moves only in the x direction, and we'll let $x = 0$ be the point where there is no force. In addition, let's assume that the force is proportional to how far the object is from the point $x = 0$:

$$F(x) = bx,$$

where b is some constant.

Now one way to analyze this problem is to dive higgledy-piggledly into Newton's laws and set this force equal to mass times acceleration; we'll do this later (although without confusion), in Chapter 20. Instead, let's investigate the object's potential energy.

If we pull the object away from $x = 0$, we must apply a force equal and opposite to F, thereby doing work. The total work we do moving the object from 0 to x is the integral

$$W(x) = \int_0^x -F(s)\, ds,$$

where we have used s as a dummy variable. Since $F(s) = bs$, this integral is equal to

$$W(x) = -b \int_0^x s \, ds = -\tfrac{1}{2} bx^2.$$

The body's potential energy changes by the same amount, so we have

$$U(x) - U(0) = -\tfrac{1}{2} bx^2.$$

If we now arbitrarily make $U(0) = 0$ (which means we calculate the potential energy with respect to the level of reference $x = 0$) we get the potential function

$$U(x) = -\tfrac{1}{2} bx^2. \tag{14.7a}$$

The graph of $U(x)$, shown in Fig. 14.4, is a parabola which opens downward if b is positive and upward if b is negative.

To avoid confusion about plus and minus signs we consider the following two cases:

$$F(x) = bx, \quad \text{for which} \quad U(x) = -\tfrac{1}{2} bx^2,$$

and

$$F(x) = -kx, \quad \text{for which} \quad U(x) = \tfrac{1}{2} kx^2,$$

where both b and k are positive constants. The graph of

$$U(x) = \tfrac{1}{2} kx^2 \tag{14.7b}$$

is shown in Fig. 14.4b.

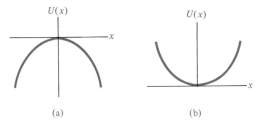

(a)   (b)

Figure 14.4 (a) Graph of the potential-energy function $U(x) = -\tfrac{1}{2} bx^2$. (b) Graph of the potential-energy function $U(x) = \tfrac{1}{2} kx^2$.

Examining the curve Fig. 14.4a, we see that at $x = 0$ it has a maximum, whereas that of Fig. 14.4b has its minimum there. You may remember that differential calculus began with Pierre de Fermat trying to locate the maximum or minimum of a curve. Fermat found that these values occurred at points where the tangent line is horizontal. We know now that the slope of a tangent to a curve is equal to the derivative at that point. Therefore, a horizontal tangent (zero slope) to a curve occurs when the derivative of the function is zero. The derivatives of the curves we are considering are

$$\frac{dU}{dx} = \frac{d}{dx}\left(-\frac{bx^2}{2}\right) = -bx$$

and

$$\frac{dU}{dx} = \frac{d}{dx}\left(\frac{kx^2}{2}\right) = kx.$$

We note two significant features from our analysis. First, the derivative $dU/dx = 0$ at $x = 0$; we already knew that this was where the maximum and minimum would be. Second, in both cases, dU/dx is the opposite of the force. In other words,

$$F = -\frac{dU}{dx}.$$

(14.8)

This result is no accident; it is a consequence of the first fundamental theorem of calculus. The potential function arises from integrating the force we apply to do work, which is the opposite of F, so F comes from differentiating the negative of the potential function $U(x)$. Putting the two facts together, we realize that *the maximum or minimum of the potential energy occurs at just the point where $dU/dx = 0$, and therefore at just the point where the object has no force on it.*

From our simple analysis, we've reached an extremely important general conclusion: a body has no force acting on it when it's at a maximum or a minimum of its potential energy. This result holds not only for the forces we have considered, but for all forces satisfying (14.8).

We can gain a physically intuitive understanding of this consequence by considering a marble and a bowl. You can think of the bowl as having the shape of the surfaces in Fig. 14.5. If you place the marble at the bottom of the bowl, it remains there; it is at the minimum of potential energy and there is no net force acting on it. Similarly, if you place the marble on top of an inverted bowl, as in Fig. 14.5a, the marble is again perfectly happy to remain at the maximum of its potential energy since there are no net forces acting on it. In each case, because the marble has no net forces acting on it, we say that it is in *equilibrium*. The place where such equilibrium occurs, where $dU/dx = 0$, is called the equilibrium position.

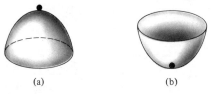

(a)   (b)

Figure 14.5 (a) Bowl corresponding to a potential $U = -\frac{1}{2}bx^2$ and unstable equilibrium. (b) Bowl corresponding to a potential $U = \frac{1}{2}kx^2$ and stable equilibrium.

Although the marble is at rest in both cases, you have a physical intuition that the situations are not really equivalent. The reason is this: If you push the marble when it's at the bottom of the bowl, it rolls around for a while, but eventually comes to rest at the equilibrium position, where it was before. If, however, you disturb in the slightest way the marble resting on top of the bowl, it rolls down the bowl and probably off the table; it will never return to its original position. The fact that an object has no force on it does not guarantee that there is stability.

Evidently, there are two types of equilibrium. When the marble is at the bottom of the bowl, it is said to be in *stable equilibrium*. When it's on top of the inverted bowl, it is said to be in *unstable equilibrium*.

Physically, the difference between the two types of equilibrium is as follows: When the equilibrium is stable, a disturbance produces a restoring force back toward the equilibrium position. This is a direct consequence of the fact that the potential energy is a minimum at that point. *If the disturbance increases the potential energy of the object, then the object will fall back to its original position, to a point of lower potential energy, when released.*

Just as the force resulting from the bowl and gravity causes the marble to move toward the bottom, we can imagine the force associated with a potential-energy curve pushing the object back toward its equilibrium position. The force, given by

$$F = -\frac{dU}{dx},$$ (14.8)

is in the opposite direction to the slope of the potential-energy curve. So, as depicted in Fig. 14.4b, if the object is moved in the positive x direction, where the slope of the potential is positive, the force is in the opposite direction, back toward the equilibrium position. Likewise, if the object is moved in the negative x direction, the force is in the opposite direction, restoring the object to its equilibrium position.

When the equilibrium is unstable, a slight push causes a force on the object directed *away* from where it was. *If the disturbance lowers the potential energy, the object will continue to move farther from its original position.* As shown in Fig. 14.4a, if the particle is moved in the positive x direction, the slope of the potential-energy curve is negative and hence the force is positive. This force in the positive direction pushes the object farther from its original position.

There is also a third kind of equilibrium called *neutral equilibrium*, which can be illustrated by a marble lying on a flat surface. If the marble is displaced slightly its potential energy doesn't change and it suffers neither a restoring force nor a repelling force. Neutral equilibrium corresponds to regions where the potential-energy curve is constant.

Pendulums, roller coasters, and guitar strings are examples of systems with points of stable equilibirum. These are cases in which something set into motion continues in motion for a while, as kinetic energy turns into potential energy and back into kinetic, and so on. As time goes on, more and more of that energy turns into heat, until finally the system has the least energy it can possibly have: no kinetic energy (it comes to rest) and its lowest potential energy. That is why stable equilibrium always occurs at the minimum of potential energy.

Unstable systems include a pencil balanced on its point, a row of dominoes, and a house of cards. All these systems have excess potential energy, which can easily be turned into other forms, and eventually will be.

Example 6

Why is the Tower of Pisa becoming less stable?

We can understand the tower's stability in terms of potential energy. Imagine the entire mass of the tower to be concentrated at one point, known as the center of mass. For uniform symmetric objects this point is at the geometric center. All we need to know about the center of mass is that it represents the point where we can consider the force of gravity as acting. The center of mass and the force of gravity on the tower in a vertical position and in a leaning position are shown below.

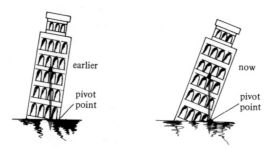

Considering the potential energy of the center of mass as being measured from the pivot point (around which the tower would fall), we see that if the tower were vertical, the center of mass would rise if a force tended to push the tower over slightly; therefore the potential energy would also increase. So when vertical, the tower is in stable equilibrium. On the other hand, in its present precarious position, the center of mass would be lowered if the tower were pushed over slightly. Since its potential energy would decrease with the push, the tower is in less stable equilibrium.

Ideas of equilibirum and stability are applied to areas besides physics as well. One example is an ecological system. Simple ecological systems, like a small jungle, tend to be stable. When there are enough little animals around for big animals to eat, the system is in equilibrium. If the predators eat too many prey, however, then there aren't enough little animals and the big animals start dying. The presence of fewer predators allows the little animals to recover; then the big animals recover too.

Another example of a stable system is a well-designed government. The term *checks and balances* is used to express the idea that the system automatically responds in a way that opposes disturbances. On the other hand, it is easy to think of political and economic systems that are unstable.

In retrospect we see that we have used a very simple example as a kind of metaphor for describing the behavior of increasingly complicated systems – a guitar string, a row of dominoes, or a jungle. The description continues to make sense, even though it becomes more and more difficult to write mathematical equations to describe the situation. In

author. At one point in his life, Boscovich was appointed chairman of a committee that was asked to investigate the stability of the dome of St. Peter's Cathedral in Rome. The committee report on that dome is still considered a minor classic in engineering analysis.

In his book *A Theory of Natural Philosophy*, published in 1763, Boscovich made use of the idea of stability to formulate a model for his interpretation of nature. According to Newton, at very great distances the gravitational force diminishes as $1/r^2$. But that was known to be true only at large distances. Boscovich wondered what happens at very small distances, distances so small that you can't see them.

He postulated that the force of gravity is alternately attractive and repulsive depending on the distance by which two masses are separated. Figure 14.6 is a diagram taken from his book illustrating this force as a function of distance. (Boscovich didn't know about potential energy.) If the force has the behavior shown with positive force indicating repulsion and negative force indicating attraction, there would be several positions where a point mass would find no force on it.

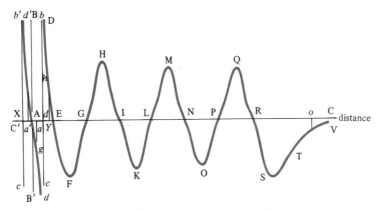

Figure 14.6 Boscovich's force of gravity alternates between attraction and repulsion depending upon the distance between two masses.

Some of the positions are unstable. For example, there is point G in Fig. 14.6. If the object is displaced slightly to the right, the force becomes positive, thereby pushing the object farther away. Other points are stable, as for example point N. If we push the object to the left, the force becomes repulsive and pushes the object back to where it was. And if we push the object to the right, the force becomes attractive (negative) and pulls the object back to its original position.

Boscovich thought that if the gravitational force acts in this way, then point masses would combine together with some point masses occupying stable positions with respect to other point masses. From this idea he imagined how atoms – point masses – acting gravitationally would comprise molecules and crystals and all the objects that make up the world.

Boscovich's idea had no scientific validity because no one could devise an experiment to discover if it was right; there was no way to test his idea.

Nevertheless, it did have a certain appeal. The basic picture he painted of intrinsic stable equilibrium in the forces that act between atoms had a profound influence on the people who followed him. John Dalton, who produced the first modern evidence for an atomic theory, and Michael Faraday, who formulated the idea of fields of force, were both among those influenced by Boscovich. Thus Boscovich's curious idea helped lay the groundwork for our modern atomic theory and the theory of fields of force.

CHAPTER 15

TEMPERATURE AND THE GAS LAWS

There are however innumerable other local motions which on account of the minuteness of the moving particles cannot be detected, such as the motions of the particles in hot bodies, in fermenting bodies, in putrescent bodies, in growing bodies, in the organs of sensation and so forth. If any one shall have the good fortune to discover all these, I might almost say that he will have laid bare the whole nature of bodies so far as the mechanical causes of things are concerned.

Isaac Newton, in *Unpublished Papers of Isaac Newton*

15.1 TEMPERATURE AND PRESSURE

Everybody talks about the weather, and that usually means the temperature, an inescapable part of our environment. Yet Newton's laws of mechanics tell us nothing about temperature. Is there any connection between mechanics and temperature?

In Chapter 13 we saw a connection. If you drop a block from above a table, its potential energy first turns into kinetic energy, and then is transformed into thermal energy when the block hits the table. After a while the only evidence that those events occurred is a slight warming of the surroundings, that is, a small increase in temperature.

What really happens is that the kinetic energy of the falling block is turned into the energy of motion of atoms and molecules. The energy is still there, but the motions are in random directions, not the organized motion of a whole block of matter. The energy of those random motions is internal energy, and the evidence of its existence is a change in temperature.

Before we go any further, let's make a short digression on temperature scales. Most of the world uses the Celsius scale, but in the United States the Fahrenheit scale is predominant. By *temperature* we mean a number assigned on a definite scale so that we can tell whether it was hotter today in Nairobi or in Calgary. Temperature scales offer a means of comparing the temperature of an object to standards set by fixed calibration points.

On the Fahrenheit scale, water freezes at 32° and boils at 212°. Temperatures on this scale are followed by units of degrees Fahrenheit, abbreviated °F. On the Celsius scale the freezing and boiling points of water are given by 0° and 100°, respectively. The abbreviation °C is used for this scale. In Section 15.5 we'll discuss how these common temperature scales came into being.

Example 1

If the temperature at a seaside resort is 86°F, what is the corresponding temperature in °C?

To find the conversion between Fahrenheit and Celsius temperature scales, let T_C denote the temperature in degrees Celsius, and T_F that in degrees Fahrenheit. The relation between T_F and T_C is linear, as illustrated in the figure on p. 289,

$$T_F = AT_C + B,$$

where A and B are constants. To find these constants, we use the fact that water freezes at $T_F = 32$ and $T_C = 0$. This immediately gives $B = 32$, so

$$A = \frac{T_F - 32}{T_C}.$$

But water boils when $T_F = 212$ and $T_C = 100$, so

$$A = \frac{212 - 32}{100} = \frac{180}{100} = \frac{9}{5}.$$

Hence

$$T_F = \tfrac{9}{5}T_C + 32.$$

We can also solve this last equation for T_C in terms of T_F to find

$$T_C = \tfrac{5}{9}(T_F - 32).$$

In particular, when $T_F = 86°F$, we get $T_C = 30°C$.

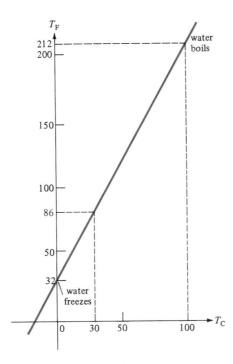

For most purposes, the best way to make this conversion is to remember a few fixed values, and interpolate between them. Unless great precision is needed, this can be done quickly without writing anything down. Here are some handy reference points.

°C	°F
0	32
10	50
20	68
30	86
40	104

Now to continue our search for a connection between mechanics and temperature. If we want to learn something about the temperature of the air, we should analyze the random motions of the molecules it contains.

The random motion of gas molecules is related not only to temperature, but also to another mechanical property, pressure. Pressure is defined as force per unit area. If an object of area A has a force F applied over that area, the corresponding pressure is

$$P = \frac{F}{A} .$$

(15.1)

In SI units, a pressure of one newton per square meter (1 N/m^2) is one pascal (1 Pa). A commonly used unit of pressure is the atmosphere (1 atm = 1.01×10^5 Pa). Other units and conversions are listed in Appendix B.

Some of the greatest minds of the eighteenth century investigated the relationship between temperature and pressure. One fruitful idea is to imagine the motion of molecules in a closed container like that shown in Fig. 15.1. The constant, rapid drumbeat of molecules bouncing off the walls of the container exerts a steady average force on the walls. Let's calculate how that force is related to quantities which specify the motion of the molecules, namely, mass and velocity.

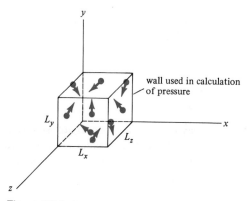

Figure 15.1 Due to collisions, the molecules of a gas exert pressure on the container walls.

Consider first a single collision. A molecule of mass m approaches the wall with velocity v, and bounces off like a rubber ball in an elastic collision. In other words, the molecule bounces off with the same speed it had before the collision. As illustrated in Fig. 15.2, the angle of incidence is equal to the angle of reflection.

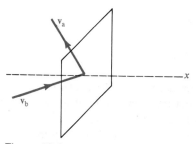

Figure 15.2 A gas molecule undergoes a change in momentum in a collision with a wall of a container.

Although the speed of the rebounding molecule is the same, its velocity is changed. As a result of the collision the molecule undergoes a change in its momentum, the quantity of motion in Newton's second law which is given by the product of its mass and velocity, $m\mathbf{v}$. The velocity $\mathbf{v}$ has components v_x, v_y, v_z, and if the direction perpendicular to the wall is the x direction, as shown in Fig. 15.2, the molecule has a positive x component of velocity before it hits the wall and a negative x component on rebound. In other words, the x component is exactly reversed as a result of the collision; the other components remain unchanged. Therefore the momentum of the molecule before collision, $\mathbf{p}_b$, is given by

$$\mathbf{p}_b = m(v_x\hat{\mathbf{i}} + v_y\hat{\mathbf{j}} + v_z\hat{\mathbf{k}}),$$

whereas the momentum $\mathbf{p}_a$ after collision is

$$\mathbf{p}_a = m(-v_x\hat{\mathbf{i}} + v_y\hat{\mathbf{j}} + v_z\hat{\mathbf{k}}).$$

According to Newton's second law the force on the molecule is equal to the rate of change of momentum, $\mathbf{F} = d\mathbf{p}/dt$, which we can approximate as

$$\mathbf{F} = \frac{\Delta\mathbf{p}}{\Delta t},$$

where $\Delta\mathbf{p}$ is the change in momentum and Δt is the change in time. By Newton's third law, this force is equal and opposite to the force the molecule exerts on the wall. Our formulas for $\mathbf{p}_a$ and $\mathbf{p}_b$ show that the change in momentum of the molecule is given by

$$\Delta\mathbf{p} = \mathbf{p}_a - \mathbf{p}_b = -2mv_x\hat{\mathbf{i}}.$$

Now suppose for a moment that the gas is extremely rarefied, so a typical molecule travels back and forth across the box many times before colliding with other molecules. Let's say a particular molecule returns and hits the wall once every t_c seconds. Then it transfers momentum to the wall at the rate $-\Delta p_x/t_c$. To calculate t_c, note that between one collision with the wall and the next, the molecule has to cross the box, hit the wall, and come back, traveling a distance $2L_x$ in the x direction (twice the length of the box in the x direction) at speed v_x. So the time between collisions is given by

$$t_c = \frac{2L_x}{v_x}.$$

All in all, then, the rate of momentum transfer from the molecule to the wall is

$$\frac{-\Delta p_x}{t_c} = 2mv_x\frac{v_x}{2L_x} = \frac{mv_x^2}{L_x}.$$

Each of the N molecules in the box has its own velocity, and therefore its own travel time and momentum transfer. Let v_{xi} denote the x component of the velocity of the ith molecule, where $i = 1, 2, \ldots, N$. Summing over all the molecules in the box, we obtain the total rate of momentum transfer to the wall, and thus the force on the wall,

$$F = \sum_{i=1}^{N} \frac{mv_{xi}^2}{L_x} = \frac{1}{L_x}\sum_{i=1}^{N} mv_{xi}^2. \tag{15.2}$$

Although the momentum transfers from a single molecule are intermittent, the total momentum transfer is a drumbeat of impulses that provides a steady push on the wall.

If the gas is not extremely rarefied, it's not quite so simple – the molecules may collide with one another before completing the round trip. But these collisions turn out to conserve both momentum and kinetic energy, so the net impact on the wall is the same as if the particles had not collided with one another.

In Eq. (15.2) we recognize the term mv_{xi}^2 as twice the contribution to the kinetic energy from the x component of velocity. Since the motion of the molecules is random, the contributions from the y and z components have the same value, so we can write

$$\sum_{i=1}^{N} mv_{xi}^2 = \frac{1}{3} \sum_{i=1}^{N} (mv_{xi}^2 + mv_{yi}^2 + mv_{zi}^2) = \frac{1}{3} \sum_{i=1}^{N} mv_i^2,$$

where

$$v_i^2 = v_{xi}^2 + v_{yi}^2 + v_{zi}^2$$

is the square of the speed of the ith molecule. But

$$\frac{1}{3} \sum_{i=1}^{N} mv_i^2 = \frac{2}{3} \sum_{i=1}^{N} \tfrac{1}{2}mv_i^2 = \tfrac{2}{3}K,$$

where K is the total kinetic energy of all the molecules. Using this in (15.2) we obtain

$$F = \frac{2}{3} \frac{K}{L_x}$$

for the total force on the wall from all the molecules in the box.

Now the pressure on the wall is the average force divided by the area. We just found the force, and we know that the area of the wall on which we are calculating the pressure is $L_y L_z$. Therefore we find the pressure to be

$$P = \frac{F}{A} = \frac{2}{3} \frac{K}{L_x L_y L_z}.$$

Since the volume V of the box is $L_x L_y L_z$, we can cast the result into the simpler form

$$P = \frac{2}{3} \frac{K}{V}. \tag{15.3}$$

Finally, since the total kinetic energy K equals the average kinetic energy $\overline{K}$ of a single molecule times the number N of molecules, we can write Eq. (15.3) as

$$P = \frac{2}{3} \frac{N\overline{K}}{V}. \tag{15.4}$$

This argument was first presented in a simplified version by James Prescott Joule. From Eq. (15.4) we see that as the random motion of the gas molecules becomes more vigorous, the average kinetic energy increases, resulting in correspondingly greater pressure on the containing walls. Furthermore, we've found a relationship between the macroscopic pressure and the microscopic average kinetic energy of the individual molecules of the gas.

Questions

1. Cite a couple of everyday examples that illustrate that heating a gas increases its pressure.

2. A spherical balloon of radius 10.0 cm contains helium gas at a pressure of 1.50×10^5 Pa. How many helium atoms are contained in the balloon if each atom has an energy of 4.2×10^{-22} J? (Helium is a monatomic gas.)

3. Compute the average kinetic energy per atom of three moles of argon gas in a cylindrical container of radius 5.0 cm and height 20.0 cm under a pressure of 3 atm. (Argon is a monatomic gas.)

4. A machine gun fires two rubber bullets per second at a speed of 5.0 m/s directly at a square plate with sides 0.10 m long. The bullets, each of mass 15.0 g, bounce back elastically along their initial direction. What is the pressure exerted on the plate by the stream of bullets?

5. What would be the relation between the pressure and average kinetic energy per molecule if the box contained a mixture of gases?

6. A mass of 5.0 kg of a gas is in a container whose volume is 0.30 m^3 at a pressure of 7.0×10^5 Pa. What is the square root of the average velocity squared of a molecule? This quantity is often referred to as the root mean square of the velocity — the rms value.

7. What temperature is the same on both the Fahrenheit and Celsius temperature scales?

15.2 THE GAS LAWS OF BOYLE, CHARLES, AND GAY-LUSSAC

By connecting two mechanical quantities, pressure and kinetic energy, we found a relationship between heat and pressure. But how does this connection aid us in finding a relationship between heat and temperature? What we need is an understanding of the behavior of gases.

A major advance in understanding gases was provided in the seventeenth century by Robert Boyle. A staunch advocate of careful, thorough experimentation, Boyle discovered experimentally a relation between the volume of a gas and its pressure: for a given sample of gas, as long as the temperature and mass remain unchanged, the pressure is inversely proportional to the volume. In other words, if you squeeze a gas, its pressure rises proportionately; gases act like springs.

Boyle's experiments suggested that for a given sample of gas at a fixed temperature, the product of the pressure P and the volume V is constant. This empirical relation,

$$PV = \text{const} \qquad \text{(at fixed temperature)}, \tag{15.5}$$

is called *Boyle's law*, and it can be compared with formula (15.4) derived in the last section. If we rewrite (15.4) as

$$PV = \tfrac{2}{3}N\overline{K} \tag{15.6}$$

it has a striking resemblance to Boyle's law. In fact, if the average kinetic energy of a gas doesn't change at a fixed temperature, then for a given sample of gas both N and $\overline{K}$ are constant, so PV is constant.

According to Boyle's law, twice as much gas at a given pressure will occupy twice the volume. Similarly, for half as much gas, the volume is also half. Equation (15.6) shows that this is reasonable, since the quantity PV is proportional to the number of gas molecules N and otherwise depends only on the average kinetic energy $\overline{K}$, which presumably depends only on temperature. This suggests that we rewrite (15.6) as follows:

$$\frac{PV}{N} = \text{a function of temperature only.} \tag{15.7}$$

It is not hard to guess something about this function. Gases expand when heated, so as temperature increases so does the quantity PV, and thus we anticipate that this function should increase with temperature.

The exact dependence on temperature was revealed by further experiments into the nature of gases carried out by Jacques Alexandre César Charles, an eighteenth-century French scientist and hot air balloon enthusiast. Charles's curiosity about the behavior of gases led him to the important discovery that *all* gases expand by the same amount with a given rise in temperature. For each degree Celsius rise in temperature, the volume of any gas expands by one-273rd of its volume at 0°C. Similarly, for each degree Celsius decrease in temperature, any gas contracts by one-273rd of its volume.

By extrapolating the behavior of a gas as the temperature is progressively decreased, we conclude that at $-273°C$ the volume would reach zero, as illustrated in Fig. 15.3. Actually, gases liquefy or solidify before they reach $-273°C$, so the volume never quite reaches zero. But if the law did hold, there could be no lower temperature than $-273°C$. This temperature, seemingly the lowest attainable, is designated as *absolute zero* on another temperature scale called the *absolute system*. The units in this system are called kelvins (K) in honor of William Thomson, Lord Kelvin:

$$T = t_C + 273.$$

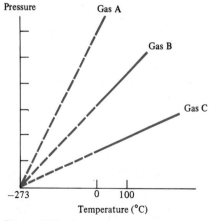

Figure 15.3 A pressure versus temperature graph of the behavior of different gases when the temperature is lowered.

Thus Charles proposed that the volume V is proportional to T, where T is the absolute temperature in kelvins. Although Charles's investigations were never published, studies made 15 years after his work by Joseph Louis Gay-Lussac, another French scientist and balloon enthusiast, confirmed that to a good approximation the volume of a given amount of any gas is proportional to the absolute temperature when the pressure is held constant. We can write this as an equation,

$$\frac{V}{T} = \text{const} \qquad \text{(for constant pressure)}, \qquad (15.8)$$

where T is the absolute temperature. This is called the gas law of Charles and Gay-Lussac.

The findings of Boyle, Charles, and Gay-Lussac amount to saying that gas behaves in much the same way as does mercury or alcohol in a thermometer. When heated, the mercury in a thermometer expands and rises up the tube. A gas also expands when heated, but with an important difference: all gases expand by approximately the same amount, one-273rd of their volume at 0°C, for each degree of rise in temperature. Often scientists conveniently speak of an *ideal* gas whose behavior is described exactly by Boyle's law and the law of Charles and Gay-Lussac. Liquids and solids, on the other hand, expand by amounts varying more noticeably with the type of substance when heated. For example, when a jar lid is hard to open, you place it in hot water. The metal lid expands more than the glass, thereby making the lid looser fitting and easier to open.

A significant conclusion drawn from Boyle's experiments was that gases must be composed of discrete particles separated by void since the gas is compressible. This interest in the behavior of gases marked the revival of the ancient Greek conjecture that matter is composed of incessantly moving particles called atoms. The original idea of atoms is credited to the fifth-century-B.C. philosopher Leucippus and his student Democritus. Although many qualitative ideas of atomic theory were developed by the ancient Greeks, controlled, quantitative investigations were not carried out until the seventeenth and eighteenth centuries.

Questions

8. If the volume occupied by a gas is decreasing, can you conclude that the temperature must be decreasing? Explain your reasoning.

9. If a helium-filled balloon is placed in a freezer, must its volume decrease? Explain.

10. The pressure in a bicycle tire is increased from 30 lb/in.^{2} to 60 lb/in.^{2}, yet the volume doesn't double. Why?

11. A gas is placed in a vessel that maintains a constant temperature at a pressure of 2.0 atm and a volume of 3.0 L. Overnight a leak develops and the pressure is found to be 1.75 atm and the volume 2.4 L. What fraction of gas has escaped?

15.3 THE IDEAL-GAS LAW

Because the quantity pressure times volume divided by the number of molecules, PV/N, increases with temperature, the easiest way to define temperature would be to set that quantity equal to the temperature. The only trouble with this scheme is that the temperature would depend on the units we choose to measure pressure and volume. To compensate for this, we insert a constant, which may be different for each set of units of pressure and volume, leaving temperature always numerically the same. Then we can write down an equation that describes the state of a gas:

$$PV = NkT,$$ (15.9)

where k is the constant. The constant universally adopted (called Boltzmann's constant) has the value

$$k = 1.38 \times 10^{-23} \text{ J/K}.$$

As we have seen, the resulting unit of temperature, called the kelvin, is the same size as 1°C. Figure 15.4 compares some temperatures on the three different scales.

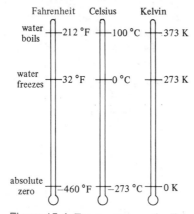

Figure 15.4 Temperature on the Fahrenheit, Celsius, and Kelvin temperature scales.

Equation (15.9), known as the *ideal-gas law*, is extremely useful in physics and chemistry. It is, of course, much more than a mere definition of the Kelvin temperature scale. Also known as an equation of state, this equation combines both Boyle's and Charles's laws, and the value of the constant k also tells us the number of molecules of gas in a given volume at known temperature and pressure.

Boyle himself had noticed small departures from the constancy of the product of pressure and volume. But, because of its simplicity and general usefulness, he ignored deviations from his law. Why don't gases follow the ideal-gas law for all ranges of temperature and pressure? One reason is that gas molecules are not point particles but have nonzero volumes. Consequently the amount of space available for each molecule is not the entire volume of the container. Instead, each molecule has a volume less than V to move around in. In addition, the forces that the molecules exert on one another are not entirely negligible. When the density of the gas is great enough for the molecules to be relatively close together, the pressure is reduced as a result of the attraction of molecules to one another.

Numerous attempts have been made to deduce a general equation holding for real gases. The most celebrated is that developed by Johannes van der Waals in 1873. Van der Waals assumed that the effect of the finite size of molecules could be taken into account by using $V - Nb$ for the volume, where b is experimentally determined for each gas. Furthermore, the attractive forces between the molecules were represented by replacing P by the quantity $P + aN^2/V^2$, where the coefficient a is also experimentally determined for each gas. Thus we have the Van der Waals equation,

$$(P + aN^2/V^2)(V - Nb) = NkT. \tag{15.10}$$

Questions

12. On a cool day when the temperature is 16°C the pressure in a car tire is 49 lb/in.^{2} (3.3×10^5 Pa). What is the pressure on a hot day when the temperature is 36°C?

13. Find the pressure of 0.70 g of argon in a volume of 5.0 L at a temperature of 25°C.

14. The state of a gas is initially specified by $T = 350$ K, $P = 6.0$ atm, and $V = 8.5$ L. If the pressure is lowered to 4.0 atm and the volume increases to 10.0 L, what is the new temperature of the gas?

15. A sample of 100 g of helium gas is at a temperature of 0°C and a pressure of 500 atm. Calculate the volume of the gas according to (a) the ideal-gas law, and (b) the Van der Waals equation if

$$a = 1.4 \text{ L}^2 \text{ atm/mol}^2, \qquad b = 0.03 \text{ L/mol}.$$

15.4 TEMPERATURE AND ENERGY

Let's summarize what we've learned so far. Heat, the random kinetic energy of molecules, is related to the pressure of a gas through

$$PV = \tfrac{2}{3}N\overline{K}, \tag{15.6}$$

which gives

$$\overline{K} = \frac{3}{2}\frac{PV}{N}. \tag{15.11}$$

On the other hand, studies of the dependence of volume and pressure of an ideal gas on temperature suggest the ideal-gas law

$$PV = NkT. \tag{15.9}$$

Solving Eq. (15.9) for PV/N and inserting that into Eq. (15.11), we are led to another result:

$$\overline{K} = \tfrac{3}{2}kT. \tag{15.12}$$

We've discovered that the average kinetic energy of an atom or molecule in a gas is directly proportional to the absolute temperature. This result agrees pleasantly with the starting point of this chapter: warmth is evidence of the disorganized motions of atoms and molecules. We have succeeded in finding the connection between heat, temperature, and energy.

One final question. Is the average kinetic energy of a gas, $\frac{3}{2}kT$, all of its energy, or can it have other forms of energy? One way to investigate this question is through an experiment. Let's theorize that the energy of a gas is $\frac{3}{2}NkT$. To test this theory, take an isolated container of gas of fixed volume and add to it a known amount of heat energy, Q (say, one joule). Because the volume is fixed, nothing moves, so no work flows in or out, and because the system is isolated, no heat, other than what we've added, can flow in or out. Since Q is the change in energy, our theory states that the rate of change in energy with respect to temperature dQ/dT must be

$$\frac{dQ}{dT} = \frac{3}{2}Nk.$$

In other words, the temperature is predicted to rise by exactly $2Q/(3Nk)$.

Experimentally, this prediction holds for certain gases, for example, the noble gases helium and argon. But for air, we would find approximately $dQ/dT = \frac{5}{2}Nk$. What does this imply? Other gases have somewhere other than kinetic energy to store energy. In fact, all gases which are composed of simple atoms are consistent with our theory, while those which are made of more complicated molecules are not; air, for example, is composed of N_2 and O_2 molecules.

Where is the extra energy stored? What happens is that molecules both vibrate and tumble, and do more of both as they heat up. And that is where the energy goes. Technically we say that the molecules are *internally excited*. Atoms are simpler and tougher structures and do not tend to get excited as easily as the more complicated molecules.

At room temperature and above, the internal excitation energy of gas molecules, like kinetic energy, is proportional to the temperature. This allows us to write down a simple equation for the total energy of a gas, U:

$$U = qNkT. \tag{15.13}$$

For simple monatomic gases like helium, q is found to be equal to $\frac{3}{2}$. For molecular gases, it is larger.

Questions

16. What is the average kinetic energy per atom of helium gas at a temperature of 23°C?

17. What is the root-mean-square speed of a helium atom in the preceding question?

18. One mole of hydrogen gas at 80°C has a total energy of 10.0 J. What is the value of q?

19. If 4.0 J of energy is added to a one mole of argon gas, then how much will the temperature of the gas increase?

20. You wake up one cold morning, turn up your thermostat, and heat the air in your house from 285 to 293 K. What is the change in the total internal energy of the air in your house? (*Hint:* The answer is zero – why?)

15.5   A FINAL WORD

A morning shower is steamy hot, a refreshing glass of lemonade is icy cold. But how hot is *hot*, or cold *cold*? What exactly do these two terms mean? How do we measure the degree of warmth, or lack of it, in a body? In other words, how can we specify the temperature of something? In order to discuss temperature, a scale on which to compare various objects is needed.

An intuitive feeling for temperature comes from the sense of touch. But whereas the length and mass of an object can be measured in terms of another object (and that measurement will be the same for every judge), the sense of touch does not provide an objective means for measuring temperature. What one might call "a comfortable shower," another might call too hot or too cold. Like force, temperature can only be measured by its effects – and a dependable, reproducible scale is needed on which changes can be measured.

One of the earliest thermometers was invented in 1602 by our old friend Galileo Galilei. Called a thermoscope, Galileo's thermometer merely consisted of a glass bulb containing air and having a long, open-ended stem, which was placed in water. When the temperature changed, the air inside the bulb expanded or contracted and, correspondingly, the water in the stem fell or rose. But Galileo's apparatus lacked one vital ingredient needed to understand the changes in temperature: a visible, fixed scale.

One of the first temperature scales was established by the Danish astronomer Olaus Roemer, who lived in the eighteenth century. Roemer chose two convenient, useful, and reliable fixed points: the melting point of snow and the boiling point of water. He designated the melting point of snow as $7\frac{1}{2}°$, and he set the boiling point of water at $60°$. The former, seemingly arbitrary point was chosen by Roemer for meteorological temperature so as to position one-eighth of his entire scale below freezing. As a result, zero degrees on Roemer's scale approximated the temperature of an ice and salt mixture, which was widely believed to be at the lowest possible attainable temperature, and so all readings on his thermometer were assumed to be positive.

Roemer did not publish anything about his thermometer, and its existence went largely unnoticed save by a very few of his contemporaries, including a young Polish scientific instrument maker by the name of Daniel Fahrenheit, who sensed a future in calibrated thermometers. In 1708, Fahrenheit watched Roemer graduate several thermometers. Later he described what he saw in a letter to Hermann Boerhaave, a colleague:

> I found that he had stood several thermometers in water and ice, and later he dipped these in warm water, which was at blood-heat and after he had marked these two limits on all the thermometers, half the distance between them was added below the point in the vessel with ice, and the whole distance divided into 22½ parts, beginning with 0 at the bottom, then 7½ at the point of the vessel with ice and 22½ degrees for that at blood-heat.

For ease of construction Fahrenheit multiplied each degree by four, so that the upper point became $90°$ and the lower one $30°$. Later he changed the upper point to $96°$ and the lower to $32°$ for a very good instrument maker's reason: the $64°$ interval between them could then be engraved into single degrees by successively dividing it in half six times. With this temperature scale, Fahrenheit (somewhat inaccurately) found the temperature of boiling water to be 212°. With 212° fixed as the boiling point of water, this scale is the one popular in the United States. As a result of the adjustment in the scale, normal body temperature became 98.6° instead of Fahrenheit's 96°. The range 0 to 100°F is the range of common weather temperatures.

In 1742 Anders Celsius, a Swedish astronomer, proposed a centigrade system in which the temperature scale is made up of 100 equal degrees spanning the liquid state of water: Celsius called the freezing point of water 100° and the boiling point of water 0°. Shortly after his death, Celsius's colleagues at the Uppsala Observatory inverted the scale to make the freezing point 0° and the boiling point 100°, giving us the scale generally used today; it is really Celsius' scale inverted, so we should call a temperature on this scale Ɔ° rather than °C as an abbreviation.

CHAPTER 16

THE ENGINE OF NATURE

Everybody knows that heat can cause movement, that it possesses great motive power; steam engines so common today are a vivid and familiar proof of it. . . . The study of these engines is of the greatest interest, their importance is enormous, and their use increases every day. They seem destined to produce a great revolution in the civilized world. . . .

Despite studies of all kinds devoted to steam engines, and in spite of the satisfactory state they have reached today, the theory of them has advanced very little and the attempts to improve them are still directed almost by chance.

Sadi Carnot, "The Motive Power of Heat" (1824)

16.1 THE AGE OF STEAM

The age of steam is past. The steam engine is a curiosity, an object of nostalgia that has been replaced by diesel engines, electric motors, turbine engines, and gasoline engines to drive the wheels of civilization. Nonetheless, steam did have its day. The steam engine not only caused the industrial revolution, which changed our lives, it also led to discoveries in physics so profound that they changed the way we think. How did investigations into the nature of steam engines lead to a deeper understanding of the universe?

First, we need to understand how a steam engine operates. In essence, a steam engine is a device which heats water in a closed container, a boiler, thereby converting it to steam. As the steam and water mixture becomes hotter the pressure rises; the steam engine controls and takes advantage of the force applied by that pressure. When the high-pressure steam is released through a valve to a cylinder where it can push a movable piston, it produces work. So a steam engine starts with heat and produces work.

More than one stroke of an engine is needed to drive a civilization. After the first stroke of the piston the low-pressure steam that remains can be expelled from the cylinder through another valve, and the process can be repeated over and over again.

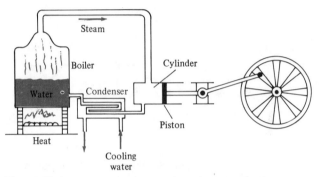

Figure 16.1 Schematic representation of a steam engine.

In the 1700s engineers were building steam engines that worked by suction rather than positive pressure, but these early devices left much to be desired. The age of steam had not yet arrived because crucial ideas were missing. The discovery that made the steam engine into a practical device was patented in 1769 by a Scottish instrument maker named James Watt. When asked to repair a model of a steam engine that was used as a lecture demonstration, Watt conducted a series of experiments that led him to realize that the engine wasted most of its heat in warming up the walls of the cylinder. Watt's idea was to condense the steam by cooling it outside of the cylinder so that the cylinder walls could always stay hot. Later engineers worked up the courage to use the potentially explosive power of positive pressure. These were among a series of developments and improvements that made the steam engine a practical source of power. The age of steam had arrived.

For the next half century, engineers devised ways to make steam engines more efficient. Their goal was to get more work out of each ton of coal used to heat the boiler. Like James Watt, these engineers were concerned with practical results, not theory. And by the 1820s steam engines worked well: there were steamships plodding across the seas and trains chugging across continents.

In 1824 a young French military engineer, Nicolas Léonard Sadi Carnot (1796–1832), published a remarkable essay on steam engines. After reviewing the industrial, political, and economic importance of the steam engine, Carnot raised the question: Is there an assignable limit to the efficiency of such engines? Although he never succeeded in making steam engines work better, his ideas had a profound influence in another direction.

James Watt, whose purpose had been to make money, invented a more efficient steam engine, and thereby revolutionized society. Sadi Carnot, whose purpose was to make steam engines more efficient, gave birth to thermodynamics, and thereby revolutionized physics. Thermodynamics is the science that deals with phenomena involving heat.

Figure 16.2 Nicolas Léonard Sadi Carnot. (Courtesy of the Archives, California Institute of Technology.)

Carnot's approach was radically different from that of steam engineers before him. Instead of tinkering with knobs, valves, and piston strokes of engines, Carnot developed an abstract theory of how engines work. He wanted to formulate the underlying principles governing the ideal engine. His question was not how to extract a little more work from a ton of coal, but what is the maximum amount of work that can be extracted from a ton of coal.

16.2 WORK AND THE PRESSURE–VOLUME DIAGRAM

Carnot undertook his quest unaware of the law of conservation of energy, which had not yet been formulated. For him, heat was not a form of energy that could be transformed into other forms, but rather a fluid, called caloric, which was conserved and not consumed by the engine. So Carnot never thought that the most efficient engine would be one that turned all the available heat energy into work. That is just as well, because it is the wrong answer.

As discussed in Section 13.5, Carnot visualized caloric running a steam engine in much the same way as water runs a water wheel. As water falls from greater to lesser height, it drives the water wheel, but is itself not consumed. The "spent" water must be carried away or the wheel will soon be flooded and stop functioning. By analogy, Carnot imagined that a steam engine runs by the flow of caloric from the higher temperature of the boiler to the lower temperature of the surroundings. In fact a steam engine would not function unless heat were extracted at the lower temperature; that was precisely the job of the condenser that James Watt invented.

That heat must be extracted doesn't depend in any way on steam. Air can similarly be used to drive an air engine. For example, air can be heated in a large can, raising it to high pressure. Then a valve opens, allowing the high-pressure air to push on a piston which does work because a force is applied through a distance. Heating air pressurizes it because as it becomes hotter, if it can't expand, its pressure rises. The reverse is also true: cooling air lowers the pressure. Expanding the volume of a gas lowers the pressure and tends to cool it, whereas compressing tends to warm it. These are the essential points in Carnot's analysis of the efficiency of an engine.

Let's express this idea quantitatively by assuming that air obeys the ideal-gas law (which it very nearly does). Then, as shown in Chapter 15, we can describe the state of the gas by

$$PV = NkT,\tag{15.9}$$

and we also can write the internal kinetic energy plus the energy stored in the molecules or atoms as

$$U = qNkT.\tag{15.13}$$

Imagine a gas confined to a cylinder which has a movable piston at one end as shown in Fig. 16.3. If we compress the gas by pushing in the piston, the volume of the gas changes and either the pressure or temperature or both must change. As a result of this compression, the gas will not be in an equilibrium state; the pressure near the piston initially will be greater than that far away from the piston. Consequently we cannot define the state of the gas, that is, specify T, P, or U, until the gas settles down. However, if we push the piston slowly in small steps, waiting for equilibrium to be reestablished after each step, we can compress the gas so that it is never too far from an equilibrium state. This kind of process is known as a *quasistatic* process, and in practice it can be achieved fairly well.

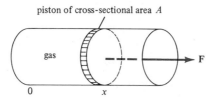

Figure 16.3 The work done by a gas depends on its pressure and volume.

With that aside, let's return to the quantitative analysis. Consider a piston in a circular cylinder, closed at the left end, as shown in Fig. 16.3. If a variable force F moves the piston from 0 to x, the work done by this force is expressed by the integral

$$W = \int_0^x F(x')\,dx'.$$

Differentiating this equation by the first fundamental theorem of calculus we obtain

$$\frac{dW}{dx} = F(x).$$

If the force F is applied by a gas inside the cylinder under pressure P, then

$$F = PA,$$

where A is the cross-sectional area of the piston, so we get

$$\frac{dW}{dx} = PA.$$

Now the state of a gas is usually described by its pressure P, its volume V, and its temperature T. In this case the volume is $V = Ax$ so $dV/dx = A$ since A is constant, and the equation for dW/dx becomes

$$\frac{dW}{dx} = P\frac{dV}{dx}.$$

We eliminate x from this equation by using the chain rule,

$$\frac{dW}{dx} = \frac{dW}{dV}\frac{dV}{dx}$$

and obtain the fundamental relation

$$\frac{dW}{dV} = P \tag{16.1}$$

which relates the work W to the pressure P and the volume V. To calculate the work W_{12} done *by the gas* in expanding the cylinder from initial volume V_1 to final volume V_2 we integrate this relation and obtain

$$W_{12} = \int_{V_1}^{V_2} P\,dV. \tag{16.2}$$

To carry out the integration, we need to know how the pressure P varies as a function of the volume V.

Note that this is the work done by an *expanding* gas. If an equal but opposite force $-F$ *compresses* the gas, then the gas gains energy (or does negative work $-W_{12}$) as a result of the compression.

Recall that the pressure, volume, and temperature of an ideal gas completely specify the state of the gas. Since these three variables are related by the ideal-gas law, knowledge of two alone specifies the state. When work is performed by a gas, P and V are the natural variables to define the state, which can be represented by a point (V, P) on a pressure – volume diagram.

Figure 16.4 illustrates this idea. As the gas is expanded from a smaller volume V_1 to a larger volume V_2, it defines a set of points (V, P), which we call a *PV* curve. Since the work done by the gas is an integral, it is equal to the area of the region under the curve and above the interval $[V_1, V_2]$.

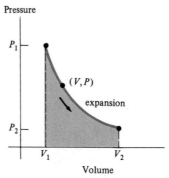

Figure 16.4 The state of a gas is represented by a point on a pressure–volume graph and the work done by the expanding gas is equal to the area of the region under the curve.

The actual shape of the PV curve will depend on how P and V are related to each other and to the temperature T. Different dependencies will produce different curves, or paths, that represent how the pressure and volume were related during the expansion or compression process. Three such paths are illustrated in Fig. 16.5. The work done by the expanding gas is different along each of these paths because the corresponding areas are different. So the work done in going from one state (V_1, P_1) to another (V_2, P_2) depends not only on the initial and final states, but also on the path taken between the states.

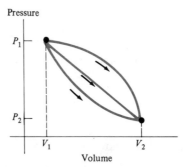

Figure 16.5 The work done by an expanding gas is equal to the area of the region under the curve in a PV diagram and depends on the path taken.

Example 1

A frictionless piston compresses a gas in a chamber that keeps the pressure constant, a process known as an isobaric compression. If (V_1, P_1) specifies the initial state of the gas and (V_2, P_1) the final state, find the work done by the gas in this compression if $P_1 = 3$ atm, $V_1 = 5$, and $V_2 = 1$ (in liters).

As the gas is compressed it gains energy and does a negative amount of work. So let's calculate instead the corresponding positive work W_{12} done by the gas if it *expands*

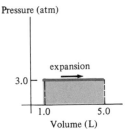

from 1 L to 5 L at a constant pressure of 3 atm. According to Eq. (16.2) this work is

$$W_{12} = \int_1^5 P \, dV = 3 \int_1^5 dV = 3 \times 4 = 12 \text{ L atm} = 1200 \text{ J}.$$

The work done by the gas in compression is -12 L atm. In this example the PV curve is a horizontal line segment and the work W_{12} is equal to the area of a rectangle.

Suppose a gas is initially compressed from a pressure P_1 and volume V_1 to another state specified by P_2 and V_2 along path A in Fig. 16.6. Next imagine that the gas is allowed to expand and follows path B back to the initial state, where B lies above path A. Has any net work been done? As indicated in Fig. 16.6 the net work done in the complete cycle is the difference between the area of the region under curve B and that under curve A. This difference is the area of the shaded region between the two paths. The work done by the gas is positive if the upper path is an expansion and the lower path a compression. In other words, if the cycle is clockwise on the PV diagram, the gas does positive work.

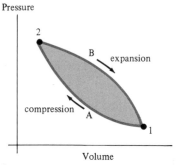

Figure 16.6 The net work done by a gas in a complete cycle is equal to the area of the region enclosed by the path.

Example 2
Suppose the gas in Example 1 is compressed from initial state (5, 3) to (1, 3) as before, and then is allowed to expand linearly to a third state (5, 4) (volume 5 L, pressure 4 atm) before being cooled at constant volume to return to its initial state (5, 3) as shown in the figure below. How much work is done by the gas in this cycle?

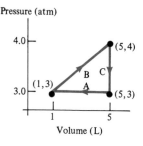

First, we recognize that the total work done is the sum of the work done along each of the three straight-line paths, A, B, and C:

$$W_{\text{tot}} = W_A + W_B + W_C.$$

From Example 1, we already know that the work done along path A is $W_A = -12$ L atm. Since path B taken from the state $(1, 3)$ to the state $(5, 4)$ is a straight line on the PV diagram, the work done along B is the area of a trapezoid,

$$W_B = \text{base} \times \text{average height} = (5 - 1) \frac{4 + 3}{2} = 14 \text{ L atm.}$$

For path C the work done is zero because the volume does not change. Thus we find the total work done by the gas to be equal to 2 L atm. Of course, this is also equal to the area of the triangle enclosed by the complete path.

Questions

1. When a cold can of soda pop is opened, a thin fog forms near the opening. Formulate a possible explanation.

2. Why do you think the valve on a bicycle pump gets hot when you pump up a tire? Once you've answered that, then explain why the valve on the compressed air at a gas station doesn't heat up.

3. Cite the underlying physics behind the formation of a mushroom cloud after a large bomb explodes.

4. At an altitude of 30,000 ft the air temperature is $-30°F$, yet passenger jets flying at this altitude use air conditioners to cool the air. Explain why.

5. For an isobaric compression, show that the work done by the gas is also equal to $Nk(T_2 - T_1)$, where T_2 and T_1 are the final and initial temperatures of the gas, respectively.

6. One mole of an ideal gas is in the state given by point A in the PV diagram.

   (a) What is the temperature of the gas at A?
   (b) If the gas expands from A to B, how much work does it do?

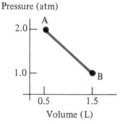

Pressure (atm)

7. Suppose N molecules of an ideal gas are compressed from a volume V_1 to a volume V_2 in such a way that the temperature varies according to $T = T_1 (V_1/V)^{3/2}$. Calculate the work done by the gas in this compression in terms of N, k, T_1, V_1, and V_2.

16.3 THE FIRST LAW OF THERMODYNAMICS

Everyday experiences show us that whenever a hot body is placed in contact with a cool body, heat flows from the hot body to the cool one until the two reach a common equilibrium temperature. Numerous ingenious experiments in the eighteenth and nineteenth centuries indicated that for two bodies in thermal contact, when heat leaves one body, an equal amount enters the other body. These experiments led to the idea of conservation of caloric – an invisible, colorless, weightless fluid. Caloric was neither created nor destroyed, but merely transferred from one body to another.

Conservation of caloric was eventually abandoned when experiments showed that it is not conserved. Late in the eighteenth century, Benjamin Thompson, also known as Count Rumford, studied the origin of heat in the production of a cannon. As director of the Bavarian arsenal, he supervised the boring of cannons for that kingdom. He observed that the cooling water had to be continually replaced during the boring process because the heat from the boring tool boiled it away. According to the caloric theory, the metal chips formed in the boring process released caloric to the water, causing it to boil. Thompson noted that even when chips were not produced but the boring tool was turning, heat was still produced. In other words, caloric could be created by friction and could be produced endlessly; evidently, caloric was not conserved. Thompson's experiments indicated a close connection between the work done by the boring tool and the heat produced.

In the 1840s a series of careful experiments performed by the British scientist James Prescott Joule once and for all destroyed the principle of conservation of caloric. As discussed in Section 13.4, Joule's experiments used a slowly descending weight that turned paddle wheels in a container of water. By precise measurements of the increase in temperature of the water and the work done by the falling weight, Joule uncovered the relationship between heat and work. In these experiments, the container of water was carefully insulated to prevent heat from entering or leaving the system. Such a system is said to be adiabatically shielded, and the transfer of work into internal energy, when no heat is allowed to enter or leave the system, is called an *adiabatic* process. In adiabatic processes, the work done depends only on the initial and final states of the system, in contrast to the more general process discussed in the previous section, in which the work done depends on the path taken from one state to another.

Work and heat are both forms of energy in the process of transfer. We define heat as that form of energy which is transferred from one body to another solely by virtue of differences in their temperature. Joule's experiments showed that neither heat nor mechanical energy is conserved independently. Rather, his experiments indicated that the mechanical energy lost from a system is converted into heat. Thus the total quantity of mechanical energy and heat is conserved.

Joule's contribution to thermodynamics was discovering the law of conservation of energy: the heat energy added to a system is equal to the sum of the work done by the system and the change in internal energy of the system. Known as the *first law of thermodynamics*, this result can be stated mathematically as

$$Q = W + \Delta U, \tag{16.3}$$

where Q is the *heat added*, W is the work done *by* the system, and ΔU is the change in internal energy of the system.

To understand the distinction between internal energy and heat and work, consider the analogy of a bank account. You can add money to your account by depositing either cash or checks. Likewise, you can withdraw money from your account in the form of cash or checks. The change in the amount of money in the account over any period is equal to the algebraic sum of all the deposits and withdrawals. The total balance shown in your monthly statement does not depend on which transactions were in cash and which were by check; only the algebraic sum of the deposits and withdrawals is reflected in the balance. So it is with the internal energy function: although heat and work contribute separately, only the algebraic sum of the two matters.

Example 3

100 g of a certain ideal diatomic gas contains 3.57 moles.

(a) What is the internal energy of the gas at 35°C?

(b) If the gas is heated from 35°C to a temperature T with the pressure and volume changing linearly as shown, find the change in internal energy for this process.

(c) Find the heat required to produce the change in internal energy in (b).

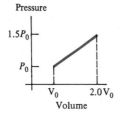

(a) From Eq. (15.13) we know that the internal energy of a gas is given by

$$U = qNkT, \tag{15.13}$$

where $q = \frac{5}{2}$ for a diatomic gas and T is the absolute temperature. Using the fact that 6.02×10^{23} atoms are in one mole, we can substitute to find

$$U = (\tfrac{5}{2}) (3.57 \text{ mol})(6.02 \times 10^{23} \text{ atoms/mol}) \times (1.38 \times 10^{-23} \text{ J/K})(308 \text{ K}),$$

$$U = 2.28 \times 10^4 \text{ J}.$$

(b) The change in internal energy is given by $\Delta U = qNk(T_f - T_i)$, so we need to find the final temperature of the gas. According to the ideal-gas law, $PV = NkT$, we can find the final temperature by forming a ratio:

$$\frac{P_i V_i}{P_f V_f} = \frac{T_i}{T_f}.$$

From the graph we know that $P_i = P_0$, $P_f = 1.5P_0$, $V_i = V_0$, $V_f = 2V_0$, and $T_i = 308$ K. Solving, we find $T_f = 924$ K. Therefore the change in internal energy is 4.57×10^4 J.

(c) According to the first law of thermodynamics, $Q = \Delta U + W$, where Q is the heat added and W is the work done by the gas. We can calculate the work done by the gas from the PV diagram since the work done is equal to the area of a trapezoid. Breaking the trapezoid into a triangle and a rectangle, we have

$$W = \tfrac{1}{2}V_0(0.5P_0) + P_0V_0 = \tfrac{5}{4}P_0V_0.$$

By the ideal-gas equation, $P_0V_0 = NkT_i$, and the work done by the gas is $W = \tfrac{5}{4}NkT_i = 1.14 \times 10^4$ J. Therefore the heat added is 5.71×10^4 J.

Questions

8. Can a gas absorb heat without any change in internal energy?

9. Since the total mechanical energy of a body is the sum of the kinetic and potential energies, why isn't the change in internal energy equal to the sum of the heat added and the work done by a gas?

10. A monatomic ideal gas starts with pressure $P_1 = 1.0 \times 10^7$ Pa, volume $V_1 = 0.1$ m^3, and temperature $T_1 = 1500$ K. First, its pressure is decreased to $P_2 = 3 \times 10^6$ Pa, with the volume kept constant. Then the gas expands to a volume $V_2 = 0.25$ m^3, with the pressure constant, as shown in the diagram.

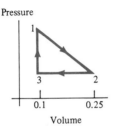

Pressure

Volume

(a) Determine the change in internal energy of the gas in going from point 1 to point 2.

(b) Calculate the work done in the process.

(c) Find the heat added to the gas in the process.

11. For the process in Question 7, determine the change in internal energy and the heat added.

16.4 ADIABATIC AND ISOTHERMAL PROCESSES

All kinds of heat engines basically work because the force of an expanding gas can be used to make something move. When that's done, gas must be compressed again to start the next engine cycle. Thus the expansion and compression of gases are the crucial elements of a working engine. To analyze engines, we need to ask how the pressure of a gas behaves while its volume is changing.

The answer depends on how the change is accomplished. For example, it's possible to keep the gas at constant pressure and make it expand by heating it. On the other hand, if the temperature is carefully kept constant, then according to $PV = NkT$ the pressure will decrease as the gas expands. In other words, the relation between P and V depends on what T is doing.

For analyzing engines, two processes are particularly important. One is the case of constant temperature, called an *isothermal* process. In the other, the system is isolated so that no heat can flow in or out. Then the behavior of the temperature, and therefore the relation between P and V, can only be deduced indirectly by invoking the first law of thermodynamics (or in other words, the conservation of energy). This second kind of process is called *adiabatic*. We discuss the adiabatic case first.

In an adiabatic process, the gas is expanded or contracted in a container which is so well insulated from its surroundings that no heat flows in or out of the system. Mathematically, this means that $Q = 0$ in the first law of thermodynamics (16.3). Hence for an adiabatic process the first law becomes

$$W + \Delta U = 0, \tag{16.4}$$

where W is the work done by the system and ΔU is the change in internal energy. If W is positive, then ΔU is negative, so the work done by the gas equals the decrease in internal energy.

We now investigate the implications that Eq. (16.4) has on the relations between pressure, volume, and temperature of an ideal gas.

Let U_0 denote the initial value of the internal energy. Then the change in internal energy is $\Delta U = U - U_0$, and Eq. (16.4) gives us

$$W = -U + U_0.$$

Differentiating with respect to volume V we obtain

$$\frac{dW}{dV} = -\frac{dU}{dV}.$$

But by Eq. (16.1) we have $dW/dV = P$, so the previous equation becomes

$$\frac{dU}{dV} = -P. \tag{16.5}$$

We also know that the internal energy is given by

$$U = qNkT, \tag{15.13}$$

which can be combined with the ideal-gas as $NkT = PV$ to give us

$$U = qPV,$$

where q is a positive constant. Differentiating this equation with respect to V, using the product rule, we get

$$\frac{dU}{dV} = qP + qV\frac{dP}{dV}.$$

Equating this to (16.5) and rearranging terms we obtain

$$\frac{1+q}{q}P\frac{dV}{dP} + V = 0,$$

a differential equation relating P and V. The factor $(1 + q)/q$ is a constant greater than 1 which we denote by γ, and the differential equation becomes

$$\gamma P\frac{dV}{dP} + V = 0.$$

To solve this equation, we first multiply each member by $V^{\gamma-1}$, which gives us

$$P\gamma V^{\gamma-1}\frac{dV}{dP} + V^\gamma = 0.$$

The reason for doing this is that now we recognize the left-hand side of the equation as simply the derivative of the product PV^γ. In other words, the differential equation now reads

$$\frac{d}{dP}(PV^\gamma) = 0,$$

which means that PV^γ must be a constant. Thus we have shown that *in an adiabatic process the pressure and volume of an ideal gas are related as follows:*

$$PV^\gamma = C, \tag{16.6}$$

where C and γ are constants, with $\gamma > 1$. Therefore the pressure as a function of volume is given by

$$P = CV^{-\gamma}, \tag{16.7}$$

whereas the volume as a function of pressure is

$$V = (C/P)^{1/\gamma} = (\text{const})\, P^{-1/\gamma}. \tag{16.8}$$

The corresponding PV diagram is given in Fig. 16.7.

We can also incorporate the ideal-gas law $PV = NkT$ to express the temperature directly in terms of P or of V. The results are

Pressure

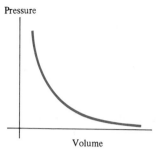

Volume

Figure 16.7 PV diagram for an adiabatic process: $PV^\gamma = C$.

$$T = \frac{PV}{Nk} = \frac{C}{Nk} V^{1-\gamma} = (\text{const}) \, V^{1-\gamma},\tag{16.9}$$

and

$$T = \frac{C}{Nk} \left(\frac{C}{P}\right)^{(1-\gamma)/\gamma} = (\text{const}) \, P^{(\gamma-1)/\gamma}.\tag{16.10}$$

Since $\gamma > 1$, the power of V in (16.9) is negative, so T is a decreasing function of V. In other words, when a gas expands adiabatically, its temperature falls as the volume increases. In (16.10) the exponent of P is postive, so T is an increasing function of P. Thus, when a gas is compressed adiabatically the temperature rises as the pressure increases.

For a monatomic ideal gas we know from Chapter 15 that $U = \frac{3}{2}PV$ so $q = \frac{3}{2}$ and $\gamma = (1 + q)/q = \frac{5}{3}$.

Example 4

Suppose that 2 moles of helium gas are compressed along the path from A to B adiabatically as shown in the PV diagram. At point B the temperature is 194 K and the pressure is 3.2 atm.

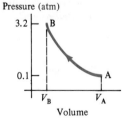

(a) Find the volume of the gas at point B.
(b) Assuming the pressure at A is $P_A = 0.1$ atm, find the volume of the gas at A.
(c) Find the change in internal energy of the gas.
(d) Find the amount of work done on the gas during compression.

(a) From the ideal-gas law, $PV = NkT$, we find the volume at B to be $V_B = 0.01$ m^3.

(b)   Since the compression is adiabatic, we know that the quantity $PV^\gamma = \text{const}$, where $\gamma = \frac{5}{3}$ for a monatomic gas. Therefore we have $P_A V_A^{5/3} = P_B V_B^{5/3}$, which tells us that $V_A = V_B (P_B/P_A)^{3/5}$, or $V_A = 0.08 \text{ m}^3$.

(c)   According to Eq. (15.13), $U = qNkT$ specifies the internal energy of a gas, where for a monatomic gas $q = \frac{3}{2}$. The change in internal energy is given by $\Delta U = U_B - U_A = \frac{3}{2}Nk(T_B - T_A)$. We are given the temperature T_B and can use the ideal-gas law to find that the temperature at A is 49 K. Substituting back into our equation, we find that $\Delta U = 3.6 \times 10^3$ J.

(d)   Since the compression is adiabatic, no heat is transferred to the gas, that is, Q, $= 0$. By the first law of thermodynamics, $W + \Delta U = 0$, where W is the work done by the gas. Therefore, the work done *on the gas* is equal to $-W = \Delta U = 3.6 \times 10^3$ J; since the gas was compressed, the work done on it is positive.

Next we consider an *isothermal* process. Here a gas is expanded or contracted in a container which is immersed in a heat reservoir, a large body whose temperature is constant. Since $U = qNkT$, if the temperature is held constant the internal energy U also remains constant, and the first law of thermodynamics states that

$$Q = W$$

because $\Delta U = 0$. The ideal-gas law $PV = NkT$ implies that in an isothermal process the product of pressure and volume is constant, so the PV curve (called an *isotherm*) is a hyperbola like the one shown in Fig. 16.8.

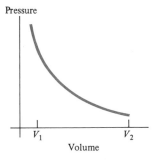

Figure 16.8 In an isothermal process the PV curve is a hyperbola, $PV = C$.

To calculate the work done by a gas in isothermal expansion, we use the integral

$$W_{12} = \int_{V_1}^{V_2} P \, dV, \tag{16.2}$$

where now $P = NkT/V$ with T constant. Therefore

$$W_{12} = NkT \int_{V_1}^{V_2} \frac{dV}{V} = NkT \ln\left(\frac{V_2}{V_1}\right). \tag{16.11}$$

For an expanding gas, $V_2 > V_1$ and W_{12} is a positive quantity, so the gas does a positive amount of work which must equal the amount of heat Q added to the system because the internal energy remains constant. To keep the temperature of the gas constant, this added heat must come from the reservoir.

On the other hand, if the gas is isothermally compressed, then the work done by the gas is negative so Q is also negative, which means that heat must leave the system. In this case the heat is absorbed by the reservoir.

Questions

12. Gently blow across your knuckles with your mouth wide open. Compare that feeling to that from blowing across your knuckles with your lips close together. Why does the air feel cooler the second time?

13. One mole of an ideal gas for which $\gamma = \frac{5}{3}$ is at 300 K and 1.0 atm. Find the initial and final energies of the gas and the work done by the gas when 800 J of energy is added at (a) constant volume, (b) constant pressure.

14. Suppose that 2.0 mol of helium gas are compressed isothermally along the path AB as shown in the PV diagram. At point A the volume is 6.0 L and the pressure is 2.5 atm.

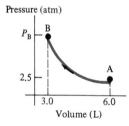

Pressure (atm)

(a) Find the temperature of the gas at point B.
(b) Assuming the volume at point B to be 3.0 L, find the pressure.
(c) Calculate the work done on the gas during the compression.
(d) How much heat was added to the gas during the compression?

15. An ideal gas at pressure P_1 and volume V_1 is compressed adiabatically to a volume V_2 and pressure P_2. Show that the work done by the gas is given by $W = (P_2V_2 - P_1V_1)/(\gamma - 1)$.

16. A monatomic gas initially at a pressure of 1.5 atm and a volume of 0.3 m³ is compressed adiabatically to a volume of 0.1 m³.

(a) Find the pressure of the gas in the final state.
(b) Determine the work done by the gas in compression.
(c) Calculate the final temperature of the gas.

16.5   THE SECOND LAW OF THERMODYNAMICS

Now that we've explored the behavior of gases, let's return to the air engine. In the cycle of an engine, one stroke of the piston allows the gas to expand and cool. This is called the power stroke. To keep the engine going, two things must be accomplished: the piston must return to where it started, and the same amount of air must replace the air in the boiler that was lost. One obvious way to achieve this would be to push the piston back in and let it recompress the air in the cylinder.

To push the piston back in would require exactly the same amount of work as the piston performed on the way out (ignoring friction), hence no net work is achieved. In the real frictional world the situation is worse: more work would have to be added to the machine than was extracted, just to keep it going.

The situation can be remedied, however, by simply cooling the air after the outward piston stroke. Cooling the air in the cylinder causes it to contract and so the piston is pulled inward. This means that less force, and hence less work, is required to return the piston to its original position. One more step also needs to be included. The cool, dense air is heated back to boiler temperature and pressure, then injected back into the cylinder. At this point the engine is back to its initial condition and a net amount of work has been done by the machine.

The crucial step that made the engine run was to cool the air after the power stroke, in other words, to remove heat at lower temperature. This is the "runoff" from Carnot's water-wheel analogy. Watt's invention, the condenser, serves precisely this purpose. That's also why your car engine must have an efficient cooling system. Figure 16.9 illustrates a schematic representation of a heat engine.

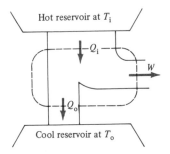

Figure 16.9 Schematic representation of a heat engine.

Carnot's water-wheel analogy, as we mentioned, is not quite accurate. Heat is not a fluid but a form of energy. What an engine does is turn heat into work. The subtle and remarkably profound insight of Carnot's analogy is that not all of the heat, but only part of it, can be turned into work. The remaining heat must be discarded.

The initial and final states of the gas in an engine are the same, so the change in internal energy must be zero. Now if the heat put into the boiler is Q_i, and the heat necessarily extracted is Q_o, the first law of thermodynamics requires that

$$W = Q_i - Q_o, \tag{16.12}$$

where W is the net work done by the engine.

The purpose of any engine is to do as much work as possible for a given expenditure of heat extracted from a source, such as burning coal or ignited gasoline. The efficiency e of a machine is defined as the ratio of the work done to the amount of heat put in:

$$e = \frac{W}{Q_i}.$$

Using Eq. (16.12), we can write the efficiency as

$$e = 1 - \frac{Q_o}{Q_i}. \tag{16.13}$$

This indicates that for optimum efficiency we want to discard as little heat as possible. However, no engine can run without discarding some heat, and therefore, even ideally, e is always less than 100%. Moreover, any real engine, having friction, will create more heat to discard, causing the efficiency to be even less than ideal.

The experimental observation that it is impossible to make an engine that is 100% efficient is known as the *second law of thermodynamics*:

A process whose *only* net result is to take heat from
a reservoir and convert it completely to work
is impossible.

We haven't yet answered the question Carnot set out to answer: Does an ideal engine exist? What is the highest efficiency a perfect, frictionless engine can possibly have? So far we've only argued that the efficiency must be less than one. But how much less?

Although the question seems to be a problem in engineering, the answer has implications far beyond the realm of engines. One of the implications concerns the properties of all matter. The other relates to the flow of time itself.

The implications for the properties of matter come from the fact that the result cannot depend on what the working fluid of an engine is. Otherwise it would be possible to design an engine that circumvents Carnot's arguments. We therefore arrive at general principles applicable to all matter.

The second point, having to do with the flow of time, stems from the fact that, once heat has flowed "downhill" from high temperature to low temperature, it is very difficult to run it "uphill" again. If some amount of heat Q_i flows into a perfect, frictionless engine, producing work W, and depositing heat Q_o at lower temperature, the very best that can ever be done is to use exactly the same amount of work W, to retrieve Q_o from the low temperature, depositing Q_o and W together as high-temperature heat Q_i. But that ideal, the net result of which is that nothing happens, is unattainable in the real world.

Once any real engine has used heat, the best any other real engine can do to reverse the process is return to high temperature a little less heat than was originally extracted, leaving little more waste at low temperature; that's what the second law of thermodynamics says. Once this has happened the universe has been irreversibly changed forever.

Questions

17. Suppose that an inventor approached you with an idea to make an engine that would extract heat from the ocean and use it to power a ship without the need to expel heat at a lower temperature. Would you invest in such an idea?

18. An engine has an output of 800 J per cycle and an efficiency of 40%. How much heat is absorbed and how much is rejected in each cycle?

19. Logically, could a device satisfy the first law but violate the second law of thermodynamics? Devise an example to explain your answer.

16.6 THE CARNOT ENGINE

Although we have tasted the flavor of Carnot's reasoning, we have not yet followed enough of it to see what it is based on, nor to understand exactly where it leads. Let's remedy that situation by examining the logic of one of his central arguments concerning the efficiency of engines. Carnot invented an engine of his own, similar to the air engine we've discussed, and no more practical, but with a few further refinements.

The Carnot engine consists of a high-temperature source of heat at temperature T_i (i for input) and a lower-temperature reservoir at temperature T_o (o for output), plus a sealed cylinder containing a gas and having no valves, but with a movable piston. In one cycle the engine uses the following processes: First, the piston is as far in as it will go, so the gas is compressed to its smallest volume of the cycle. This corresponds to point 1 on the *PV* diagram in Fig. 16.10. At this point the temperature of the gas is T_i, the same as that of the high-temperature reservoir. Next the cylinder is placed in contact with the heat source – the boiler – and the gas expands, pushing the piston partially out. Although the gas would tend to cool in the expansion, its temperature is held constant since the cylinder is in contact with the heat reservoir. In other words, the gas moves along an isotherm from point 1 to point 2 in Fig. 16.10. As we discussed in Section 16.5, an amount of heat is added to the gas during the isothermal expansion. So in this first step, some heat Q_i is extracted from the reservoir at T_i and some work is done by the engine.

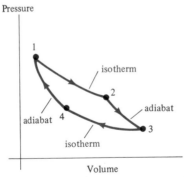

Figure 16.10 The Carnot cycle illustrated on a *PV* diagram.

In the second step of the Carnot cycle, the cylinder is removed from the heat source and the gas is allowed to expand further, doing more work. This time, however, the cylinder is not connected to any source of heat, so the gas cools as it adiabatically expands. In Fig. 16.10 this corresponds to the adiabat from point 2 to point 3. The gas expands until its temperature drops to T_o, the temperature of the cool reservoir.

For the next step in the cycle, the cylinder is placed in contact with the cool reservoir and the cylinder is pushed in partially. This compression would tend to heat the gas, but since the gas is in contact with the cool reservoir, its temperature remains constant at T_o, while heat flows out of the gas and into the reservoir. The path from point 3 to point 4 in Fig. 16.10 illustrates this isothermal compression. During this step some work is done on the gas, but since the temperature is held constant, the pressure remains lower than it could otherwise have been. This is accomplished by allowing some amount of heat Q_o to flow out into the cool bath.

During the final step, the cylinder is removed from contact with the cool bath, and work is done on it to push the piston further in. This pressurizes the gas, and because heat can't escape, the gas warms. At the end of this stroke, represented by the adiabat from point 4 to point 1 in Fig. 16.10, the compressed gas reaches the initial temperature T_i. And the cycle is ready to begin again. From the discussion of Section 16.2, the amount W of net work done is equal to the area of the region enclosed by the cycle.

The entire cycle can operate in reverse without violating any physical principles because each individual step has been designed to work equally well backwards as forwards. If we run the Carnot cycle forward, an amount of heat Q_i is extracted at high temperature T_i, an amount of net work W is done, and the remaining heat $Q_i - W = Q_o$ is rejected at the low temperature T_o. If we run the cycle backwards, net work must be put into the machine. Heat is extracted from the low temperature, and energy equal to the heat in plus the work is discarded at high temperature. That's the kind of machine we use to cool our houses and food on a hot day. It is the principle of a refrigerator. Carnot prescribed a machine that can run equally well backwards or forwards. In other words, the Carnot engine is a reversible engine.

Carnot's result for an ideal engine states that no engine that works between two heat reservoirs can be more efficient than a reversible engine that works with the same reservoirs. We can prove this statement as follows: Suppose you have a Carnot engine that can run forwards or backwards with an efficiency

$$e = 1 - \frac{Q_o}{Q_i} = \frac{W}{Q_i}. \tag{16.14}$$

However, knowing that Carnot engines are not really practical, you obtain another device from a competing company. The manufacturer claims that this engine has an even better efficiency:

$$e^* = 1 - \frac{Q_o^*}{Q_i} = \frac{W^*}{Q_i}.$$

In other words, for a given amount of heat Q_i extracted from the boiler, this engine will do more work, $W^* > W$, and dump less heat, $Q_o^* < Q_o$, than the impractical Carnot device.

How can we test the manufacturer's claim? Suppose we use the practical engine to run the Carnot device backwards as a refrigerator. The practical engine extracts heat Q_i, does work W^*, and deposits heat Q_o^*. The Carnot device uses up W, less than the available W^* to extract heat Q_o, which is more than Q_o^* from the low-temperature bath, and returns the same amount Q_i to the boiler. The net result is that the boiler doesn't need any fuel, since we return to it all the heat that we used. The combined machine still does usable work $W^* - W$ at the expense of the cool bath, which is losing heat $Q_o - Q_o^*$. If we install the engines in a ship, we could use them to drive the ship across the ocean by extracting heat from the ocean, which we use as the low-temperature reservoir. We've invented a device to make use of the limitless energy stored in the ocean.

That's impossible. The starting point of Carnot's reasoning is that work cannot be done by extracting heat over and over again from a single temperature source; that was in his water-wheel analogy. But that is just what we've done. Making heat run "uphill," with no other net effect, or running an engine from a single temperature source, are logically equivalent, and both equally impossible. That is the second law of thermodynamics, and it cannot be proved or deduced from other laws. Its logical consequences are Sadi Carnot's great gift to us. In the view of many physicists, the second law is perhaps the most profound of all the laws of physics.

An engine more efficient than the Carnot engine would violate the second law of thermodynamics. In other words, no engine can be more efficient than the Carnot engine. The crucial property of the Carnot engine is that it is reversible. In fact, any engine operating between temperatures T_i and T_o that can run with equal efficiency backwards and forwards must by the same arguments have the same efficiency as the Carnot engine.

What is the ideal value of the efficiency of a Carnot engine? Since the efficiency is the same for different kinds of machines with different working fluids, it cannot depend on any of the details of the machine. The efficiency can only depend on the things that are the same for all, namely, the temperatures of the reservoirs, T_i and T_o.

Let's calculate the efficiency of the Carnot cycle illustrated in Fig. 16.11. During the isothermal expansion from V_1 to V_2, the heat extracted is equal to the work done by the gas. By Eq. (16.11), we know that this amount of heat is given by

$$Q_i = W_{12} = NkT_i \ln(V_2/V_1). \qquad (16.11)$$

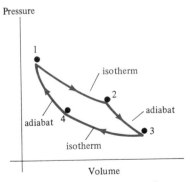

Figure 16.11 The Carnot cycle.

The second stroke expands the cylinder further, from V_2 to V_3. Then the third stroke, this time with the gas in contact with the bath at T_o, squirts out heat by contracting from V_3 to V_4. Along the isotherm from point 3 to point 4, the heat extracted is given by

$$Q_o = NkT_o \ln(V_4/V_3).$$

Now we could find the ratio Q_o/Q_i and therefore the efficiency if we knew how the volumes were related to each other.

Since strokes 2 and 4 are adiabats, we know that $PV^\gamma = $ const, and because $P = NkT/V$, we have

$$NkTV^{\gamma-1} = \text{const}$$

for adiabats. Since Nk doesn't change, it follows that

$$T_i V_2^{\gamma-1} = T_o V_3^{\gamma-1} \qquad \text{(from 2 to 3)},$$

$$T_o V_4^{\gamma-1} = T_i V_1^{\gamma-1} \qquad \text{(from 4 to 1)}.$$

Taking the ratio of these equations, we get

$$\frac{V_3}{V_4} = \frac{V_2}{V_1}.$$

Using this result for the ratio of the volumes and inserting it into the expressions for Q_i and Q_o, we find

$$\frac{Q_o}{Q_i} = \frac{T_o}{T_i}. \qquad (16.15)$$

Therefore the efficiency of a Carnot engine is given by

$$e = 1 - \frac{T_o}{T_i}. \qquad (16.16)$$

This astonishingly simple result is one of the most important in all of thermodynamics and raises momentous issues for both engineering and physics.

Example 5

A reversible engine operates with an efficiency of 40% and rejects 150 cal of heat to a reservoir at 300 K each cycle.

(a) What is the temperature of the hot reservoir?
(b) How much work is done each cycle?

(a) For a reversible or Carnot engine the efficiency is given by Eq. (16.16), $e = 1 - (T_o/T_i)$. Solving this equation for T_i we find $T_i = 500$ K.
(b) Since the efficiency is defined as $e = W/Q_i$, and $Q_o/Q_i = T_o/T_i$, we have $W = eQ_oT_i/T_o$, and we find $W = 100$ cal.

Questions

20. Why do designers of power plants try to increase the temperature of the steam fed into engines as much as possible?

21. An engine works between reservoirs at 450 K and 300 K, extracting 100 J from the hot reservoir during each cycle.

(a) What is the maximum possible efficiency for this machine?

(b) What is the greatest amount of work it can perform during each cycle?

22. An ideal gas having $\gamma = \frac{5}{3}$ follows the cycle shown in the PV diagram; the temperature at point 1 is 300 K. Determine the efficiency of this cycle.

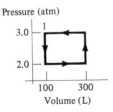

Pressure (atm)

3.0

2.0

100   300

Volume (L)

23. A diatomic gas is used in a heat engine. Starting at point 1 with $P_1 = 1.0 \times 10^5$ N/m^2, $V_1 = 1.0 \times 10^{-3}$ m^3, and $T_1 = 400$ K, the cycle follows that for an idealized gasoline engine, as shown in the diagram. For the compression 1 to 2, the volume decreases to 10^{-4} m^3, and for the pressure increase in 2 to 3, $P_3 = 2P_2$.

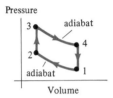

Pressure

adiabat

3

4

2

adiabat   1

Volume

(a) Fill in the following table with the values of pressure and temperature for the various points indicated:

	1	2	3	4
P	1.0×10^5 N/m^2			
V	1.0×10^{-3} m^3	1.0×10^{-4} m^3	1.0×10^{-4} m^3	1.0×10^{-3} m^3
T	400 K			

(b) Calculate the heat added and indicate in which part of the cycle this occurs.

(c) Determine the heat extracted and indicate in which part of the cycle this occurs.

(d) Find the work done in one cycle.

(e) Calculate the efficiency of the engine.

24. An ideal monatomic gas initially at 300 K and occupying a volume of 2.0×10^{-4} m^3 is allowed to expand to a volume of 4.0×10^{-4} m^3, so the pressure (in Pa) is given by

$$P = 1.0 \times 10^5 - (1.0 \times 10^{-3})/V^2,$$

where V is measured in m^3.

(a) Find the initial and final pressures of the gas.
(b) Determine the final temperature of the gas.
(c) Calculate the work done by the gas during the expansion.
(d) Find the amount of heat added to the gas during the expansion.

25. A certain monatomic gas is taken through the cycle shown in the *PV* diagram. At point A, under a pressure of 4.0×10^5 Pa and a temperature of 1200 K, the gas expands adiabatically from a volume of 2.56×10^{-4} m^3 to 5.00×10^{-4} m^3 at point B. From B to C the pressure is constant, and from C to A the volume is constant.

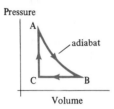

(a) Find the temperature of the gas at point B.
(b) Calculate the pressure at point B.
(c) Determine the temperature at point C.
(d) Calculate the heat added to the gas during the cycle.
(e) Find the net heat extracted during the cycle.
(f) Calculate the work done and the efficiency of the cycle.

16.7   A FINAL WORD

The efficiency of real engines is an important problem, intimately related to the so-called energy crisis, problems of the environment, and so on. Every analysis of those questions leads us back to Sadi Carnot. Nevertheless, we will be more interested in yet other consequences of his reasoning that have implications for the flow of time and the properties of matter. How did the focus of our discussion shift from grubby, clanking steam engines to the universe itself? Carnot's reasoning gently led us from arguments about engines to arguments about principles. His imaginary devices are no longer idealized steam engines, they are idealizations of the human will, the magical but physical law-abiding machines that allow us to find the outer limits of the extent to which we can ever aspire to alter the world and by extension the universe.

Within Carnot's own brief lifetime he was respected, but his work was largely ignored. One reason was that it was abstract and difficult to understand. Another was that he was an engineer and his ideas had no practical engineering significance at the time. His work, however, was rescued from obscurity by theoretical physicists who discovered it decades after it was done.

The first of these was a German physicist named Émile Clapeyron, who cleverly simplified Carnot's arguments by doing what Carnot did not: using the "law" of conservation of caloric. Because this idea was wrong, many believed that Carnot's deductions were based on faulty arguments. They were not; only Clapeyron's arguments were. The next physicist who realized that there was something entirely new at the heart of Carnot's arguments was Rudolph Clausius. The conservation of energy, although true and profoundly important, just wasn't enough to explain a world in which every machine would eventually grind to a halt if it weren't wound up or supplied with new fuel. Another idea was needed, something even more subtle and profound than the conservation of energy. Clapeyron and Clausius found that fruitful idea in Carnot's work, and Clausius called it *entropy*. That is what the next chapter of our story is about.

CHAPTER  17

ENTROPY

For the present I will limit myself to quoting the following result: if we imagine the same quantity, which in the case of a single body I have called its entropy, formed in a consistent manner for the whole universe (taking into account all the conditions), and if at the same time we use the other notion, energy, with its simpler meaning, we can formulate the fundamental laws of the universe corresponding to the laws of the mechanical theory of heat in the following simple form:

1. The energy of the universe is constant.
2. The entropy of the universe tends to a maximum.

Rudolph Clausius, in *Annalen der Physik*, Vol. 125 (1865)

17.1 TOWARD AN UNDERSTANDING OF ENTROPY

In this chapter we turn our attention to the *entropy principle*, a concept which, like Newton's second law, is an organizing principle for understanding the world. The principle is relatively simple to state, but understanding its meaning is more challenging.

Through theoretical studies of Carnot's work in 1865, the German physicist Rudolph Clausius introduced a new physical quantity closely linked to energy. He called it *entropy* a word which sounds like *energy* and comes from the Greek word for *transformation*. The use of entropy provides a way to analyze the behavior of energy in transformation.

To obtain an intuitive feeling for the concept of entropy, let's start with a familiar mechanical system we've used many times: Galileo's experiment with a ball rolling down and up two inclined planes. If friction is ignored, the ball rolls down one plane and back up the other, conserving mechanical energy. But if friction is *not* ignored, as time goes on the ball loses energy. This energy is transformed to heat which warms the ball and the material on the inclined planes. Without friction, all of the energy in the system can be accounted for by describing the motion of a single object, the ball itself. But with friction, the energy is shared among all the atoms and molecules that have been warmed by friction. As time goes on, even remote parts of the apparatus are warmed through conduction of heat. As a result, the number of objects sharing the energy continues to grow. When the ball finally comes to rest at the lowest point of its travels, a huge number of atoms have gained roughly the same fraction of the energy originally available. The number of objects sharing this energy has increased dramatically, from one object (the ball) to a number on the order of 10^{26} (the number of atoms and molecules in the ball, the material of the planes, their supports, and everything else in the system). Consequently, everything in the system is slightly warmer. Presently (in Example 2) we shall show that everything warms up by about 0.002°C, not enough for anyone to notice.

When all the energy resides in the ball, the ball can perform a positive amount of work. But at the end of the experiment, when the energy has been distributed among 10^{26} atoms and molecules in the form of heat, no useful work can be extracted. The entropy of the system is a measure of the amount of energy *unavailable* for work. For this particular experiment the amount of energy transformed to heat keeps increasing as more and more subunits share the energy. Very crudely, the entropy tends to increase in proportion to the number of subunits sharing the energy.

The example of the rolling ball illustrates two important points. First, the transfer of energy in this process in *irreversible*. We cannot make the ball gain energy and roll back up the inclined planes by cooling the system. Second, the entropy S increases as more and more subunits come to share the available energy. But once all the atoms have roughly equal shares of the energy, the entropy can increase no further.

These two points are contained in the following statement of the entropy principle: *In an irreversible process, the total entropy of a system always increases until it reaches a maximum value.* After that, nothing else happens.

Example 1

If the experiment just described is carried out with a 0.50-kg ball which is initially lifted to a height of 1.0 m, estimate the final change in the average energy per atom in the system.

Initially the system has potential energy

$$E = mgh = (0.50 \text{ kg})(9.8 \text{ m/s}^2)(1.0 \text{ m}) = 4.9 \text{ J}.$$

The average energy per atom is roughly $E/N \approx 4.9 \times 10^{-26}$ J.

Example 2

By the end of the foregoing experiment, everything in the system is slightly warmer. How much warmer?

For this estimation, we treat the system as though it were an ideal gas consisting of $N = 10^{26}$ atoms, and we assume that all the initial potential energy E (calculated in Example 1 to be 4.9 J) is equally distributed as heat among the N atoms. Then the increase in energy of each atom is E/N. Now in Chapter 15 we found that the average kinetic energy $\overline{K}$ of an atom in an ideal gas at absolute temperature T is given by

$$\overline{K} = \tfrac{3}{2}kT, \tag{15.12}$$

so the change in absolute temperature ΔT caused by a change $\Delta \overline{K}$ in the kinetic energy is

$$\Delta T = \frac{\Delta \overline{K}}{3k/2} .$$

In this example, $\Delta \overline{K} = E/N$ so

$$\Delta T = \frac{E/N}{3k/2} .$$

Substituting the values $E/N = 4.9 \times 10^{-26}$ J and $k = 1.38 \times 10^{-23}$ J/K we find the temperature of the system rises by a meager 2.4×10^{-3} K, or about 0.002°C. (Most solid materials have more energy per atom than an ideal gas at the same temperature so the temperature change is actually smaller – perhaps half this size.)

Example 3

For the system in Example 1, what is the average speed of the center of mass of the ball at the end of the experiment?

The ball doesn't actually come to rest; it continues to jiggle around with the same fraction of the original energy of the system that each atom ends up with, namely E/N. The average speed $\overline{v}$ of any particle of mass M in the system at the end of the experiment satisfies

$$\tfrac{1}{2}M\overline{v}^2 = E/N.$$

Hence for the center of mass of the ball we have

$$\overline{v} = \sqrt{\frac{2E}{NM}} = \sqrt{\frac{2(4.9 \text{ J})}{(10^{26})(0.50 \text{ kg})}} = 4.4 \times 10^{-13} \frac{\text{m}}{\text{s}} .$$

Now that's splitting hairs!

Questions

1. Is there any change in the entropy of a ball that is strictly obeying the law of inertia?

2. Can you think of a process for which the entropy decreases?

17.2 ENGINES AND ENTROPY

There is a quantitative meaning for entropy that arises out of an analysis of reversible
processes. In studying the efficiency of a Carnot engine we learned that if an ideal engine
absorbs an amount of heat Q_i at an absolute temperature T_i and then discards heat Q_o at
absolute temperature T_o, the quantities are related by Eq. (16.15), $Q_o/Q_i = T_o/T_i$, or

$$\frac{Q_o}{T_o} = \frac{Q_i}{T_i}. \tag{17.1}$$

In other words, in a reversible process, the quantity Q/T is conserved, and therefore has
physical significance. The ratio

$$\Delta S = \frac{Q}{T} \tag{17.2}$$

is also called *the change in entropy* and has units joules per kelvin (J/K).

In a reversible process Q/T stays constant, but in an irreversible process the ratio
Q/T increases. Increasing Q/T has the same effect as friction increasing the entropy of a
system such as the rolling ball described earlier. To better understand the relation between
this meaning of entropy and the qualitative description in Sec. 17.1 we compare an ideal
Carnot engine with a nonideal engine.

At a high temperature T_i a Carnot engine has entropy change $\Delta S_i = Q_i/T_i$, and at
a low temperature T_o its entropy change is $\Delta S_o = Q_o/T_o$. For the ideal Carnot engine,
Eq. (17.1) states that

$$\Delta S_o = \Delta S_i \qquad \text{(ideal)}.$$

But a less efficient (nonideal) engine starting with the same Q_i and T_i produces less work
and therefore deposits more heat Q_o' at temperature T_o. Hence its entropy change at the
low temperature T_o is

$$\Delta S_o' = \frac{Q_o'}{T_o},$$

a quantity larger than the ratio $\Delta S_o = Q_o/T_o$. In other words,

$$\Delta S_o' > \Delta S_i \qquad \text{(nonideal)}.$$

So ideal engines keep Q/T constant, and nonideal engines increase it.

Now we can see the analogy to friction. The ideal Carnot engine doesn't create any
entropy. That's why a Carnot engine connected to a Carnot refrigerator could operate
forever, but do nothing besides run itself. It would be exactly analogous to a frictionless
ball in Galileo's experiment that rolls up and down the inclined planes forever. In the

real world there is always friction and rolling balls always come to rest. Likewise, in the real world there are no ideal Carnot engines and entropy is not conserved.

What property of the Carnot cycle makes the change in entropy zero? The key is found in the reversibility of the cycle. In order for a process to be reversible, friction must be eliminated and the process must be quasistatic. Most processes in nature, however, are irreversible. To prove that the entropy increases in irreversible processes, let's develop a theoretical method to handle such processes.

If we now adopt the convention that the heat extracted is negative and heat input is positive, so that $Q_o = -|Q_o|$ and $Q_i = |Q_i|$, then the relation for the Carnot engine,

$$\frac{|Q_i|}{T_i} = \frac{|Q_o|}{T_o}, \tag{17.1}$$

becomes

$$\frac{Q_i}{T_i} + \frac{Q_o}{T_o} = 0. \tag{17.3}$$

This was the starting point of Clausius's derivation of the entropy principle: *In a reversible Carnot cycle, the entropy of the system does not change.* What about *any* reversible engine?

Clausius realized that any reversible engine operating between two heat reservoirs can be approximated as accurately as desired by a series of alternating quasistatic adiabatic and isothermal paths. Here's how: first a number of adiabats are drawn, then adjacent adiabats are connected with two isotherms. The temperatures of the isotherms correspond to the temperatures at the top and bottom of the strip. This idea is illustrated in Fig. 17.1.

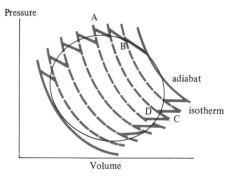

Figure 17.1 Any reversible cycle on a *PV* diagram can be approximated as well as desired by a series of alternating adiabats and isotherms.

The lines A, B, C, D, for example, constitute a Carnot cycle in Fig. 17.1. In other words, an arbitrary cycle can be approximated by a series of Carnot cycles. Since no heat is absorbed or rejected in the adiabatic parts of the cycle, all the isothermal parts may be paired off into Carnot cycles. And we know how to analyze Carnot cycles. The approximation can be made as close to the actual cycle as we wish by making the mesh of adiabatic and isothermal lines still finer. For each of the small Carnot cycles, we know that

$$\frac{\Delta Q_i}{T_i} + \frac{\Delta Q_o}{T_o} = 0. \tag{17.3}$$

Summing over all the isothermal paths along which heat ΔQ_j is either absorbed or rejected at temperature T_j, we obtain

$$\sum_j \frac{\Delta Q_j}{T_j} = 0. \tag{17.4}$$

By letting the heat ΔQ_j become arbitrarily small and the number of such subcycles become arbitrarily large, the sum approaches an integral:

$$\oint \frac{dQ}{T} = 0, \tag{17.5}$$

where the symbol $\oint$ is that for a line integral taken around a complete cycle. In other words, the change in entropy for *any* reversible system is zero. This implies that for a reversible cycle, the entropy difference between two states depends only on those states and not on the particular path connecting them – a result of Clausius first published in 1854.

To understand why the entropy is independent of the path, consider the system moving reversibly along path 1 from A to B as shown in Fig. 17.2, then back to A along a different reversible path. Equation (17.5) implies that

$$\int_A^B \frac{dQ}{T} + \int_B^A \frac{dQ}{T} = 0.$$
path 1   path 2

Because each path is reversible,

$$\int_B^A \frac{dQ}{T} = -\int_A^B \frac{dQ}{T},$$
path 2   path 2

so that

$$\int_A^B \frac{dQ}{T} = \int_A^B \frac{dQ}{T}.$$
path 1   path 2

Since this is true of an arbitrary cycle through points A and B, the change in entropy is independent of the path taken, provided that path is reversible. This result is surprising because the heat entering or leaving the substance does depend on the path, yet the entropy change doesn't.

From this analysis it seems reasonable to define the change in entropy between any two states A and B connected by a reversible process as

$$\Delta S = S_B - S_A = \int_A^B \frac{dQ}{T}. \tag{17.6}$$

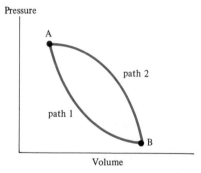

Figure 17.2 A reversible process from a state A to a state B made along two different paths in the *PV* diagram.

The change in entropy depends *only* on the initial and final equilibrium states of a system and not on the path joining them.

What is the change in entropy for an irreversible process? An irreversible process cannot be represented by a continuous path on a *PV* diagram because it doesn't move through a series of equilibrium states. However, we can join the ends of an irreversible process on a *PV* diagram by a reversible process, as shown in Fig. 17.3. Then we can approximate the entire cycle by a sequence of small Carnot cycles. Consider a portion of the cycle operating between T_{ij} and T_{oj} in which heat ΔQ_{ij} is extracted from the hot

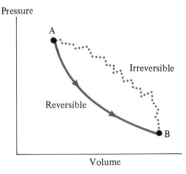

Figure 17.3 The final equilibrium states of an irreversible process can be connected by a reversible process between the same points.

reservoir and an amount ΔQ_{oj} is discarded at low temperature. The irreversible cycle is less efficient than a reversible cycle operating between the same two temperatures, so we have

$$1 - \frac{-\Delta Q_{oj}}{\Delta Q_{ij}} < 1 - \frac{T_{oj}}{T_{ij}} \,,$$

or by rearranging terms,

$$\frac{\Delta Q_{ij}}{T_{ij}} + \frac{\Delta Q_{oj}}{T_{oj}} < 0.$$

Summing over all such cycles, and passing to the limit we obtain

$$\oint \frac{dQ}{T} < 0, \tag{17.7}$$

for an irreversible cycle. What has happened is that the irreversible part of the cycle has generated more heat than it should have, and that extra heat has been extracted from the system to bring it back to its starting point. That's why the integral $\oint dQ/T$ is negative. The result (17.7) is known as the *Clausius inequality* and it allows us to prove that in any irreversible process the entropy of an isolated system always increases.

Here's how: Imagine that the irreversible process shown in Fig. 17.3 proceeds from the equilibrium state A to B. For the complete cycle, first going from A to B in an irreversible process, then returning from B to A by a reversible path, Clausius inequality implies

$$\underbrace{\int_A^B \frac{dQ_I}{T}}_{\text{irreversible}} + \underbrace{\int_B^A \frac{dQ_R}{T}}_{\text{reversible}} < 0,$$

or

$$\int_A^B \frac{dQ_I}{T} < -\int_B^A \frac{dQ_R}{T},$$

or

$$\int_A^B \frac{dQ_I}{T} < \int_A^B \frac{dQ_R}{T}.$$

From the definition of entropy for the reversible path, we know

$$\Delta S = S_B - S_A = \int_A^B \frac{dQ_R}{T}, \tag{17.6}$$

so

$$\int_A^B \frac{dQ_I}{T} < S_B - S_A.$$

This last inequality implies that as a result of the irreversible process taking the isolated system from equilibrium state A to B, the entropy of the final state is greater than the initial state. This is true even for an adiabatic process occurring in an isolated system. For an adiabatic process, no heat is exchanged, so the left-hand side of the equation is zero, implying that $S_B > S_A$. In other words, even for an isolated system, the entropy of the system increases after any irreversible process.

This result makes a profound statement about the behavior of the universe. Since the universe is an isolated system, the entropy of the universe must increase in time. As a consequence, the universal usefulness of energy decreases.

Example 4

An ideal gas is allowed to expand freely and adiabatically from a volume V_1 to a volume V_2. What is the change in entropy for this irreversible process? (A free expansion is one in which no work is done, as when a balloon bursts in a vacuum.)

Since no heat is exchanged with the surroundings, we might think that the change in entropy is zero. However, something irreversible does happen. We never observe the reverse process in which the air in a room freely contracts. On the other hand, since neither work nor heat is exchanged the temperature of the gas does remain constant, because the temperature of an ideal gas depends only on its energy, U. We compute the change in entropy by considering a *reversible* process connecting the two states: a quasistatic isothermal expansion from V_1 to V_2 at constant T. In this case, the change in entropy will be given by Q/T, where Q is the heat absorbed by the gas. Since the internal energy is constant in this process, the work done by the gas arises from heat absorbed by it. From Eq. (16.2) we calculate the work done by the gas as

$$Q = W = \int_{V_1}^{V_2} P \, dV = NkT \int_{V_1}^{V_2} \frac{dV}{V} = NkT \ln\left(\frac{V_2}{V_1}\right).$$

Therefore, the change in entropy for the irreversible process is

$$\Delta S = Q/T = Nk \ln(V_2/V_1).$$

Questions

3. Find the increase in the entropy of the system described in examples 1 and 2.

4. Construct an argument explaining why a gas by itself never freely contracts.

5. Suppose a vessel consists of two chambers, each of the same volume. In one chamber there is helium gas, and in the other there is argon gas at the same temperature and pressure. The partition between the chambers is suddenly removed. Does the entropy of the system increase? Explain your answer.

6. Calculate the change in entropy for the case of Question 5.

7. One mole of an ideal gas expands reversibly and isothermally from a volume of 15 L to a volume of 45 L. (a) Find the change in entropy of the gas. (b) Determine the change in entropy of the universe for this process.

8. A system absorbs 200 J from a heat bath at 300 K, does 50 J of work and rejects 150 J of heat at a temperature T, then returns to its initial state.

(a) Calculate the change in entropy of the system for a complete cycle.
(b) If the cycle is reversible, what is the temperature T?

9. A 2.0-kg block is dropped from a height of 3.0 m above the ground, strikes the ground, and remains at rest. If the block, air, and ground are all initially at a temperature of 300 K, what is the change in entropy for the universe in this process?

17.3 ENTROPY AND THE SECOND LAW OF THERMODYNAMICS

Clausius's indelible contribution to thermodynamics was his analysis of irreversible processes. He derived his inequality (17.7) from the observation that heat never flows by itself from low temperature to high temperature. For example, an ice cube placed in water never causes the water to boil. He concluded that a negative value of the change in entropy for an isolated system would correspond to heat flowing from cold to hot. Consequently, the change in entropy for any cycle must be zero if it is reversible, and positive if it is irreversible. Later he capsulized his ideas in the lines quoted at the beginning of this chapter: "The energy of the universe is constant. The entropy of the universe tends to a maximum."

Let's look into another simple, but crucial, example of how the entropy principle works when there is no friction involved. Suppose a body consists of two parts that are not at the same temperature. One body is at temperature T_i and the other is at T_o. Now imagine momentarily connecting the two parts by a piece of copper wire, as in Fig. 17.4.

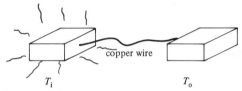

Figure 17.4 Heat conduction from a hot body to a cool body.

Since copper is a good conductor of heat, heat flows from the hot piece at T_i to the cold piece at T_o. The entropy change of the hot piece is

$$\Delta S_i = -\frac{Q}{T_i}. \tag{17.2}$$

No work is done, and all the heat flows into the cool body at T_o. The entropy of the cooler body is increased by

$$\Delta S_o = \frac{Q}{T_o}.$$

The process has thereby increased the entropy of the two pieces combined by

$$\Delta S = \Delta S_o + \Delta S_i = \frac{Q}{T_o} - \frac{Q}{T_i},$$

which is positive because $T_o < T_i$.

The flow of heat warmed the cooler piece and cooled the warmer piece. The two temperatures are now closer together, but so long as the temperatures are not equal, we can reconnect the copper wire and repeat the process. Each time we do so, the entropy of the combined body will increase slightly, and the two temperatures will become closer to each other. We can repeatedly reconnect the two bodies until the two temperatures become equal. Then no heat will flow, and the entropy will no longer increase. At that point, without changing the mass, volume, or energy of the system, we have made the entropy as large as it can possibly be. And at that point the combined body has reached thermal equilibrium. This situation is closely analogous to the ball at the end of Galileo's experiment, which is in a state of mechanical as well as thermal equilibrium.

Example 5

A copper wire connected between two large pieces of metal conducts 40 J of heat from one piece at 400 K to the other at 350 K. What is the change in entropy for this process?

The hot piece of metal loses an amount of energy $Q = 40$ J. Since this occurs at a constant temperature $T_i = 400$ K, the change in entropy of this body is

$$\Delta S_i = -Q/T_i = -(40 \text{ J})/(400 \text{ K}) = -0.10 \text{ J/K}.$$

Now the other metal piece gains 40 J at a temperature of 350 K, so its change in entropy is

$$\Delta S_o = Q/T_o = (40 \text{ J})/(350 \text{ K}) = 0.11 \text{ J/K}.$$

Therefore, the overall change in entropy is

$$\Delta S = \Delta S_o + \Delta S_i = 0.01 \text{ J/K}.$$

This simple and obvious argument has breathtaking consequences. As we've seen, the entropy of a body can decrease: in our example, the hot body kept losing entropy. However, the *combined entropy* of both pieces increased. If energy can flow in or out of a body, its entropy can increase or decrease. But if the total energy of a system is fixed, then the entropy can only increase until it reaches a maximum value. At that point the system has reached thermal equilibrium, and nothing more will happen.

What we've seen now is that a rolling ball eventually coming to rest at its lowest point and two pieces of matter placed in contact eventually reaching the same temperature are both consequences of the law of increase of entropy. In both cases, of course, energy is conserved, but in neither instance does simple conservation of energy help predict the result that the entropy principle gives us. That is why Clausius chose the name *entropy*. Entropy is something that tags along with energy, keeping track of how useful or well organized the energy is. As time goes on, entropy tends to increase, meaning that energy tends to more random, disorganized, useless forms.

The association of entropy with usefulness of energy is obvious in our analysis of Galileo's experiment, and we can also see the connection in the example we just discussed. Before the combined body reached thermal equilibrium, there were still two pieces at temperatures T_i and T_o; instead of connecting a copper wire between them, we could have connected an engine, and thereby extracted work. In other words, a copper wire is just an example of the most inefficient possible "engine," since it merely transforms heat from high to low temperature, doing no work at all. Once the process is finished, entropy has increased to the maximum value, everything is at the same temperature, and no work can be obtained. An increase in entropy means a loss of the ability to do work. Although energy is conserved, its usefulness is destroyed.

Also from examples, we see that the entropy of a system can change in two distinct ways: Another form of energy may turn into heat, as in Galileo's experiment, or heat may flow from higher temperature to lower, as in the case of the two bodies connected by a copper wire. But whatever the cause, the entropy of the universe, that is, of a system and its surroundings, always increases up to a maximum value. The energy of a system becomes less organized, less able to do useful work.

The entropy of the universe increases with time as the energy of the universe becomes less useful. Ordered mechanical energy eventually and inexorably is converted into the unordered, random motion of atoms. This conclusion is a consequence of the second law of thermodynamics and allows the second law to be cast into another form:

The entropy of the universe always increases toward a maximum.

As we shall see, this law makes a profound statement about the fate of the universe.

Questions

10. A 1200-kg car traveling at 80 km/h crashes into a brick wall. If the temperature of the air is 25°C, calculate the entropy change of the universe.

11. Suppose 300 J of heat is conducted from one reservoir at a temperature of 500 K to another reservoir at a temperature T. Calculate the change in entropy of the system if T equals (a) 100 K, (b) 200 K, (c) 400 K, (d) 490 K. What can you conclude about the change in entropy as the reservoirs are closer in temperature?

12. One mole of an ideal gas undergoes a free, adiabatic expansion from $V_1 = 12$ L, $T_1 = 400$ K to $V_2 = 24$ L, $T_2 = 400$ K. Afterward it is compressed isothermally back to its original state. (a) Compute the change in entropy of the universe in this process. (b) Show that the work made useless is given by $T \Delta S$.

13. Which process is more wasteful: (a) a 1.0-kg ball starting from a height of 1.5 m and rolling down and up inclines until it comes to rest at its lowest point, where the temperature of everything is 300 K; or (b) the conduction of 150 J of heat from a reservoir at 350 K to one at 300 K?

17.4   AN IMPLICATION OF THE ENTROPY PRINCIPLE

We know that entropy depends in some way on how many parts of a system have a share of the energy, and that when heat flows, an amount of entropy Q/T accompanies the flow. Entropy has one more very important quality: it is associated with the internal order or configuration of a body. For example, under otherwise identical conditions (total volume, temperature, etc.) we might imagine organizing a certain large number of molecules into either a liquid or a solid. The difference between these two states is that in the solid state, the molecules are arranged in a neat crystal lattice. If we form both states, the liquid will turn out to have more entropy then the solid.

We can see this clearly by noting what happens when an ice cube melts. At constant temperature (0°C) as heat flows into the ice from the warm drink it is immersed in, instead of warming up, the ice melts. An amount of heat Q flows in at temperature T, meaning that the entropy increases by Q/T. That entropy indicates the change in the internal structure of the ice from solid to liquid. In other words, the liquid has a higher entropy than the solid at the same temperature. So entropy is a measure not only of uselessness of energy, but also of disorder.

The question of why ice melts in the first place raises a paradox. If equilibrium is always a state of maximum entropy, and liquid is a state of higher entropy than solid, why do solids, like ice, ever exist in equilibrium at all? The crux of the problem is this: a body of a given *energy* reaches equilibrium when it has maximized its entropy. We don't know yet what constitutes the condition for equilibrium of a body of a given *temperature*. Let's now find the answer to that problem.

Consider something we'll call our system, which is divided into two parts. One part, which is very small compared to the system (but still large enough to be macroscopic) is called the sample. Everything else in the system is called the bath. The total energy of the system will always remain constant. We can therefore apply the entropy principle to the system: the entropy of the system tends to a maximum value. Moreover, the bath is always assumed to be in equilibrium; its entropy is always as large as it can be for the amount of energy it has. Being in equilibrium, the bath has a definite temperature T. Finally, there is the sample that is not necessarily in equilibrium. In fact the sample could be in any state at all. Heat is free to flow either way between the bath and the sample. However, the sample is so small compared to the bath that whatever heat flows in or out doesn't appreciably affect the temperature of the bath. Therefore we know that when the sample finally reaches equilibrium, its temperature will be T. But we don't know whether it will be a solid or a liquid, or more importantly what decides its fate when it reaches equilibrium.

When we put our sample in contact with the bath, everything that happens tends to increase the entropy of the entire system. Now suppose some heat Q flows into the sample and increases the energy of the sample by $\Delta E = Q$. That action also decreases the entropy of the bath by

$$\Delta S_b = -\frac{Q}{T} = -\frac{\Delta E}{T},$$

where the minus sign reflects the decrease. Since the sample is not in equilibrium, we don't know what happens to its entropy, but the event must increase the entropy of the system, or at least leave it constant – that's the second law of thermodynamics. If we call S_s the entropy of the system and S the entropy of the sample,

$$\Delta S_s = \Delta S_b + \Delta S \geq 0.$$

It follows that

$$\Delta S - \frac{\Delta E}{T} \geq 0.$$

In other words, the entropy of the sample must increase by at least as much as or more than the entropy of the bath decreased. Thus, as time goes on, the quantity ΔS –

$\Delta E/T$ continually increases, until it reaches its maximum value. These quantities, the entropy, energy, and final equilibrium temperature, are properties of the sample only. Therefore we can now ignore the bath and make a statement about the sample only.

Let's write the above result in a slightly different way by multiplying by T:

$$T \Delta S - \Delta E \geq 0,$$

where all the changes refer to those that occurred when heat flowed into the sample. Since the temperature doesn't change, we can write the left side as $\Delta(TS - E)$, and because it increases, its negative obeys

$$\Delta(E - TS) \leq 0. \tag{17.8}$$

In other words, whereas the entropy of the *system* increases to a maximum, the quantity $E - TS$ *of the sample* tends to a *minimum* when the sample is kept in a bath at constant temperature T.

The quantity $E - TS$ is called the Helmholtz free energy, or simply the free energy, and is written

$$F = E - TS. \tag{17.9}$$

The quantity F plays a role in thermodynamics somewhat analogous to that of potential energy in mechanics: the system is in a state of stable equilibrium whenever F is at a minimum. The crucial idea is that whenever a sample can lower its free energy, it does. Although that process might decrease the entropy of the sample, the action always increases the entropy of the universe.

Now we can understand why H_2O is sometimes solid and sometimes liquid. We need to compare the energies and entropies of the two states. And the state of the H_2O will be that for which the free energy is lower.

We already know that water, being a more disorganized state than ice, has more entropy at the same temperature. But what about its energy? The energy consists of the kinetic plus potential energies of all the atoms and molecules. At a given temperature, both states have the same kinetic energy. In a gas, liquid, or solid, each atom generally has $\frac{3}{2}kT$ of kinetic energy in equilibrium. The difference lies in the potential energy. In solids, unlike liquids, the molecules arrange themselves into stable configurations of the lowest possible potential energy. Of course, the arrangement that minimizes the potential energy of a bunch of molecules, fitting one together with another in just the right way, is the same for any small set of molecules, which is exactly why solids are made up of identical building blocks that repeat endlessly in a rigid lattice. The liquid, on the other hand, is disordered; it does not have all of its molecules in this optimal configuration. Therefore, the liquid has higher potential energy than the solid.

Now, the H_2O molecules must find the arrangement that will minimize the free energy. For this purpose, the solid has smaller energy but also smaller entropy. The liquid, on the other hand, has larger E and also larger S. What combination wins? The decision is made by the temperature. When T is very small, the negative term TS is small and unimportant, and the lower E state wins out: at low temperature H_2O is solid. But when the temperature is high, the term TS becomes more important, so the high S state wins and the H_2O is liquid. At some temperature in between, the winner switches from solid to liquid. That's the melting point. At that point the ice cube would rather melt completely than just warm up a little bit.

Questions

14. Develop an argument as to why different metals have different melting points.

15. In Chapter 14 we saw that nature tends to seek states of lowest possible potential energy. Does the behavior of ice at temperatures greater than the melting point contradict that idea? Explain.

16. Extend the discussion presented in the text to explain why H_2O is a gas and not a liquid above the vaporization temperature.

17. A tree takes unorganized molecules and organizes them into branches and leaves. Do you think living organisms violate the second law of thermodynamics? Explain your reasoning.

17.5 A FINAL WORD

Entropy is a measure not only of uselessness, but also of disorder. As time goes on, entropy increases. Energy is degraded to more useless forms, and matter into less ordered states. Of course, a given bit of matter might temporarily decrease its entropy, but that always means that something else nearby is increasing its entropy by at least as much, and usually more. Strictly speaking, the principle of increasing entropy only applies to systems of conserved total energy. The universe itself is such a system, so the universe appears to be headed for a state of thermal equilibrium, after which nothing else will happen. This cheerfully optimistic view of the future is generally referred to as the heat death of the universe.

There is another equally extravagant extrapolation of the entropy principle which turns around the sentence, ''As time goes on, entropy increases'' to read ''As entropy increases, time goes on.'' In other words, the increase in entropy is the very arrow of time; all other physical laws would work equally well if time ran backwards instead of forwards.

We have seen that all systems, including the universe, tend to evolve in an irreversible way. Despite the action of men and women, the inexorable law of nature is for energy to become less useful and the universe more disorganized. This tendency of the flow of time was pointed out by an eleventh-century Persian poet-mathematician, Omar Khayyám, who wrote

> The Moving Finger writes; and having writ,
> Moves on: not all thy Piety nor Wit
> Shall lure it back to cancel half a line,
> Nor all thy Tears Wash out a Word of it.

There have been many other statements of the second law of thermodynamics, but none so elegant.

CHAPTER 18

THE QUEST FOR LOW TEMPERATURES

We have seen that the gaseous and liquid states are only distant stages of the same condition of matter, and are capable of passing into one another by a process of continuous change. A problem of far greater difficulty yet remains to be solved, the possible continuity of the liquid and solid states of matter. The fine discovery made some years ago by James Thomson, of the influence of pressure on the temperature at which liquefaction occurs, and verified experimentally by Sir. W. Thomson, points, as it appears to me, to the direction this inquiry must take; and in the case at least of those bodies which expand in liquefying, and whose melting-points are raised by pressure, the transition may possibly be effected. But this must be a subject for future investigation; and for the present I will not venture to go beyond the conclusion I have already drawn from direct experiment, that the gaseous and liquid forms of matter may be transformed into one another by a series of continuous and unbroken changes.

Thomas Andrews, *Philosophical Transactions of 1869*

18.1 COOLING OFF

How do you make something colder? Making something hotter is easy. For example, if you need to warm yourself on a chilly night, you can build a fire with little or no technology. But to cool yourself on a hot day is quite another matter. The difference between heating and cooling is reflected in our history: we've had the use of fire ever since Prometheus let the secret slip long before the dawn of history. But the secret of making things colder – refrigeration – is barely older than the oldest living person.

Nonetheless, some techniques of cooling are somewhat old. The ancient Sumerians used porous jugs for cooling household water. As some water seeped out it evaporated and cooled the rest, a technique used today in bottled water dispensers. Later in history, slaves were sent to bring snow from the mountains in summer. And later still, Leonardo da Vinci invented a form of air conditioning using air blown over a block of ice that had been stored since winter.

As we turn to the past to see how low temperatures were reached, we find that the quest for low temperatures is closely tied to the history of understanding the basic states of matter – solid, liquid, and gas.

As it turns out, the Sumerians had the right idea: evaporation, that is, the change of a liquid into its gaseous form, cools the liquid. That's not the only method, but it is powerful and convenient for most purposes. However, to make something very cold – temperatures in the range 10–100 K (-263 to $-173°C$) – further knowledge of the behavior of matter is needed. With the quest for low temperature came the discovery that all elements can exist in each of the basic states under the right conditions of temperature and pressure. We turn now to find what conditions govern the states of matter.

Questions

1. Explain why you feel cool when you step out of the shower and into another room to dry yourself.

2. Explain why hot, humid weather is more uncomfortable than hot, dry weather.

3. When a frost is expected in Florida, why do citrus growers often spray their trees with water?

18.2 THE STATES OF MATTER

That all matter can exist in three basic states – solid, liquid, and gas – is a discovery less than a century old. But the idea of the states of matter can be traced back to ancient Greece. The elements of Aristotle's universe – earth, air, water, fire – also demonstrated the basic states of matter. The element earth reflected the solid appearance of matter, water generalized the liquid state, and fire and air represented the gaseous state. What was unknown to Aristotle and to others until the late nineteenth century was that any particular substance can exist in each of the three basic states. What conditions produce different forms of the same substance?

The physical state or phase of a substance depends on both its temperature and its pressure and can be illustrated visually by a *phase diagram* such as that in Fig. 18.1. For example, at low temperatures and high pressures a substance such as H_2O tends to be solid ice. At high temperatures and low pressures it tends to be gaseous, such as steam or water vapor, and somewhere in the intermediate range of temperatures and pressures it tends to be liquid.

The curve that separates the solid and liquid regions on the phase diagram is called the *melting curve*. The curve between the liquid and gaseous states is the *vapor-pressure curve*, and that between the solid and gaseous states is the *sublimation curve*. Only on these curves can the respective phases coexist in equilibrium, and if two phases are in

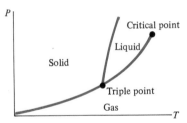

Figure 18.1 A typical pressure–temperature (phase) diagram illustrating regions in which a substance exists as either solid, liquid, or gas.

equilibrium, then specifying either the temperature or the pressure alone is sufficient to define the state. When a substance crosses one of these curves it is said to undergo a phase transition.

When the phase transition is from solid to liquid, the substance melts. The structure changes from an ordered, crystalline array of molecules and atoms to a less-ordered configuration. To achieve this change of phase a certain amount of heat must be added which goes entirely into changing the phase and does not raise the temperature. If heat Q is required to melt a substance of mass m, then the ratio Q/m is called L_f, the *latent heat of fusion*. Hence

$$Q = mL_f. \tag{18.1}$$

The latent heats of fusion for several substances are listed in Table 18.1.

As the phase diagram indicates, a substance can go directly from solid to gas without passing through the liquid state. This process, known as *sublimation*, also is accompanied by a latent heat. The most common example is the sublimation of dry ice: solid carbon dioxide sublimes at atmospheric pressure. That's why dry ice is never wet.

When a substance crosses the vapor-pressure curve, boiling occurs as liquid is changed into vapor. In this case the structure of the substance does not change radically as in the

Table 18.1 Phase Transitions and Their Temperatures and Latent Heats for Various Substances at 1 atm

Substance	Melting point (K)	L_v (kJ/kg)	Boiling point (K)	L_v (kJ/kg)
Hydrogen	14	59	20	452
Nitrogen	63	26	77	200
Oxygen	55	14	90	213
Ammonia	—	—	240	1369
Water	273	334	373	2257
Lead	600	25	2023	858

solid–liquid transition. Instead the molecules and atoms increase their separation distance. Just as in the case of melting, once a substance is at the boiling point, a certain amount of heat must be added to change the phase from liquid to vapor. If Q is the amount of heat necessary to vaporize a mass m of liquid at the boiling point, the ratio Q/m is called L_v, the *latent heat of vaporization*. Thus we have

$$Q = mL_v. \tag{18.2}$$

The boiling points and latent heat of vaporizations of a few substances are also listed in Table 18.1

Two points of special interest on the phase diagram are the triple point and the critical point, which are shown in Fig. 18.1. The triple point is a unique pressure and temperature where the three coexistence curves meet and at which all three phases can coexist. By international agreement, the triple point of water, occurring at 273.16 K and a pressure of 0.006 atm, is used as the fixed point in the absolute temperature scale. Triple-point temperatures and pressures for several other substances are listed in Table 18.2.

The critical point signals the end of the vapor-pressure curve and it also occurs for a particular pressure P_c and temperature T_c. At any higher pressure and temperature, there is no distinction between liquid and gas; there is only a homogeneous fluid phase.

A thorough study of the critical point of carbon dioxide was made by Thomas Andrews in the mid-nineteenth century. In an attempt to solve one of the great problems of his time – the liquefaction of gases – Andrews discovered what was to be a general property of all matter: above a critical temperature, no distinction between gas and liquid exists. This discovery, among others of the time, prompted physicists to investigate what distinguishes one phase from another. Table 18.3 lists the critical temperatures and pressures for several substances.

In Fig. 18.2, the same phases as in Fig. 18.1 are shown on a PV diagram. Here we find the solid state at high pressure and low volume and the gas at low pressure and large volume. In the shaded regions of the graph more than one phase can coexist. The triple point of the pressure–temperature diagram corresponds to the single, horizontal isotherm

Table 18.2 Triple-point Temperatures and Pressures for Several Gases

Substance	Temperature (K)	Pressure (10^5 Pa)
Hydrogen	13.84	0.070
Nitrogen	63.18	0.125
Oxygen	54.36	0.002
Ammonia	195.40	0.061
Carbon dioxide	216.55	5.17
Water	273.16	0.006

Table 18.3 Critical Temperatures and Pressures for
Several Gases

Substance	Temperature (K)	Pressure (10^5 Pa)
Hydrogen	33.3	13.0
Nitrogen	126.2	33.9
Oxygen	154.8	50.8
Ammonia	405.5	112.8
Carbon dioxide	304.2	73.9
Water	647.4	221.2

labeled T_p. On the other hand, the critical point is also a single point on this diagram that occurs at the maximum of the gas–liquid coexistence curve, indicating that there is a critical volume as well as a critical temperature and pressure.

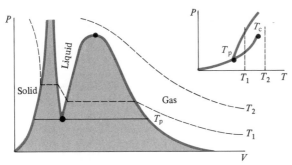

Figure 18.2 Phases of a given amount of a substance on a pressure–volume diagram.

The states of a substance in equilibrium depend on three quantities – pressure, temperature, and volume; the mathematical relationship between these variables can be extremely complicated. Nonetheless we can represent the state graphically by a surface on a pressure–volume–temperature plot. Such a representation is shown in Fig. 18.3. The region under the PV curve shown in Fig. 18.2 is a projection of this surface on a plane $T = $ const.

Questions

4. Can a liquid boil and freeze at the same time? Support your answer by reference to a phase diagram.

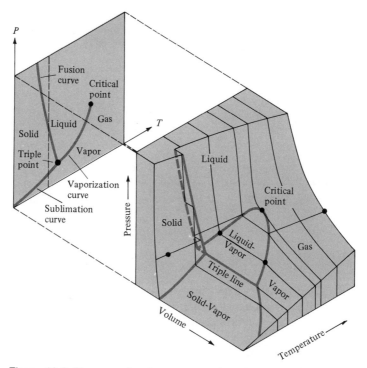

Figure 18.3 The state of a given amount of a substance can be represented by a surface in a plot of temperature, volume, and pressure.

5. For the phase diagram shown below, explain what happens as the substance changes states from (a) A to B, (b) C to D, (d) E to F, (e) G to H.

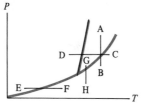

6. Can liquid become gas without passing through a phase transition? Can solid change into gas without passing through a phase transition? Explain your reasoning.

7. To raise the temperature of 1 kg of ice 1°C requires 2.0 kJ of heat; to raise the temperature of 1 kg of water 1°C requires 4.2 kJ of heat. Using this information, plot a graph of temperature versus heat added for 1 kg of ice taken from $-10°C$ to vapor at 100°C.

18.3 BEHAVIOR OF WATER

Let's consider the states of a substance without which life on earth would not exist – water. The phase diagram for H_2O is illustrated in Fig. 18.4. Imagine an ice cube inside

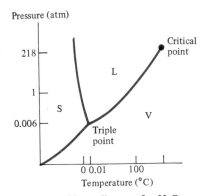

Figure 18.4 Phase diagram for H_2O.

a sealed container that is much larger than the cube and from which the air has been evacuated. Even at very low temperature, sufficient water vapor exists inside the container to create the pressure called for by the sublimation curve. As we heat the container, the ice cube warms up, and the gas pressure increases slightly as water molecules evaporate from the surface. As this process occurs, the ice remains on the sublimation curve with the solid and gas states in coexistence.

As we continue to heat the container, the temperature eventually reaches that of the triple point. At this point no further increase in temperature occurs until all the ice has melted. The added heat goes entirely into changing the phase from ice to water.

Once all the ice has melted, the water can be heated further, thereby creating more water vapor. Since the water remains on the vapor-pressure curve, just enough water evaporates with every increase in temperature as called for by the curve (the amount depends on the size of the container). If we continue to warm the container, it eventually reaches 100°C, at which point the vapor pressure is equal to one atmosphere. If you opened the container at this point, the pressure of water vapor inside and the air outside would just balance. If you were to open the container and continue to heat it, you would have the same situation as a pot of water on a kitchen stove. Whereas before, the pressure could change because the can was sealed, now the pressure is kept constant and equal to that of the atmosphere. So heating will make the water evaporate without any change in temperature – that's boiling.

On the other hand, if you kept the container sealed, the water would pass through 100°C without any apparent effect, and the pressure would continue to rise as well. This is what happens inside a pressure cooker. There is nothing special about the boiling point, 100°C for water, except the accidental fact that at that point the vapor pressure of water happens to be equal to atmospheric pressure on the planet Earth, at sea level, on an average day. Table 18.4 lists the boiling point for water at various pressures.

The phase diagram for water is slightly different from the typical diagram we saw earlier in that the melting curve bends backward. This peculiarity of water makes it possible to ski or ice skate because, at a temperature below 0°C, ice can be melted by simply applying pressure. The force of the skis or ice skates melts the ice at that particular place and the liquid lubricates the path. Otherwise, skiing on snow would be like skiing on concrete.

Table 18.4 Vapor Pressures of Water for
Various Temperatures

Temperature (°C)	Pressure (10^5 Pa)
0	0.006
20	0.023
40	0.073
60	0.199
80	0.473
100	1.01
120	1.99
140	3.61
160	6.17
180	10.0
200	15.5

Questions

8. In a car radiator the coolant water is under pressure. When the radiator cap is removed after the engine has been running for some time, why does the water boil and create a miniature geyser?

9. On hot, humid days why does a cold glass of lemonade ''sweat''? That is, why do water droplets form on the outside of the glass?

10. A real-estate developer proposes to create an artificial lake on the moon by filling a large crater with water. Would you invest in such a venture? Explain why or why not.

11. Can it be too cold to ice skate? Explain.

12. During extremely cold winters in places like Chicago, snow ''vanishes'' even though the temperature never rises above 0°C. What happens to the snow?

13. Use the phase diagram for water to explain how a snowball is formed.

14. In the ''mile high'' city of Denver (1 m = 1.6 km), a 3-min egg takes longer to cook. Explain why.

18.4 LIQUEFACTION OF GASES

At the turn of the nineteenth century, a handful of scientists may have understood the phases of water, but even fewer saw a generalization to the states of matter. The idea crystallized through research into the liquefaction of gases. Earlier a Dutch scientist, Martin Van Marum, had accidentally liquefied ammonia gas under pressure while using it to test Boyle's law; however, he didn't grasp the significance of this first liquefaction. Until 1823, the liquefaction of gases remained tenuous.

In that year, at the Royal Institution of London, a young chemist made an explosive discovery. While heating a compound in a sealed glass tube, Michael Faraday produced an oily-looking substance. When he filed the tube open to investigate the substance, the tube promptly exploded. But Faraday realized that he had liquefied chlorine gas.

This great intuitive genius of the nineteenth century would later become famous for his researches in electromagnetism, but he revealed his talents in this early episode. Unlike Van Marum, Faraday understood perfectly the significance of his discovery, and immediately undertook a quest to liquefy other gases. Using an inverted U tube of glass, in one leg of which he could heat reagents to evolve the gas in question, while the other leg, under pressure, could be cooled, if necessary, to form the condenser, he succeeded in liquefying many gases. With that success, Faraday reached the plausible conclusion that all forms of matter could exist in each of the fundamental states, given the right conditions of temperature and pressure.

The next great advance in understanding the states of matter came in 1835 when the French scientist Thilorier managed to solidify carbon dioxide (CO_2). The phase diagram of carbon dioxide is shown in Fig. 18.5. Although the phase diagram appears like the ones we discussed earlier, peculiar things happen because of the relation between the characteristic temperatures and pressures of CO_2 and the temperature and presssure at which human life on earth flourishes.

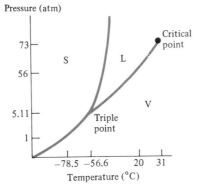

Figure 18.5 Phase diagram for carbon dioxide, CO_2.

At atmospheric pressure and temperature, carbon dioxide is commonly a gas. However, if it is compressed without allowing its temperature to rise above room temperature, carbon dioxide will condense into a liquid. On the phase diagram this compression corresponds to moving from room temperature and atmospheric pressure directly up the line to the coexistence curve, where liquid CO_2 coexists with gaseous CO_2 at high pressure.

Liquefying gases is what Faraday had already accomplished. The advance made by Thilorier comes when, starting from these conditions, the pressure is reduced too quickly for heat from the surroundings to keep the temperature constant. We can imagine opening a valve on the container of liquid plus gaseous CO_2 that leads directly into a room at atmospheric pressure. And whoosh, a high-pressure jet of CO_2 blows forth into the room.

Inside the container liquid CO_2 still coexists with gaseous CO_2, but now at reduced pressure. Since the two phases coexist, the carbon dioxide must be on the coexistence curve of the phase diagram. And to be on that curve at reduced pressure means that it must also be at reduced temperature. In other words, the liquid and gas in the container are *cooled* by evaporation. The process works for CO_2, or for that matter any other substance, just as it does for water.

We can summarize this observation as follows: To stay on the coexistence curve, temperature must decrease when pressure is allowed to decrease, just as the pressure must rise when a substance is heated. As gas escapes from a container, liquid evaporates in order to try to replace the gas. The result of the evaporation is cooling.

We can describe the same process in a more subtle way. Gas is a state of higher entropy than liquid. The entropy is higher not because the gas has a different internal structure from the liquid but because the gas is less dense. Each molecule of the gas has more space in which to move around, and that in itself is a form of disorder. In any case, when liquid evaporates to become a gas, the entropy per molecule increases. If this occurs at temperature T, then there is a corresponding transfer of heat equal to the heat of vaporization,

$$Q = mL_v \tag{18.2}$$

and the change in entropy in going from liquid to gas is

$$\Delta S = \frac{Q}{T}. \tag{17.2}$$

Example 1

Compare the change in entropy for 0.1 kg of ice melting at the triple point with that of the same mass of liquid H_2O vaporizing at the boiling point.

As Table 18.1 indicates, when ice melts it releases 334 kJ/kg of heat. Since $\Delta S = Q/T$ the change in entropy for this process is

$$\Delta S = (0.1 \text{ kg})(334 \times 10^3 \text{ J/kg})/(273 \text{ K}) = 122 \text{ J/K}.$$

On the other hand, vaporization releases 2257 kJ/kg and the associated change in entropy is

$$\Delta S = (0.1 \text{ kg})(2257 \times 10^3 \text{ J/kg})/(373 \text{ K}) = 6050 \text{ J/K}.$$

The entropy change is much greater in the liquid–vapor transition than in the solid–liquid transition. This comparison indicates that for water the increase in distance between molecules accompanying the liquid–vapor transition is more effective in increasing the entropy than the change of form from crystalline array to liquid, which occurs at nearly constant density.

There is yet another way to describe this process. At any given temperature, the CO_2 molecules have the same kinetic energy, on the average, no matter whether they are in the liquid or the gas. According to Eq. (15.12) this energy is equal to $\frac{3}{2} kT$ for each atom making up each molecule. The molecules of liquid, however, have lower potential energy than those of the gas; it is precisely this potential energy that binds the molecules together, causing the liquid to form. Condensation is a consequence of the electric force of attraction between the molecules when they are close together in the liquid state.

Therefore, when molecules evaporate from liquid to gas, they are increasing their potential energy by escaping the attraction of their neighbors. If the gas plus liquid are isolated from contact with the rest of the world, so that overall energy must be conserved, the increase in potential energy must be balanced by a decrease in kinetic energy. But we've already seen that kinetic energy is equivalent to temperature, and so the temperature falls. On a molecular level this is why evaporation causes cooling of the gas. In the next section we'll return to this explanation.

Escape from the attractive forces of other molecules explains why the gas is cooled, but what cools the liquid? Inside the liquid, the average kinetic energy per molecule is $\frac{3}{2} kT$, but in fact, some molecules have more kinetic energy than that, and some less. The fastest-moving molecules are the ones most likely to escape into the gas. That leaves behind the slower, cooler molecules. That's why the liquid cools.

Of course, all of these different versions are not opposing views, but rather alternative descriptions of a single coherent understanding of the states of matter. The change in potential energy and the latent heat of fusion are really the same quantity viewed on different levels.

Any liquid, not just CO_2, can coexist with its own vapor. But carbon dioxide differs from most other substances in one peculiar respect: its vapor pressure at the triple point is above 1 atm. When Thilorier opened his valve and allowed liquid CO_2 to cool by evaporation, as the pressure dropped, the mixture froze before it reached atmospheric pressure. CO_2 sublimes directly from the solid at a pressure of 1 atm. That's why dry ice never becomes a liquid under ordinary circumstances.

Thilorier's discovery led him to a method of producing even lower temperatures. By mixing ether with carbon dioxide and pumping away the vapor to reduce the pressure below one atmosphere, Thilorier produced a temperature of $-110°C$. Armed with this new advance in technology, Michael Faraday returned to the problem of liquefaction in 1844. Within a year he succeeded in liquefying all the known "permanent gases" except those he really wanted. Three elements – oxygen, nitrogen, and hydrogen – and three compounds – carbon monoxide, nitric oxide, and methane – failed to yield to his techniques.

The quest to liquefy these permanent gases now took on the character of an open scientific challenge. To succeed where the great Faraday had failed was no small matter. The story of that quest is filled with twists and ironies. For example, the liquefaction of oxygen was accomplished in 1877 by Louis Paul Cailletet and Raoul Pierre Pictet, who achieved success within a few days of each other 33 years after Faraday's failure set the stage. The basic process was a kind of cascade in which one bath was used to liquefy another bath under pressure. This second bath in turn was further cooled by released pressure, and so on. In the final step, the pressurized oxygen gas (Pictet achieved a pressure of 250 atm at a temperature of $-130°C$) did not liquefy. The critical point of oxygen hadn't been reached. However, when the pressure was subsequently released, both Pictet and Cailletet found that a cloud of partially liquefied oxygen was produced. The liquefaction had occurred because expansion of a gas by itself produces cooling.

Questions

15. Explain why the mist coming from an aerosol spray can is cool to the touch.

16. Formulate a simple theory of fog formation by citing the conditions that induce fog and the physical process that occurs.

17. Using the data in Table 18.1, calculate the entropy change associated with (a) lead melting, (b) ice melting, (c) nitrogen vaporizing, (d) carbon dioxide vaporizing.

18. Offer an explanation for the comparative difference in your answers to parts (c) and (d) in Question 17.

18.5 THE JOULE–THOMSON EFFECT

Beginning in 1845 and culminating in 1862, Joule, and later Joule and Thomson (the same Thomson who later became Lord Kelvin), conducted a series of experiments to test the interchangeability of work and heat.

Their investigations showed that under certain circumstances the expansion of a gas into a chamber held at a constant pressure would produce cooling. The physical reason turns out to be the same one that causes a liquid to cool when it evaporates.

The cooling effect they discovered, known as the Joule–Thomson effect, provided experimental evidence for the existence of intermolecular forces between gas molecules. It is these forces that account for the structural changes in a substance as it passes through different phases as well as for the deviations of real gases from the ideal-gas law (which were discussed in Chapter 15). Further research revealed that intermolecular forces consist of two different interactions between pairs of molecules. One force is a weak, relatively long-range attractive force known as the *Van der Waals force*. The second force is a strong, short-range repulsive force which prevents molecules from coming too close together. The regions in which each force dominates are indicated in Fig. 18.6 where the potential energy of a molecule is plotted as a function of the distance between two interacting molecules. These forces are electrical in origin and depend on the structure of the particular molecule.

In their experiments, Joule and Thomson kept a gas at constant pressure and allowed it to flow continuously through a porous plug of tightly packed cotton, as shown in Fig. 18.7. As the gas flows from the left chamber at pressure P_1 into the right chamber at lower pressure P_2, the pistons move, changing the volume, so as to keep the pressures constant. The porous plug, or nozzle, merely supports the pressure difference and the entire apparatus is isolated so that the process is adiabatic.

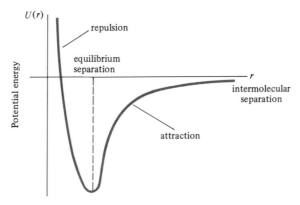

Figure 18.6 Potential-energy diagram for a molecule in a real gas. The force is repulsive at small r where $U(r)$ slopes down, and attractive at larger r where $U(r)$ slopes up.

Let's discuss why the Joule–Thomson effect indicates the existence of intermolecular forces. As the gas is pushed from the left chamber of Fig. 18.7, decreasing its volume from V_1 to zero, the work done *by* the gas is simply $W_1 = -P_1V_1$ because the pressure is held constant. On the other side of the nozzle the piston does an amount of work $W_2 = P_2V_2$ as the volume expands from zero to V_2. According to the first law of thermodynamics,

$$Q = W + \Delta U, \tag{16.3}$$

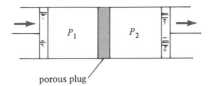

Figure 18.7 Schematic of the Joule–Thomson porous plug apparatus.

and since the process is adiabatic, $Q = 0$. Therefore, the change in internal energy of the gas is equal to the negative of the net work done by the gas:

$$\Delta U = U_2 - U_1 = -W = -(-P_1V_1 + P_2V_2) = P_1V_1 - P_2V_2. \tag{18.3}$$

From this we see that there is a conserved quantity for the process:

$$U_1 + P_1V_1 = U_2 + P_2V_2. \tag{18.4}$$

Now if we assume an ideal gas, then both the internal energy and PV are functions only of temperature:

$$U = qNkT, \tag{15.13}$$

$$PV = NkT. \tag{15.9}$$

Consequently, the temperature of the gas should remain constant during the process. But the observation is that the Joule–Thomson process produces a cooling of the gas. This departure from ideal gas behavior is attributed to intermolecular forces.

The intermolecular forces contribute potential energy, so the internal energy of the gas is not simply due to the kinetic energy of individual molecules. Let u denote the potential-energy contribution due to intermolecular forces. Then the internal energy of the gas is given by

$$U = qNkT + u. \tag{18.5}$$

According to Eq. (18.3), we should have

$$\Delta U = qNk\,\Delta T + \Delta u = P_1V_1 - P_2V_2. \tag{18.6}$$

When the gas cools, the right-hand side of this last result is found to be positive, that is,

$$P_1V_1 - P_2V_2 > 0,$$

which in turn implies

$$\Delta u > -qNk\,\Delta T. \tag{18.7}$$

Since ΔT is negative, this means Δu is positive.

As the gas expands, the potential energy of the molecules increases, because the molecules are pulling away from each other's attractive potentials. This is exactly what happens when a liquid evaporates. The increased potential energy is balanced by a decreased kinetic energy, or slower atoms, so the gas is cooler.

If the initial temperature and pressure are high enough, the repulsive part of the intermolecular force may be dominant. In this case, Δu is negative, so the gas should actually warm. Warming of the gas is not only expected, but also observed under these conditions.

Of course, intermolecular forces are responsible for change of phase as well. The difference between a gas and a liquid is that the molecules of a gas on the average are too far apart to have much influence on each other, while in a liquid, they are close enough for the Van der Waals forces between them to cause them to attract and condense. Expressed in energy terms, liquid molecules have more negative potential energy. This potential energy is the latent heat.

The Joule–Thomson effect creates a decrease in temperature that is found to be nearly proportional to the difference between the pressures of the two chambers and provides a practical method for liquefying gases. In 1895 the German chemist Karl von Linde adapted the process for the large-scale liquefaction of gases. His process did away with the cascade effect and makes use only of the gas to be liquefied, for example, nitrogen. Cooled, compressed nitrogen passes through an inside tube and through a Joule–Thomson expansion valve and back through an outer tube. The gas that is not liquefied is returned to the compressor by means of a heat exchanger, where the incoming compressed gas is cooled to a lower temperature by the outgoing cold product.

Questions

19. Using the potential-energy curve in Fig. 18.6, indicate how the potential energy decreases as molecules of a gas come closer together.

20. Use a potential-energy curve like the one in Fig. 18.6 to explain why a gas cools when it expands.

21. In the Joule–Thomson process, determine whether net work is done on the gas or by the gas, if

(a) the gas is ideal,

(b) cooling occurs,

(c) warming occurs.

What is the connection between net work and intermolecular forces?

18.6 A FINAL WORD

After nitrogen and oxygen had been liquefied, only one permanent gas remained – hydrogen. Scientists knew that if it could be liquefied at all, doing so would be a great achievement. And so two men set out in a race to liquefy hydrogen. They were Sir James Dewar, who was Faraday's successor as Professor at the Royal Institution in London, and Heike Kammerlingh-Onnes at the University of Leiden, in the Netherlands.

Each of these two great scientists spent decades pursuing the quest. In the end, Sir James Dewar won the race. He gained the lead by inventing the double-walled evacuated flask to retain cold liquids. It is known in scientific circles as the Dewar flask and in popular terms as a thermos bottle. However, in a final twist of irony, the rules of the race changed just before the finish line.

In 1895, just before Dewar succeeded in producing liquid hydrogen, his countryman, William Ramsay, isolated terrestrial helium, an element previously known only to exist in the sun. Kammerlingh-Onnes, knowing that he had lost the race for liquid hydrogen, immediately abandoned the pursuit and laid plans for an assault upon helium. That siege lasted another thirteen years. To undertake it made sense only because of the work of one of his countrymen, Johannes Diderik van der Waals.

In the 1870s, Van der Waals developed a theory of the fluid state that took into account the fact that a compressed gas, especially above its critical point, could have potential energy. Such a gas, he argued, would not obey the ideal-gas law, $PV = NkT$. [See Eq. (15.10).] In fact, by careful observation of the deviations from that law, he concluded that it would be possible to deduce the strength of the forces between the molecules. But those forces were responsible for the Joule–Thomson cooling, and in fact, liquefaction itself. Thus, by careful measurements of the properties of helium gas, even well above its critical temperature, it would be possible to predict both the conditions under which the Joule–Thomson cooling would be effective, and the temperature and pressure that had to be reached in order to liquefy it.

The results of this analysis, which Kammerlingh-Onnes carried out, gave him three indispensable pieces of information. First, helium did behave like other gases; in other words, it could be liquefied. Although the effort was daunting, it would not be wasted. Second, when liquefied, the temperature of helium would be much lower than that of liquid hydrogen. So if he succeeded, he and not Dewar would be the winner of the quest in history's eyes. And finally, by giving him in advance the values of temperature and pressure he would have to attain, the analysis made it possible to undertake the job in a rational and orderly way.

Kammerlingh-Onnes and his research group succeeded in liquefying helium on 10 July 1908. The critical temperature turned out to be, as predicted, only 5 K – just five degrees above absolute zero. The last "permanent" gas had been liquefied. For his research into the properties of matter at extremely low temperature, Kammerlingh-Onnes was awarded the Nobel prize in physics in 1913.

Today, liquid helium is made in huge commercial plants for use in scientific laboratories and industry as is liquid nitrogen and oxygen. The cost is low: a liter of liquid nitrogen costs about as much as a liter of milk, and a liter of liquid helium costs about as much as a liter of vodka.

CHAPTER 19

THE CONSERVATION OF MOMENTUM

And with respect to the general cause, it seems manifest to me that it is none other than God himself, who, in the beginning, created matter along with motion and rest, and now by his ordinary concourse alone preserves in the whole the same amount of motion and rest that he placed in it. For although motion is nothing in the matter moved but its mode, it has yet a certain and determinate quantity, which we easily see may remain always the same in the whole universe, although it changes in each of the parts of it.

René Descartes, *Principles of Philosophy* (1644)

19.1 THE UNIVERSE AS A MACHINE

In the seventeenth century science assumed its modern form and the scientific spirit infected Europe. It was then that Aristotle's view of nature was rejected and Galileo's great book of the universe was adopted. The new science was nourished by an optimism that mankind could discover the laws of nature.

One of the most significant and influential figures in seventeenth-century natural philosophy was René Descartes. Early in his life, Descartes rebelled against the traditions in which he had been thoroughly educated. He sought new foundations for knowledge, foundations which could underpin confidence in our understanding of nature. Convinced of the indubitable logic of mathematics, Descartes chose to identify mathematics with physics.

Descartes is credited with having been the first person to state the law of inertia correctly. Unlike Galileo, who claimed that an inertial motion was horizontal (always parallel with the surface of the earth), Descartes realized that in the absence of forces, a body would continue to move in a straight line in any direction.

Descartes admired and at the same time criticized Galileo's efforts at resolving what he considered to be mundane questions, such as free fall and the motion of the pendulum. Descartes had greater designs; he wanted to create a system of the world as a whole. He saw the universe as a machine, set into motion by the Creator, who created matter and the laws of motion. The mechanical universe of Descartes is no different from artifacts produced by skillful artisans; it is only a matter of scale. Like clockwork, the universe inexorably follows purely mechanical laws.

His idea of a mechanical universe, however, presented a problem. What keeps the machine going? Like clocks, machines run down and need to be wound up from time to time. Descartes had to confront the question of whether the universe needs periodic intervention from the Creator to keep it going.

To resolve this dilemma, Descartes invented a new law. Writing hypothetically, so as not to arouse the ire of the church, which had silenced Galileo only years earlier, Descartes proposed that the total quantity of motion in the universe is constant, preserved by God who created it in the beginning and constantly sustains it. The "quantity of motion" is something we now call momentum $\mathbf{p}$, a vector quantity which is the product of the mass of an object and its velocity:

$$\mathbf{p} = m\mathbf{v}. \tag{19.1}$$

Mathematically, we write Descartes's law as

$$\sum_{i=1}^{n} \mathbf{p}_i = \mathbf{p}_1 + \mathbf{p}_2 + \mathbf{p}_3 + \cdots + \mathbf{p}_n = \text{const},$$

where $\mathbf{p}_i$ is the momentum of the ith body, and the summation of the momenta of all the bodies (here there are n of them) is the total momentum. Since the total momentum is constant, if one object in the mechanical universe slows down, another must speed up.

In Chapters 13 and 14 we encountered our first example of a conservation law: conservation of energy. Descartes's law of the conservation of "quantity of motion" is another such important law distinct from conservation of energy. Because energy changes into heat, causing machines to run down, Descartes had to postulate a new law to explain the mechanics of the universe and to prevent his clockwork from eventually stopping. To find the origin of Descartes's ad hoc law, we turn once again to the laws that irrevocably altered the way mankind viewed the universe – the laws of Sir Isaac Newton.

19.2 NEWTON'S LAWS IN RETROSPECT

In Chapter 6 we introduced Newton's three laws, upon which mechanics is based, yet in subsequent chapters we haven't actually used all three laws. We have only used one

RENÉ DESCARTES.

Figure 19.1 René Descartes. (Courtesy of the Archives, California
Institute of Technology.)

law – the second law. The second law, $\mathbf{F} = m\mathbf{a}$, governs the motion of objects in Newton's
clockwork universe. In fact, we've only used this one law because for most purposes the
other two laws are contained in it.

To recount briefly, Newton's three laws are

1. The law of inertia.
2. The law, $\sum \mathbf{F} = d(m\mathbf{v})/dt$.
3. The law of action and reaction.

Suppose we have a single particle of mass m and momentum $m\mathbf{v}$. According to the
second law,

$$\mathbf{F} = \frac{d(m\mathbf{v})}{dt} = m\frac{d\mathbf{v}}{dt}.$$

Now if there are *no forces* acting on this particle, then this expression is equal to zero:

$$m\frac{d\mathbf{v}}{dt} = \mathbf{0}.$$

We know how to solve this equation. If the derivative of something is zero, then that
something must be constant. That something here is velocity. The velocity $\mathbf{v}$ must be
constant if no force acts on the particle, and constant velocity means constant speed and
constant direction. But this is the law of inertia: a body acted upon by no forces will
remain moving along a straight line with constant speed. Therefore the first law is
contained in the second law.

What about the third law? Let's imagine an object either at rest or moving at constant
velocity. In this case, no external forces act on the object. Imagine further that you take
a closer look and notice that the object is in fact a compound body, composed of two

particles, which aren't necessarily touching each other, as illustrated in Fig. 19.2a. Now we have two particles, which we'll unimaginatively label particles 1 and 2.

By the second law, zero acceleration implies that no net force acts on the two particles. Whatever forces do act between particles 1 and 2, they must add up to zero, as shown in Fig. 19.2b:

$$\mathbf{F}_{12} + \mathbf{F}_{21} = \mathbf{0},$$

$$\mathbf{F}_{12} = -\mathbf{F}_{21}.$$

If the internal forces did not add up to zero then (blurring our vision again) the compound body would have a net acceleration. But by our assumption it does not. Therefore, if particle 1 applies a force on particle 2, particle 2 must apply *an equal and opposite force* on particle 1. This is precisely what the third law states. At least in the case of two bodies moving without external force, Newton's third law is contained within the second.

Newton's laws are most easily applied to point particles, but by imagining any compound body to be made up of parts, we can extend the laws to describe the motion of real physical objects. Every object in nature is really a compound body composed of smaller parts, ultimately made up of atoms. Our description of a single body must apply to descriptions of compound bodies as well.

(a)   (b)

Figure 19.2 (a) A compound body composed of two separate particles. (b) Internal action–reaction forces on individual particles.

As briefly mentioned in Chapter 14, the point of a body where we can think of all the mass of the object as being concentrated is called its *center of mass*. This point is often identical with the geometric center of an object. Figure 19.3 illustrates the center of mass of a few objects.

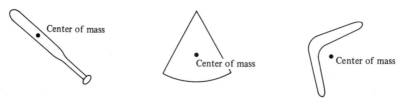

Figure 19.3 The center of mass of a few objects.

When we talk about the velocity and acceleration of a compound body, we are referring to that of its center of mass. So if, for example, no net force acts on a compound system at rest, the center of mass of the system remains at rest even though individual parts of the system may move.

Example 1

We all know that you can't pick yourself up by the seat of your pants, but perhaps you could get a wagon to move by hanging a huge magnet in front of it as shown. Will it move?

We know that in order for a system to accelerate, there must be a net *external* force acting on it. Here the system consists of you, the wagon, and the magnet. Is there a net force acting on the system from outside it? No. You might think that the magnet exerts a force on the wagon, which it does, *but* the wagon, by the third law, exerts an equal and opposite force on the magnet. The result is that *center of mass* of the system remains at rest, and you, the wagon, and the magnet could never move by this method of propulsion.

Questions

1. A spring between two masses resting on a frictionless horizontal air table is compressed and tied. If the string is cut, what can you say about the motion of the center of mass of the system?

2. A mouse is at one end of a box resting on a frictionless horizontal surface. If the mouse scrambles to the other end of the box, in which direction does (a) the box move, (b) the center of mass of the box plus mouse system move?

3. Suppose in a nightmare you find yourself locked in a light cage on rollers on the edge of a rapidly eroding cliff. Assuming that no net external forces act on the system consisting of you and the cage, is there anything you could do to move the cage away from the edge?

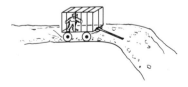

4. A projectile explodes while in flight. Fragments are blown in all directions as shown in the sketch. What can you say about the motion of the center of mass of the system after the explosion?

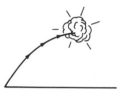

19.3 THE LAW OF CONSERVATION OF MOMENTUM

Now that we understand that Newton's second law is the guiding principle for the workings of the universe, let's use this law to deduce the law of conservation of momentum. Imagine a collection of objects that are moving through space and are not necessarily touching one another, as shown in Fig. 19.4. Suppose that when viewed from afar, the whole bunch of particles is observed not to be accelerating. We can conclude that no net external forces act on these objects. However, there may be internal forces acting between the various objects. For example, gravity acts between each pair of bodies; electric or magnetic forces as well may act between the objects, but all these forces are internal. We know from the third law (or now the second, if you like) that the sum of all the internal forces must be zero. Why? The collection of bodies is not accelerating; the center of mass of the system moves with a constant velocity.

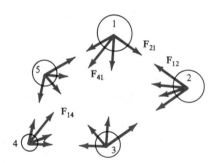

Figure 19.4 Internal forces acting on a collection of objects.

Let's call the net force acting on any one body $\mathbf{F}_i$, where the index i can be 1, 2, 3, whatever the number of the body is. The condition placed on these internal forces by the third law is

$$\sum_i \mathbf{F}_i = \mathbf{0},$$

where the sum runs over all the bodies comprising the system. Now by the second law, the force on *each* body is equal to the rate of change of *its* momentum, that is,

$$F_i = \frac{d}{dt}(m_i v_i).$$

Because the sum of all such internal forces is zero, we have

$$\sum \frac{d}{dt}(m_i v_i) = \frac{d}{dt}(m_1 v_1) + \frac{d}{dt}(m_2 v_2) + \cdots + \frac{d}{dt}(m_n v_n) = 0.$$

Since the sum of derivatives is the derivative of the sum, we can write

$$\frac{d}{dt}(m_1 v_1 + m_2 v_2 + m_3 v_3 + \cdots + m_n v_n) = 0.$$

We know how to solve this equation: if the derivative of something is zero, that something must be constant. Thus the momentum of the system, identified as the sum of the individual momenta of each of the bodies, is equal to a constant:

$$m_1 v_1 + m_2 v_2 + m_3 v_3 + \cdots + m_n v_n = \text{const.} \tag{19.2}$$

Although the individual momenta of the bodies may change, the *total* momentum of the system remains constant. Conservation of momentum guarantees that if you know the total momentum of a system at one instant, then you know it all times because that quantity never changes.

Since momentum is a vector quantity, Eq. (19.2) implies three equations when written out in components:

$$m_1 v_{1x} + m_2 v_{2x} + m_3 v_{3x} + \cdots + m_n v_{nx} = \text{const}, \tag{19.3a}$$

$$m_1 v_{1y} + m_2 v_{2y} + m_3 v_{3y} + \cdots + m_n v_{ny} = \text{const}, \tag{19.3b}$$

$$m_1 v_{1z} + m_2 v_{2z} + m_3 v_{3z} + \cdots + m_n v_{nz} = \text{const.} \tag{19.3c}$$

Example 2

Imagine a howitzer (mass $M = 300$ kg) floating somewhere out in space. As called for in the script of a science-fiction western, it fires a shell (mass $m = 20$ kg) at 150 m/s.
    (a)   With what velocity does the howitzer recoil?
    (b)   How much chemical energy is transformed into kinetic energy?

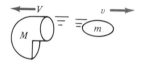

    (a)   Initially, the momentum of the system is zero, because both the howitzer and shell are at rest. After the explosion, the shell moves off at speed $v = 150$ m/s whereas the howitzer recoils with a speed we'll label V in the opposite direction. The sum of the two momenta must remain zero. Taking the direction of the velocity of the shell as the positive direction, we have by conservation of momentum, Eq. (19.3a),

$$mv - MV = 0.$$

Solving for V, we find

$$V = (m/M)v = (20/300)(150 \text{ m/s}) = 10 \text{ m/s}.$$

(b)   We know that energy is always conserved and that here chemical energy is transformed into kinetic energy of the shell and howitzer. Therefore, this amount of energy is

$$K = \tfrac{1}{2}MV^2 + \tfrac{1}{2}mv^2,$$

which turns out to be 2.4×10^5 J.

Note that we first used conservation of momentum rather than conservation of energy to find the recoil speed of the howitzer. The reason for this is that initially we did not know the exact amount of chemical energy transformed into kinetic energy.

Example 3

A firecracker at rest explodes into three fragments. Two fragments, each having the same mass, fly off perpendicular to each other with the same speed $V = 8$ m/s. The third fragment has half the mass of either of the other fragments. Find the velocity $\mathbf{v}$ of the third fragment.

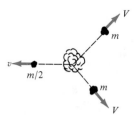

Before the explosion, the firecracker was at rest and had zero momentum. So after the explosion, the momenta of the fragments must add up to zero. Let's introduce the coordinate system shown below in which the momenta of the fragments are indicated. (We've indicated that the unknown momentum of the third fragment must be in the opposite direction to the vector sum of the momenta of the other two fragments.)

Calling m the mass of one of the identical fragments, we have by conservation of momentum in the x direction

$$0 = mV_x + mV_x - \tfrac{1}{2}mv_x,$$

or

gravitational, or otherwise, and it would make no difference whatsoever; momentum is conserved regardless of the nature of the participating forces. The law of conservation of momentum, like the conservation of energy, is a vast and powerful principle. It makes an overall statement about nature without fussing over the details of the forces involved.

We now have two conservation laws, which may be applied together as a potent aid to understanding certain kinds of problems. These are problems in which two objects initially moving in some direction briefly interact, as in a collision, and move off in different directions. The word *interact* in physics means that some force is applied between the objects. We may not know the details of the force, but we can conclude that for a brief time the particles exert forces on each other because the motion of each object is changed. The laws of conservation of energy and momentum can help us analyze the subsequent motion of particles in this type of encounter.

Although it might not appear that the world is simply a collection of bodies colliding with each other, this type of problem is important in modern physics. Much of twentieth-century physics is an exploration of quantum mechanics – the physics of atoms and elementary particles. What physicists want to know in quantum mechanics is the hidden internal structure of atoms, nuclei, and even protons and neutrons themselves. But there are very few ways to find out what goes on inside a nucleus.

One way to study the nucleus is to accelerate it to a large momentum, smash it into another nucleus, and observe what happens. Such experiments are undertaken in colossal particle accelerators, such as the one at Fermilab, where protons collide with other protons. The debris of collisions can be seen as tracks of tiny bubbles produced inside chambers of liquid hydrogen. Figure 19.5 is a photograph of tracks made in a bubble chamber.

Figure 19.5 Particle tracks produced in a bubble chamber as result of a high-energy collision.

From such tracks, momenta can be measured and particles identified. Although this procedure lacks a certain delicacy, no one has found another kind of experiment to probe the inside of a nucleus. Someone once described it as similar to attempting to learn music by listening to a piano falling down a flight of stairs.

In preparation for application of the equations of conservation of energy and momentum, let's first write energy and momentum in terms which will allow us to compare them easily. We know that the momentum of a body is given by

$$\mathbf{p} = m\mathbf{v}, \tag{19.1}$$

and the kinetic energy is given by

$$K = \tfrac{1}{2}mv^2.$$

Recalling that $v^2 = \mathbf{v} \cdot \mathbf{v}$, we can use the result $\mathbf{v} = \mathbf{p}/m$ from Eq. (19.1) and substitute into the kinetic-energy expression:

$$K = \frac{1}{2}m \left(\frac{p}{m}\right)^2,$$

$$K = \frac{p^2}{2m}. \tag{19.4}$$

This result clearly shows the connection between kinetic energy and momentum and allows us to simplify many equations.

Suppose we have two objects, each of mass m, on a horizontal plane. One object, which we'll call the target, is at rest; the second object hits the target, as shown in Fig. 19.6. The result of this interaction is what we'll study. Although we don't know anything about the nature of the forces that operate between the bodies when they're close together, we nonetheless will discover properties of the encounter by applying the law of conservation of energy and momentum.

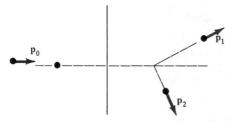

Figure 19.6 Collision of two objects of equal mass.

Initially the target has zero momentum; the momentum of the other object we'll call $\mathbf{p}_0$. After the encounter, the bodies will move off with momenta $\mathbf{p}_1$ and $\mathbf{p}_2$ (for the target). What can we say about $\mathbf{p}_1$ and $\mathbf{p}_2$? The conservation of momentum implies that the total momentum of the system before the encounter is equal to the total momentum of the system after the encounter:

$$\mathbf{p}_0 = \mathbf{p}_1 + \mathbf{p}_2. \tag{19.5}$$

Since all motion takes place on a horizontal plane, the potential energy of the bodies doesn't change and we need not worry about it. Furthermore, if no energy is dissipated as heat in a collision, we say that the collision is *elastic*. If some kinetic energy is transformed into heat, we call that collision *inelastic*. In this problem, we'll assume that the collision is elastic. The conservation of energy implies that the kinetic energy before the encounter is equal to the kinetic energy of the objects after the encounter. Using Eq. (19.4), we equate kinetic energies and get

$$\frac{p_0^2}{2m} = \frac{p_1^2}{2m} + \frac{p_2^2}{2m},$$

because the mass is the same for both objects. Multiplying by $2m$ we get

$$p_0^2 = p_1^2 + p_2^2. \tag{19.6}$$

Equation (19.5), being a vector equation, represents three equations, which we could proceed to solve analytically. But before solving the problem in that way, let's understand what that vector equation means. It says that the vector $\mathbf{p}_0$, whatever it is, is the sum of two pieces, $\mathbf{p}_1$ and $\mathbf{p}_2$. In other words, the vectors form a triangle.

In addition, Eq. (19.6) tells us that the square of the length of $\mathbf{p}_0$ is equal to the sum of the squares of the lengths of $\mathbf{p}_1$ and $\mathbf{p}_2$. The vector triangle obeys the Pythagorean theorem so $\mathbf{p}_1$ is perpendicular to $\mathbf{p}_2$. Already we have a prediction: the two objects will fly off at right angles to each other, as shown in Fig. 19.6.

We can also solve the problem analytically, instead of geometrically. Let's take p_0^2 and work out the dot product:

$$p_0^2 = \mathbf{p}_0 \cdot \mathbf{p}_0 = (\mathbf{p}_1 + \mathbf{p}_2) \cdot (\mathbf{p}_1 + \mathbf{p}_2),$$

$$p_0^2 = p_1^2 + p_2^2 + 2\mathbf{p}_1 \cdot \mathbf{p}_2.$$

Using Eq. (19.6), which tells us

$$p_0^2 = p_1^2 + p_2^2, \tag{19.6}$$

we find that

$$\mathbf{p}_1 \cdot \mathbf{p}_2 = 0. \tag{19.7}$$

There are two ways in which this last result can hold true. One solution is to say that the dot product of the two vectors is zero because the vectors are perpendicular to each other, a conclusion reached earlier. The other solution, however, didn't appear when we solved the problem geometrically: either $\mathbf{p}_1$ or $\mathbf{p}_2$ can be zero. This second solution solves Eq. (19.7) as well and is consistent with the conservation of energy.

If $\mathbf{p}_2 = \mathbf{0}$, then the target remains still, indicating that the other object missed the target completely, an unexciting prospect. If, however, $\mathbf{p}_1 = \mathbf{0}$, then we had a head-on collision in which particle 2 (the target) flew off and particle 1 was left standing still. You can easily verify this conclusion by making two pennies collide head on.

The two laws of conservation of momentum and energy working together give us a powerful means of analyzing collisions. These laws transcend the details of the forces involved; they are a bookkeeping device that tells us what is the same before and after the collision. The following examples illustrate this approach to understanding collisions.

Example 4

A 0.03-kg ball traveling at 0.08 m/s collides head on with a 0.05-kg ball which is initially at rest. If the collision is elastic, find the speed of each ball after the collision.

Let's call the momentum of the incident ball mv_0. After the collision, we can assume that both balls move off in the same direction with momenta mv and MV. Since all motion is along one direction, conservation of momentum implies

$$mv_0 = mv + MV.$$

Conservation of energy requires that

$$\tfrac{1}{2}mv_0^2 = \tfrac{1}{2}mv^2 + \tfrac{1}{2}MV^2.$$

We need to solve these two equations for our unknowns v and V. To accomplish this, let's solve the first equation for v,

$$v = v_0 - (M/m)V,$$

and substitute it into the second equation:

$$\tfrac{1}{2}mv_0^2 = \tfrac{1}{2}m[v_0 - (M/m)V]^2 + \tfrac{1}{2}MV^2.$$

Multiplying out the square and then solving for V, we find

$$mv_0^2 = mv_0^2 + m(M/m)^2V^2 - 2Mv_0V + MV^2,$$

$$V = 2v_0/(1 + M/m).$$

Substituting numbers, we find that $V = 0.06$ m/s.

To find v, we use the result $v = v_0 - (M/m)V$ and obtain $v = -0.02$ m/s. The minus sign here tells us that the ball reverses its direction as a result of the collision.

Example 5

A glider of mass $M = 0.60$ kg has a piece of clay at one end. Initially it is traveling with a speed $V_0 = 0.20$ m/s along an air track when it collides with and sticks to a glider of mass $m = 0.40$ kg which is moving at $v_0 = 0.10$ m/s. It is observed that the two gliders move off together at a speed V.

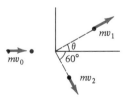

(a) Determine the speed V.

(b) What amount of kinetic energy is dissipated in the collision?

(a) Since the two gliders stick together, we know that the collision is inelastic; some amount of energy must be dissipated by friction in order to have the two gliders move off as one. Therefore we first apply conservation of momentum:

$$MV_0 + mv_0 = (M + m)V,$$

where after the collision both gliders have the same speed V. Solving for V, we find

$$V = (MV_0 + mv_0)/(M + m),$$

which numerically is 0.16 m/s.

(b) The energy dissipated is equal to the difference between the initial and final kinetic energy of the system:

$$\text{energy dissipated} = (\tfrac{1}{2} MV_0^2 + \tfrac{1}{2} mv_0^2) - \tfrac{1}{2}(M + m)V^2,$$

which turns out to be 1.2×10^{-3} J.

Example 6

A proton traveling with a speed $v_0 = 600$ m/s interacts with another proton initially at rest. From particle tracks, we know that one proton moves off at 60° from the initial direction.

(a) What is the direction of the other proton after interaction?

(b) What are the speeds of the two protons after the interaction?

(a) Since we have a collision between two identical particles, we can make use of the discussion in the text. From that discussion, we know that the particles will move off at right angles to each other. Therefore the angle θ in the diagram above is equal to 30°.

(b) To find the speeds of the protons, let's use the conservation of momentum equation in component form. The initial momentum in the x direction is mv_0, so we have

$$mv_0 = mv_1 \cos 30° + mv_2 \cos 60°,$$

where v_1 and v_2 are our unknowns. In the y direction, the initial momentum is zero (by the way we chose our coordinate axes), so conservation of momentum for that component implies

$$0 = mv_1 \sin 30° - mv_2 \sin 60°.$$

From this last equation we can solve for v_1,

$$v_1 = v_2 \sin 60°/\sin 30°,$$

and substitute this value into our momentum equation for the x direction:

$$v_0 = v_2 \sin 60° \cot 30° + v_2 \cos 60°.$$

Solving for v_2, we find

$$v_2 = v_0/(\sin 60° \cot 30° + \cos 60°),$$

which tells us that $v_2 = 300$ m/s. Once we have this value, it is an easy matter to substitute back into a momentum equation and find that $v_1 = 520$ m/s.

Example 7

A compact car, $m_A = 1300$ kg, and a sports car, $m_B = 1000$ kg, approach an intersection, each traveling at 14 m/s. They collide and move off at an angle θ as indicated in the diagram.

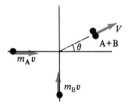

Find

    (a)   the speed of the entangled cars after the collision,
    (b)   the angle θ,
    (c)   the amount of energy dissipated in the collision.

We know that the total momentum before the collision is equal to the total momentum after the collision. Let's write this out in components. For the x direction, we have

$$m_A v = (m_A + m_B)V \cos \theta,$$

where both cars have the same speed after the collision. Because we have two unknowns, V and θ, we need another equation before we can solve for them. In the y direction, conservation of momentum implies

CHAPTER 20

HARMONIC MOTION

Another question concerns the oscillations of pendulums, and it falls into two parts. One is whether all oscillations, large, medium, and small, are truly and precisely made in equal times. The other concerns the ratio of times for bodies hung from unequal threads; the times of their vibrations, I mean. . . . As to the prior question, whether the same pendulum makes all its oscillations – the largest, the average, and the smallest – in truly and exactly equal times, I submit myself to that which I once heard from our Academician [Galileo]. He demonstrated that the moveable which falls along chords subtended by every arc [of a given circle] necessarily passes over them all in equal times. . . .

As to the ratio of times of oscillations of bodies hanging from strings of different lengths, those times are as the square roots of the string lengths; or should we say that the lengths are as the doubled ratios, or squares, of the times.

Galileo Galilei, *Two New Sciences* (1638)

20.1 FINDING A CLOCK THAT WOULDN'T GET SEASICK

Navigation has provided one of the most persistent motives for measuring time accurately. All navigators depend on continuous time information to find out where they are and to chart their course. But until about two centuries ago, no one was able to make a clock that could keep time accurately at sea.

Early travelers noticed that the North Star, unlike other stars, does not change its position with respect to the earth; it appears to be suspended in the northern sky. The farther northward they traveled, the higher in the sky the North Star appeared; at the North Pole it would be directly overhead. By measuring the elevation of the North Star above the horizon with a sextant, a navigator can determine the distance from the North Pole and the latitude, as Fig. 20.1a illustrates.

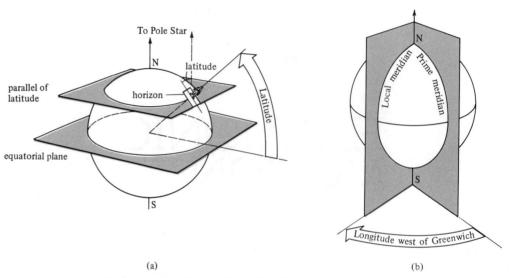

(a)   (b)

Figure 20.1 (a) Latitude determined by position of the North Star.
(b) Longitude requires knowing a reference time.

On account of the earth's rotation, however, charting a course east or west presented a more complex problem. Only by knowing the time very accurately was a navigator able to calculate longitude. Because of the earth's rotation, the sun appears to travel across the sky from east to west at the rate of one degree in four minutes. (It takes the sun 24 hours to move once around the earth, so it must move through 360° in 24 hours, which is 360/24, or 15° in 1 hour, or 1° in 4 minutes.) Now if a navigator determines the local time from the position of the sun and has a clock which very accurately tells the time in Greenwich, England (through which, by international agreement, the zero longitude line runs), he can easily determine his longitude. For every four minutes his clock, showing Greenwich time, differs from the local time, he is one degree of longitude away from Greenwich. Even at night, a navigator can determine the longitude by using star charts to determine the local time and by comparing that time to Greenwich time. What is essential is an accurate timepiece.

The earliest time-keeping devices, built by the ancient Egyptians, consisted of an alabaster bowl, wide at the top and narrow at the bottom, which had horizontal markings on the inside to tell the time. As water dripped out through a hole in the bottom of the bowl, successive lines were exposed. For centuries the basic design of the water clock remained unchanged; Galileo used a water clock in his fertile experiments with balls rolling down inclined planes.

Sometime between the eighth and eleventh centuries, Chinese artisans constructed a water clock that had the characteristics of a mechanical clock. Falling water powered a wheel that contained small cups around its rim. As a cup filled with water, it became heavy enough to trip a lever which allowed the next cup to move into place and advance the wheel by a step. In thirteenth-century Europe, variations of the Chinese water clock became popular. Aside from the fact that these clocks did not keep very good time, they also tended to freeze in the European winters.

Sand clocks (hourglasses) introduced in the fourteenth century avoided the problem of freezing, but because of the weight of sand, they were limited to measuring short intervals of time. One of the chief uses of the hourglass was to determine a ship's position by "dead reckoning." Sailors would throw a log overboard with a long rope attached to it, and then count knots, which were tied in the rope at equal intervals, as the rope played out for a specific amount of time. In this way sailors could crudely estimate the speed, or knots, at which the ship was moving. By knowing their speed and how long they had traveled in a certain direction, they could track their position.

The first truly mechanical clocks were built in the fourteenth century and consisted of pulleys and weights with escapements, similar to present-day cuckoo clocks. The accuracy of these early mechanical clocks depended on the friction between parts, the driving weights, and the skill of the craftsman constructing it. No two clocks would show the same time let alone keep accurate time. What was needed was some sort of periodic, repeating device whose frequency was essentially a property of the device itself.

20.2 SIMPLE HARMONIC MOTION

The event that led to accurate timepieces was the analysis of *periodic* or *harmonic motion* – motion that repeats itself in equal intervals of time. When an object moves back and forth over the same path in harmonic motion, we say that it is *oscillating*. We now explore this type of motion.

In Chapter 14, we discussed the stability of an object subject to a force

$$F = -kx,$$

which acts on the object whenever it moves away from the equilibrium point $x = 0$. This force, for example, describes the behavior of a spring. Associated with this force is the potential energy

$$U = \tfrac{1}{2}kx^2.$$

We used this potential energy as a model to study stability and discovered that it typified many systems that oscillate: a marble rolling in the bottom of a bowl, a mass oscillating at the end of a spring, a swinging pendulum, a vibrating guitar string, and even a complex political or ecological system fluctuating. All these systems have the property that if they are disturbed from equilibrium, the restoring force that acts on them tends to move them back into equilibrium.

These systems have another property in common too: when disturbed from equilibrium, the system tends to return to equilibrium by the action of the restoring force, but they overshoot that point because of inertia. After the overshoot the restoring force again acts to return the system to equilibrium. The result is that the system winds up oscillating back and forth, like a marble in a bowl, a mass on a spring, and a guitar string. Figure 20.2 shows "snapshots" of a mass oscillating on the end of a spring. The horizontal displacement of the mass, plotted as a function of time, traces out a path resembling a sine or cosine curve, as suggested by Fig. 20.2.

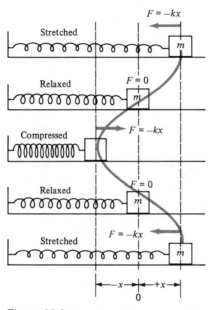

Figure 20.2 Snapshots of a mass oscillating on the end of a spring.

To determine this curve mathematically we apply Newton's second law, $F = ma$. Since the acceleration is the second derivative of the displacement, $a = d^2x/dt^2$, and we are considering the spring force $F = -kx$, we have

$$m\frac{d^2x}{dt^2} = -kx.$$

Dividing by the mass and explicitly denoting x as a function of time, we obtain

$$\frac{d^2}{dt^2}x(t) = -\frac{k}{m}x(t). \qquad (20.1)$$

We already believe that the solution of this differential equation, $x(t)$, whose second derivative is proportional to the negative of the function itself, will be a function that oscillates back and forth with time.

In Chapter 3, we encountered two functions which have this property: the sine and cosine functions. To recapitulate,

$$\frac{d}{d\theta}\sin\theta = \cos\theta, \qquad (3.6)$$

and

$$\frac{d}{d\theta} \cos \theta = -\sin \theta. \tag{3.7}$$

From these derivatives we can verify that

$$\frac{d^2}{d\theta^2} \sin \theta = -\sin \theta,$$

$$\frac{d^2}{d\theta^2} \cos \theta = -\cos \theta.$$

The sine and cosine functions satisfy the type of differential equation we have, but how are they and θ related to $x(t)$? As time goes on, the oscillating mass traces out a sine curve on moving paper, as Fig. 20.2 illustrates. This suggests that

$$x(t) = A \sin \theta(t), \tag{20.2}$$

where A is some constant and $\theta(t)$ some angle that depends on time t. The cosine function is also a solution, but we'll discuss it later. In Example 1 below we use the differential equation (20.1) to show that we can arrange matters so that $\theta(t)$ increases uniformly with time,

$$\theta(t) = \omega_0 t,$$

where ω_0 is positive constant (called the *angular frequency*) given by

$$\omega_0 = \sqrt{\frac{k}{m}}, \tag{20.3}$$

and that A (called the *amplitude*) is the largest positive value $x(t)$ can have. Thus a solution of the differential equation (20.1) is

$$x(t) = A \sin \omega_0 t,$$

where $\omega_0 = \sqrt{k/m}$, and simple harmonic motion is a result of a linear restoring force ($F = -kx$).

In our discussion of uniform circular motion in Chapter 9, we had precisely the same connection between the angle which locates an object in such motion and time. This is no accident. As Figure 20.3 illustrates, the shadow of an object undergoing uniform

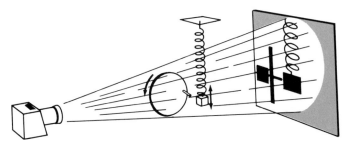

Figure 20.3 The shadow of a peg on an object executing uniform circular motion exhibits simple harmonic motion.

circular motion executes exactly the same motion as the oscillating mass and spring system. The oscillating mass is one component of uniform circular motion.

In Chapter 9, we called ω_0 the angular speed. When we use it to describe harmonic motion, we call it the *angular frequency*. The angular frequency is the angle (in radians) an object in circular motion moves through per unit time. This angular frequency is intimately related to the number of oscillations a corresponding object in harmonic motion makes per second for the reason we saw earlier. Here's the connection: An object in circular motion moves through 2π radians before it returns to where it started; in other words, it moves through 2π radians in one cycle. The frequency f is the number of cycles an object in harmonic motion completes per second. Since 2π radians corresponds to one cycle,

$$f = \omega_0/(2\pi). \tag{20.4}$$

In SI units, frequency is measured in hertz, abbreviated Hz, where 1 Hz = 1 cycle/second.

The time required to complete one oscillation, known as the period T, is simply the reciprocal of the frequency:

$$T = 1/f. \tag{20.5}$$

Figure 20.4 illustrates the period of a sinusoidal function.

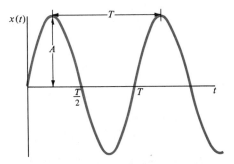

Figure 20.4 Period and amplitude of a sinusoidal function.

Example 1

Show that if a function

$$x(t) = A \sin \theta(t) \tag{20.2}$$

satisfies the differential equation

$$\frac{d^2}{dt^2} x(t) = -\frac{k}{m} x(t), \tag{20.1}$$

then with a suitable choice of initial conditions we have $\theta(t) = \omega_0 t$, where $\omega_0 = \sqrt{k/m}$ and A is the largest value of $x(t)$.

Differentiate (20.2) with the help of the chain rule to get

$$\frac{dx}{dt} = \frac{dx}{d\theta}\frac{d\theta}{dt} = A \cos \theta(t)\frac{d\theta}{dt} \ . \tag{20.6}$$

Differentiate again using both the product rule and the chain rule to obtain

$$\frac{d^2x}{dt^2} = -A \sin \theta(t)\left(\frac{d\theta}{dt}\right)^2 + A \cos \theta(t)\frac{d^2\theta}{dt^2} \ . \tag{20.7}$$

To satisfy (20.1) we want the right-hand side of (20.7) to reduce to

$$-\frac{k}{m}x(t) = -\frac{k}{m}A \sin \theta(t).$$

Therefore it suffices to choose $\theta(t)$ so that

$$-A\left(\frac{d\theta}{dt}\right)^2 = -\frac{k}{m}A \quad \text{and} \quad A\frac{d^2\theta}{dt^2} = 0.$$

The first of these implies $(d\theta/dt)^2 = k/m$, so $d\theta/dt = \pm\omega_0$, where $\omega_0 = \sqrt{k/m}$. If we want θ to increase with time we choose the plus sign and obtain $d\theta/dt = \omega_0$, which, incidentally, implies $d^2\theta/dt^2 = 0$. Therefore $\theta(t) = \omega_0 t + C$, where C is a constant. If the spring is in the relaxed position (as shown in Fig. 20.2) at time $t = 0$, then $\theta = 0$ when $t = 0$ and we must choose $C = 0$. Thus

$$x(t) = A \sin \omega_0 t.$$

Since the largest value of $\sin \theta$ is 1, the factor A represents the largest value of the displacement $x(t)$.

Now to interpret the results physically. The angular frequency depends on the physical characteristics of the system, namely, the spring constant k and the mass m. The stiffer the spring, the larger the value of the spring constant, and by Eq. (20.3), the larger the number of oscillations in one second. In other words, stiffer springs make the system oscillate more rapidly. This makes sense because a stiffer spring exerts a greater force and tends to accelerate the mass more. Equation (20.3) also tells us that the greater the mass, the slower the oscillations. We expect that larger values of m would lead to slower oscillations because of inertia. Because the frequency depends only on the physical characteristics of a particular mass and spring system, we refer to ω_0 as the *natural angular frequency*; it is the angular frequency at which the system will naturally oscillate. The frequency of the oscillations is independent of the amplitude A.

Once we have $x(t)$ for a particular system, we can find everything there is to know about the motion of that system. For example, to find the velocity of the mass, we need only take dx/dt; we already did this in Eq. (20.6), which now becomes

$$\frac{dx}{dt} = A\omega_0 \cos \omega_0 t. \tag{20.8}$$

Similarly, the acceleration at any instant is found by differentiating the velocity,

$$\frac{d^2x}{dt^2} = -\omega_0^2 A \sin \omega_0 t = -\omega_0^2 x. \tag{20.9}$$

The type of oscillatory motion we are studying is called *simple harmonic motion*. And an object which executes this type of motion is called a *simple harmonic oscillator*, abbreviated as SHO. The term *simple* refers to the absence of external forces such as friction or viscosity. The word harmonic refers to music: musical instruments generally vibrate harmonically. As we've already seen, the world is full of simple harmonic oscillators. Once we understand completely the SHO, we can put it in our pocket, and explore the world of problems it solves.

Example 2

A particle of mass 0.25 kg undergoes simple harmonic motion with an amplitude of 0.15 m and a frequency of 100 Hz. What is

    (a)   its angular frequency?
    (b)   the spring constant?
    (c)   its maximum velocity?
    (d)   its maximum acceleration?

    (a)   Using $\omega_0 = 2\pi f$, we have $\omega_0 = 630$ rad/s.
    (b)   Knowing that the natural angular frequency $\omega_0^2 = k/m$, we have

$$k = m\omega_0^2 = (0.25 \text{ kg})(630 \text{ rad/s})^2 = 9.9 \times 10^4 \text{ N/m}.$$

(To end up with units of N/m, we drop the radians because radians are dimensionless.)
    (c)   From Eq. (20.8),

$$v(t) = A\omega_0 \cos \omega_0 t, \tag{20.8}$$

we see that because the maximum value of the cosine is 1, the maximum value of the velocity is

$$v_{max} = A\omega_0 = (630 \text{ rad/s})(0.15 \text{ m}) = 94 \text{ m/s}.$$

    (d)   For the same reason as in (c), we see that Eq. (20.9) implies that the maximum acceleration is

$$a_{max} = A\omega_0^2 = (630 \text{ rad/s})(0.15 \text{ m}) = 6.0 \times 10^4 \text{ m/s}^2.$$

Questions

1. Give some additional examples of harmonic motion.

2. For a particle undergoing simple harmonic motion, where (at either the equilibrium point or the endpoints) does the maximum value occur in the
    (a) force acting on it,   **(b)** speed,   **(c)** acceleration?

3. Any real spring, of course, has mass. If the mass of a spring were taken into account, how would this change the frequency of a mass-and-spring system? Give a qualitative argument.

4. What changes could you make in a simple harmonic oscillator to double its

(a) maximum speed,   (b) maximum acceleration?

5. A small (0.10-kg) mass is attached to the bottom of an unstretched vertical spring and set into motion. If the maximum speed of the mass is 0.20 m/s and its maximum acceleration is 0.5 m/s^2, find

(a) the spring constant,
(b) the amplitude of the oscillations,
(c) the frequency of the oscillations.

6. A 2.5-kg block hangs from a spring. If a 0.5-kg mass is attached to the block, the spring stretches an additional 0.05 m. Determine the frequency of oscillations if only the 2.5-kg block is attached to the spring.

7. A mass is attached to a spring of spring constant k. If the spring is cut in half and the same mass suspended from one of the halves, how are the frequencies of oscillation related, after and before the spring is cut?

8. Two springs of spring constants k_1 and k_2 are attached to a block of mass m as shown. What will be the frequency of oscillations?

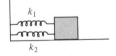

20.3 ENERGY CONSERVATION AND SIMPLE HARMONIC MOTION

In Chapter 14, we argued that an object near a potential-energy minimum would oscillate because potential energy would turn into kinetic energy, which in order to be conserved would turn back into potential energy. That was how we explained why the object doesn't stop at the equilibrium point where there's no force acting on it. Conservation of energy led us to expect oscillations in the first place. Now we've obtained the same result without mentioning energy at all. To connect the two approaches, let's examine the energy of a simple harmonic oscillator.

We already know that the potential energy is given by

$$U(x) = \tfrac{1}{2}kx^2. \tag{14.1}$$

From the displacement $x(t)$ given by Eq. (20.2) we can express the potential energy as a function of time:

$$U(t) = \tfrac{1}{2}kA^2 \sin^2 \omega_0 t. \tag{20.10}$$

The kinetic energy of the system is $\tfrac{1}{2}mv^2$, where the velocity v is related to $x(t)$ through $v = dx/dt$. We already found this derivative:

$$\frac{dx}{dt} = A\omega_0 \cos \omega_0 t, \tag{20.8}$$

so the kinetic energy as a function of time is

$$K(t) = \tfrac{1}{2}mv^2 = \tfrac{1}{2}m\left(\frac{dx}{dt}\right)^2,$$

$$K(t) = \tfrac{1}{2}m\omega_0^2 A^2 \cos^2 \omega_0 t. \tag{20.11}$$

Therefore the total energy of the system is

$$E = K + U,$$

$$E = \tfrac{1}{2}m\omega_0^2 A^2 \cos^2 \omega_0 t + \tfrac{1}{2}kA^2 \sin^2 \omega_0 t.$$

At first glance it appears as if the total energy is not constant because it seems to depend on t. To show that the total energy is indeed constant, we recall that $\omega_0^2 = k/m$, and substitute this into the expression for E:

$$E = \tfrac{1}{2}m(k/m)A^2 \cos^2 \omega_0 t + \tfrac{1}{2}kA^2 \sin^2 \omega_0 t,$$

$$E = \tfrac{1}{2}kA^2(\cos^2 \omega_0 t + \sin^2 \omega_0 t).$$

The term in parentheses, according to trigonometry, is always equal to one, so we have the result

$$E = \tfrac{1}{2}kA^2, \tag{20.12}$$

which is a constant. Figure 20.5 shows graphs of $U(t)$, $K(t)$, and the total energy E.

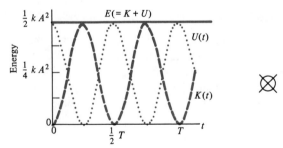

Figure 20.5 Graphs of potential, kinetic, and total energy for a simple harmonic oscillator.

Example 3
A spring of spring constant 20 N/m hangs unstretched. A 0.20-kg mass is attached to the free end and released. Find (a) how far below the initial position the mass descends, (b) the energy of the oscillator.

When the mass is released, its gravitational potential energy is converted into potential energy stored in the spring, and it momentarily stops descending when the two energies are equal. Let d be the distance the mass falls. Then its initial potential energy is mgd and its final energy is $\frac{1}{2}kd^2$. Equating these two energies

$$\frac{1}{2}kd^2 = mgd,$$

we can solve for d:

$$d = 2mg/k = 2(0.20 \text{ kg})(9.8 \text{ m/s}^2)/(20 \text{ N/m}) = 0.20 \text{ m}.$$

This result makes sense because the greater the mass the more inertia the object has, the greater the force needed to stop it, and the farther it stretches the spring; the stiffer the spring, the less it will stretch.

The amplitude of the oscillations will be half the distance it falls because it oscillates equally up and down about that point. Therefore, substituting $A = 0.10$ m into Eq. (20.12), we find

$$E = \frac{1}{2}kA^2 = \frac{1}{2}(0.20 \text{ N/m})(0.10 \text{ m})^2 = 0.001 \text{ J}.$$

Equation (20.12) represents the initial amount of energy put into the oscillator by an outside agent performing work, as when you stretch a spring, or pluck a guitar string. As a result of friction that energy is gradually transformed to heat, so the energy of a real harmonic oscillator decreases with time and the amplitude of successive oscillations diminishes. But the frequency of the oscillations, being independent of the amplitude, remains constant; only the amplitude of successive oscillations decreases.

Robert Hooke, after whom the spring force law is named, understood the essential feature of spring oscillations – that even in the presence of friction, the frequency remains constant. In the 1650s he experimented with the idea of using a metal spring to regulate the frequency of a clock. The first spring-controlled clock was built, however, by Christian Huygens, a Dutch physicist. His idea was to use a spiral spring – the type still used today in mechanical watches.

Questions

9. When a simple harmonic oscillator is at one-third of its maximum displacement, what fraction of its total energy is in kinetic energy?

10. A simple harmonic oscillator of mass 0.40 kg has a frequency of 15 Hz and a total energy of 30 J. What is the amplitude of its oscillations?

11. Express the total energy of a simple harmonic oscillator, Eq. (20.12), in terms of its maximum velocity.

12. From conservation of energy, show that the instantaneous velocity of a simple harmonic oscillator is given by

$$v(t) = \pm\omega_0\sqrt{A^2 - x^2},$$

where $x = x(t)$ is the instantaneous displacement.

20.4 INITIAL CONDITIONS

When we searched for a solution to the simple harmonic oscillator, we found one of the form $x(t) = A \sin \omega_0 t$. But is it the *only* solution? It is easy to verify that another function,

$$x(t) = B \cos \omega_0 t,$$

satisfies the same differential equation of simple harmonic motion,

$$\frac{d^2}{dt^2}x(t) = -\frac{k}{m}x(t), \tag{20.1}$$

where B is any constant. And we can form another solution to the equation by taking the sum of the two solutions:

$$x(t) = A \sin \omega_0 t + B \cos \omega_0 t. \tag{20.13}$$

To make matters worse, there are other possibilities, like $x(t) = C \sin(\omega_0 t + \phi)$, where ϕ is a constant angle. However, it can be shown that this and any other possibility can be converted into a sum of sine and cosine functions by a suitable choice of A and B. For this reason we say that Eq. (20.13) represents the general solution.

Example 4

Verify that $x(t) = A \sin \omega_0 t + B \cos \omega_0 t$, where $\omega_0^2 = k/m$, satisfies Eq. (20.1).
   Calculating the first derivative, we have

$$\frac{dx}{dt} = A\omega_0 \cos \omega_0 t - B\omega_0 \sin \omega_0 t,$$

and taking the derivative again, we get

$$\frac{d^2 x}{dt^2} = -A\omega_0^2 \sin \omega_0 t - B\omega_0^2 \cos \omega_0 t.$$

Writing $\omega_0^2 = k/m$ we have

$$\frac{d^2 x}{dt^2} = -\frac{k}{m}(A \sin \omega_0 t + B \cos \omega_0 t).$$

Recognizing that the term inside the brackets is $x(t)$, we see that this solution also satisfies Eq. (20.1).

   Now that we have many solutions to the equation of simple harmonic motion, how do we choose one which describes a particular case? One way to find out is to ask where the oscillator is, and how fast it is moving at some time; a convenient choice of time is $t = 0$ (when we start our clock). This will identify a particular solution. To illustrate how this works, let's begin with the general solution

$$x(t) = A \sin \omega_0 t + B \cos \omega_0 t. \tag{20.13}$$

The two constants A and B are yet undetermined. (Remember that the natural angular frequency ω_0 is determined by the mass-and-spring constant of the oscillator.) But by specifying the displacement and velocity of the oscillator at time $t = 0$, we will determine the constants A and B.

Now suppose that, as Fig. 20.6 illustrates, at $t = 0$ the oscillator has a displacement $x(0)$ and zero velocity, $v(0) = 0$. According to Eq. (20.13), we have

$$x(0) = A \sin 0 + B \cos 0.$$

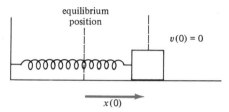

Figure 20.6 Initial conditions for a simple harmonic oscillator.

Since $\sin 0 = 0$ and $\cos 0 = 1$, this condition requires that $B = x(0)$, so we've determined one of the constants, B.

Before we impose the second condition, that involving the initial velocity, we need to calculate the velocity at any instant:

$$v(t) = \frac{dx}{dt} = A\omega_0 \cos \omega_0 t - B\omega_0 \sin \omega_0 t. \tag{20.14}$$

Setting the velocity equal to 0 at $t = 0$, we get

$$v(0) = 0 = A\omega_0 \cos 0 - B\omega_0 \sin 0,$$

which implies

$$A\omega_0 = 0,$$

hence $A = 0$. So we've determined A as well. Therefore, the position at *any* time for an oscillator that starts out in *this* particular way is

$$x(t) = x(0) \cos \omega_0 t.$$

By specifying the *initial conditions* – the displacement and velocity of the oscillator at time $t = 0$ – we can determine the constants that specify the simple harmonic motion for a particular oscillator. We've seen another example of this process in Chapter 6, when we found the general description of a projectile and imposed initial conditions on position and velocity to determine a specific trajectory.

In some respects the example we just solved is typical of what happens in general. Somewhere in the process of solving the differential equation, two integrations are required (one to go from acceleration to velocity, the second to go from velocity to position). Accompanying the two integrations are two constants, which here appear in the solution as A and B. By imposing initial conditions, we determine those constants and specify the particular solution.

There is no need to memorize the specific solution for every simple harmonic oscillator. All you need to know is $x(t)$, Eq. (20.13), and how to apply the specific initial conditions for a particular problem.

Example 5

A 0.5-kg mass is attached to a spring with spring constant $k = 120$ N/m and set into motion. A clock is started when the oscillator has a displacement $x(0) = 0.3$ m and a velocity $v(0) = 6.0$ m/s. What is the displacement of the oscillator at time t?

Starting with the general solution, Eq. (20.13),

$$x(t) = A \sin \omega_0 t + B \cos \omega_0 t,$$

we set $x(0) = 0.3$ m, which implies

$$0.3 \text{ m} = A \sin 0 + B \cos 0,$$

or that $B = 0.3$ m.

Taking the derivative of $x(t)$ and setting it equal to $v(0)$, we have

$$v(0) = 6.0 \text{ m/s} = \omega_0 A \cos 0 - \omega_0 B \sin 0.$$

Solving for A and using $\omega_0 = \sqrt{k/m} = 15$ rad/s, we find $A = 0.4$ m. Now that we know the two constants A and B, we can describe the displacement of the oscillator at any instant as

$$x(t) = (0.4 \text{ m}) \sin(15t) + (0.3 \text{ m}) \cos(15t),$$

where t is in seconds.

By understanding general principles and working out specific examples, we gain not only additional insights into the way things work, but often obtain technological improvements as well. In 1713 the British government offered a prize of £20,000 to anyone who could build a clock accurate enough to enable a seafarer to determine longitude to within one-half of a degree (about 35 miles). Among the many dexterous craftsmen who sought to win the ample award was an English clockmaker, John Harrison. For 40 years he struggled to construct a spring-driven clock which could cope with rolling seas, temperature-induced expansion and contraction, and the corrosive salt spray. Finally in 1761 he sent his son on a voyage to Jamaica to test his clock, but only after the government forced him to build an identical model lest the original be lost at sea. His masterpiece was a technological triumph – it allowed the navigator to determine longitude to within one-third of a degree.

Questions

13. At time $t = 0$ an oscillator having a natural frequency of 35 Hz has a displacement $x(0) = 0$ and a velocity $v(0) = 20$ m/s.

   (a) What is the displacement of the oscillator at any time?
   (b) Find the maximum velocity of the oscillator.

14. Suppose that initially an oscillator has a displacement of 0.25 m, a velocity of -10 m/s, and a frequency of 10 Hz.

(a) What is its amplitude?

(b) Find the displacement at any instant.

15. Show that $x(t)$ as given in Eq. (20.13) leads to energy conservation.

20.5 THE SIMPLE PENDULUM

A very special and important aspect of simple harmonic motion is that the angular frequency ω_0 does not depend on the amplitude (A or B) of the motion. This independence means that the time for each complete cycle, the period T, also does not depend on the amplitude; if the oscillator makes large oscillations, it moves rapidly, and if it makes smaller oscillations it moves more slowly. Even in the real world, where friction makes oscillations die down, the oscillator always takes the *same* amount of time for each cycle. This is the characteristic that allows a simple harmonic oscillator to be used as a timing device. The discovery of this fact led immediately to the invention of the first accurate clocks. Even today, wristwatches which are accurate to within a few seconds per month use as timekeeping devices a kind of harmonic oscillator – a quartz crystal.

But earlier clocks used a different oscillator – the pendulum. Galileo made the crucial discovery that a pendulum takes the same time per swing, even as its motion dies down, and he thereby laid the groundwork for improved timekeeping. In the *Two New Sciences* he eloquently summarized his observations.

Folklore places Galileo's discovery in the Duomo, or Cathedral, in Pisa. The famous Leaning Tower of Pisa is actually the bell tower of the magnificent, high-ceilinged cathedral. Hanging from the ceiling on a long cable is a lamp, which one day Galileo supposedly noticed swinging back and forth, probably just after it had been lit. Timing the swings by comparing them to his own pulse, Galileo realized that they always took the same time even as the swings became smaller and smaller.

This famous lamp, called Galileo's lamp, still hangs in the Cathedral at Pisa. However, there's one thing wrong with the tale: the church's records show that the lamp was installed in the 1650s, ten years after Galileo's death.

Setting fables aside, let's analyze the motion of a simple pendulum to find out precisely what factors determine the period. The idea is to use Newton's second law to find a differential equation that describes the motion of a pendulum and cast it into the form of Eq. (20.1),

$$\frac{d^2x}{dt^2} = -\omega_0^2 x. \tag{20.1'}$$

In other words, we need to find an equation which says that the second derivative of the oscillating quantity is proportional to the negative of the quantity. Then, without any further trouble, we'll know that the quantity multiplying $-x$ (the position occupied by ω_0^2) is the square of the angular frequency with which a pendulum swings. In this case it will turn out that the oscillating quantity is not the displacement but rather the angle of the pendulum cord with respect to the vertical.

Let's consider the simple pendulum, like the one shown in Fig. 20.7a. We call it simple because we are idealizing the pendulum as a point of mass m at the end of a massless string of length L. An analysis based on Newton's second law, however, is not limited to this case; it's only easier. Suppose we start the pendulum in motion by pulling it aside and releasing it. When the pendulum moves through an angle θ from the vertical, it goes through a distance θL along a circular arc from its equilibrium position. The force responsible for tending to restore the pendulum to its equilibrium position (hanging straight down) is its weight mg. We can resolve this force into components parallel and perpendicular to the string, as illustrated in Fig. 20.7b. The perpendicular component, the one that is always tangent to the circular arc, causes the pendulum to accelerate back to its equilibrium position; from Fig. 20.7b we see that this component is $-mg \sin \theta$. The acceleration along this path is the second derivative of the displacement along the circular arc:

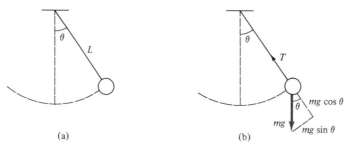

Figure 20.7 (a) The simple pendulum. (b) Force diagram for pendulum.

$$\frac{d^2}{dt^2}(L\theta) = L\frac{d^2\theta}{dt^2} .$$

Therefore, along the arc Newton's second law implies

$$mL\frac{d^2\theta}{dt^2} = -mg \sin \theta.$$

Canceling the mass and dividing through by L, we get

$$\frac{d^2\theta}{dt^2} = -\frac{g}{L} \sin \theta.$$

This last result is not the equation for a simple harmonic oscillator; the second derivative of the displacement (here θ) is not proportional to $-\theta$ but to $-\sin \theta$. Furthermore, it can be shown that no elementary function will satisfy this differential equation.

The intrepid physicist, however, is not daunted by such minor obstacles. If physicists only solved problems that they knew how to solve exactly, they would accomplish very little. The essence of practical physics is to ignore what is unimportant and to approximate. As Table 20.1 indicates, when expressed in radians, θ is approximately equal to $\sin \theta$; the smaller the angle, the closer the agreement. Even for an angle of $\pi/4$ rad ($45°$) the difference between θ and sin θ is only about 10%. Therefore, as long as we only consider small oscillations, we can safely replace sin θ by θ in our equation:

$$\frac{d^2\theta}{dt^2} = -\frac{g}{L}\theta. \tag{20.15}$$

Now this equation is exactly like the simple harmonic oscillator equation; the variable now is θ, but that doesn't matter. We know that the solution is

$$\theta(t) = \theta_0 \cos \omega_0 t, \tag{20.16}$$

where θ_0 is the amplitude determined from the initial conditions. This solution is reasonably good if the amplitude of the swings is small.

Table 20.1 Comparison of θ (in rad) with sin θ for Small Angles

θ (deg)	θ (rad)	sin θ
0	0	0
0.5730	0.0100	0.0100
5.730	0.1000	0.0998
11.459	0.2000	0.1987
17.189	0.3000	0.2955
45.000	$\pi/4 = 0.7854$	0.7071

By comparing Eq. (20.15) with the simple harmonic oscillator equation, we see that the angular frequency of the oscillations is

$$\omega_0 = \sqrt{g/L}, \tag{20.17}$$

which means that the period is

$$T = 2\pi/\omega_0 = 2\pi\sqrt{L/g}.$$

Consequently, on any given planet, the frequency of a simple pendulum depends only on its length. Unlike a mass on a spring, for which the frequency $\sqrt{k/m}$ does depend on the mass, the frequency of a pendulum is independent of its mass.

The reason the natural frequency is independent of the mass is exactly the same reason that the acceleration of a falling body on the surface of the earth doesn't depend on its mass: through Newton's second law, $F = ma$, and the universal law of gravity, $F = GmM_e/R_e^2$, the mass m cancels. The ingenious Isaac Newton used pendulums of different masses to test this cancelation with a precision of one part in a thousand. Because pendulums of identical length but different mass have equal frequencies, this proves exactly the same law as dropping a penny and a feather in a vacuum. Newton realized that the pendulum experiment works without being in a vacuum and is easier to observe.

Example 6

A pendulum has an amplitude of 20° and a length of 2.0 m. Find (a) its natural frequency, (b) its maximum velocity.

Using Eq. (20.17), we find

$$\omega_0 = \sqrt{g/L} = \sqrt{(9.8 \text{ m/s}^2)/(2.0 \text{ m})} = 2.2 \text{ rad/s}.$$

Differentiating Eq. (20.16), we obtain

$$\frac{d\theta}{dt} = -\theta_0 \omega_0 \sin \omega_0 t.$$

The velocity of the pendulum along the circular arc is given by $v = L \, d\theta/dt$, so the maximum speed is

$$v_{max} = \theta_0 \omega_0 L = (20°)(\pi \text{ rad}/180°)(2.2 \text{ rad/s})(2.0 \text{ m}),$$

which turns out to be 1.5 m/s. Note that one must express θ_0 and ω_0 in radians, not degrees, to get the right answer.

Questions

16. What must be the length of a pendulum to have a period of 1.0 s?

17. A pendulum has a period of 4.0 s on the surface of the earth. What will be its period on the surface of the moon?

18. A certain pendulum has a period of 3.0 s. How will the period change if the length is increased by 60%? If the length is decreased by 60%?

19. Qualitatively argue, taking into account the mass of the rod on which a pendulum swings, how the frequency would change from that of a simple pendulum.

20. A simple pendulum of length 1.50 m makes 80 oscillations in 200 s. What is the local acceleration due to gravity?

20.6 A FINAL WORD

We started out to study harmonic oscillation, a motion executed by various things, like pendulums and guitar strings. Understanding such periodic motion was crucial to the development of accurate timepieces. In our analysis we had to ignore air resistance and friction – idealizations we have often made. Yet for the pendulum, we approximated even further when we found out that it is not quite a harmonic oscillator, because its motion is along a circular arc. Is physics imprecise?

Many people believe that physicists seek the most fundamental and precise equations that govern the behavior of the universe. But in fact, physicists don't have completely universal equations at their disposal. Newton's laws aren't such principles; his laws don't accurately describe objects as small as atoms or as large as galaxies. And although we understand atoms (quantum theory) and galaxies (general theory of relativity), we don't have one fundamental set of laws that explains both at the same time. Many physicists, though, search for such a law and believe that it soon will be within their grasp.

Suppose, however, that we already knew the fundamental laws that govern the universe. What would we do then? The obvious approach would be to write down those equations and find all the solutions. This would be excruciatingly difficult, since the laws presumably would be expressed as differential equations. But in principle it would seem it could be done; a differential equation can be solved numerically by a sufficiently powerful computer even if it is impossible to express the solution analytically by formulas. So, if we found all the solutions, we would unlock all the secrets of the universe. Or would we?

Solving the equations of the universe numerically is something we would *not* want to do even if we could. The reason is very simple. The computer printout would be as complicated as the universe itself – and we already have the universe! What we want from physics is not the precise numerical results that describe exactly how everything behaves. Instead, what we seek is something much more subtle. We want understanding, insight, and at best a kind of trained and dependable intuition about why things work the way they do.

In studying the differential equation of the simple harmonic oscillator we gained an understanding of how some things work, even though we don't know of any physical system that precisely satisfies this equation. Yet, as we look at the world around us, mentally armed with this equation and its solutions, we begin to see everywhere examples of things that we know have this equation buried somewhere deep in their behavior. Our understanding of how things work has been inexpressibly enriched once we grasp the idea of extracting from complicated phenomena simple, underlying elements. Harmonic motion is often one of those elements. The road to insight often does not go through meticulous, complete, and precise description. It usually starts out in quite a different direction, passing first through crude but clever estimations and approximations.

CHAPTER

21

RESONANCE

First of all, it is necessary to note that each pendulum has its own time of vibration, so limited and fixed in advance that it is impossible to move it in any other period than its own unique natural one. Take in hand any string you like, to which a weight is attached, and try the best you can to increase or diminish the frequency of its vibrations; this will be a mere waste of effort. On the other hand, we confer motion on any pendulum, though heavy and at rest, by merely blowing on it. This motion may be quite large if we repeat our puffs; yet it will take place only in accord with the time appropriate to its oscillations. If at the first puff we shall have moved it half an inch from the vertical, by adding the second when, returned toward us, it would commence its second vibration, we confer a new motion on it; and thus successively with more puffs given at the right time (not when the pendulum is going toward us, for thus we should impede the motion and not assist it), and continuing with many impulses, we shall confer on it impetus such that much greater force than a breath would be needed to stop it.

Galileo Galilei, *Two New Sciences* (1638)

21.1 FORCED OSCILLATIONS

Galileo was not the only famous member of the Galilei family; his father Vincenzo was an accomplished and articulate musician. Understandably, Vincenzo was interested in how sound was produced. In 1589 he published his work on the relationship of the lengths and tensions of strings to the tones they produced. This study may have been the first experimentally derived law ever to have been discovered to replace a rival law. In the sixteenth century, music was considered a branch of mathematics. The Pythagorean idea that harmonious tones are produced by strings whose lengths are in definite ratios dominated music theory. Vincenzo argued that the complex sounds of musical instruments had to be determined by ear, rather than by mathematics alone. His ideas instilled a keen ear and insatiable curiosity into the eldest of his seven children, Galileo.

Not only did Galileo uncover the factors that determine the frequency of a pendulum, but he also understood the phenomenon of resonance. As the lines opening this chapter reveal, he noted that the swings of a pendulum can be made increasingly large by repeated, timed applications of a small force, like a puff of air. This method of making a pendulum swing is an example of *forced oscillations* – vibrations induced by an external driving force. Galileo further realized that if the frequency of the external driving force exactly matches the natural frequency ω_0 of the system, a spectacular effect takes place: the amplitude of the vibrations becomes exceedingly large. When a vibrating system is driven by a periodic force at the natural frequency of the system, we say that *resonance* occurs. Galileo knew that the phenomenon of resonance lay at the heart of the sounding boards of his father's clavichords and violins, and even of the power of a singer's voice.

Sometimes resonance can cause an oscillating system literally to break apart. Television commercials and movies have capitalized on this dramatic effect by showing wine glasses shattering when a singer hits a certain note. Is this impressive effect possible, or is it just Hollywood trickery?

Before we can answer, we need first to understand how and why resonance occurs. Perhaps the most familiar example of forced oscillations is a child's swing. Everyone knows how to push a swing to make it oscillate with a large amplitude. If you want a child to swing high, you push in step with the motion of the swing, as Fig. 21.1a illustrates. By applying a small force at the same point of each swing, you are timing your pushes to the natural frequency of the swing, which is a type of pendulum. The oscillations become larger and larger because you are adding energy to the system with each push. If, however, the pushes are not in step with the motion, as in Fig. 21.1b, the driving force opposes the motion and can cause the amplitude to diminish.

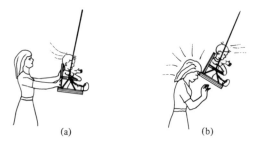

(a)   (b)

Figure 21.1 Forced oscillations of a swing by a force (a) in phase and (b) out of phase with the natural frequency.

The repeated application of a small force can create large-amplitude vibrations if (a) the force is in step (or in phase) with the oscillating system, and (b) the driving force repeats with the same frequency as the natural frequency of the system. Under these conditions, resonance occurs.

will be a solution of Eq. (21.1). We now show this is so by substituting $x = A \sin \omega t$ into (21.1) and determining A to satisfy the equation.

The second derivative of $x(t)$ is proportional to $x(t)$ itself: you can verify this through differentiation:

$$\frac{d^2x}{dt^2} = -\omega^2 x.$$

Substituting this result into Eq. (21.1), which we write as

$$\frac{d^2x}{dt^2} + \omega_0^2 x = a_0 \sin \omega t, \tag{21.1}$$

we obtain

$$-\omega^2 A \sin \omega t + \omega_0^2 A \sin \omega t = a_0 \sin \omega t$$

or

$$(-\omega^2 A + \omega_0^2 A - a_0) \sin \omega t = 0.$$

The only way this can be true for all times is if

$$-\omega^2 A + \omega_0^2 A - a_0 = 0.$$

Let's now pause and recall for a moment what we are looking for. We want to understand how and when resonance occurs. More dramatically, we want to know if it is possible to break a wine glass by singing the right note. The wine glass is represented by our basic equation, Eq. (21.1), with ω_0 its natural frequency (which you hear if you tap the wine glass), and $x(t)$ the distortion of the shape of the wine glass. The singer's (live or taped) voice causes the air to push the glass with a driving force $F_0 \sin \omega t$, leading to a disturbance of the glass, $A \sin \omega t$. The size of the resulting disturbance is proportional to A; if A becomes too large, the glass will shatter. From our last equation we already know what the amplitude of the disturbance will be:

$$A = \frac{a_0}{\omega_0^2 - \omega^2} \tag{21.3}$$

provided that $\omega \neq \omega_0$.

Let's interpret this equation for A. The wine glass rings with a definite frequency ω_0. If the sound waves striking the glass are from a bass note, that is, a low frequency ω, which is much less than ω_0, $\omega \ll \omega_0$, then we can ignore ω^2 compared to ω_0^2 and Eq. (21.3) gives us

$$A \approx a_0/\omega_0^2.$$

For sound waves a_0 is very small, consequently A is also very small. This means that the glass vibrates only slightly at the bass note frequency ω. In other words, nothing spectacular happens.

On the other hand, if the sound comes from a soprano, it has a high frequency, $\omega \gg \omega_0$. This time we ignore ω_0^2 compared to ω^2 in (21.3) and the resulting amplitude of the glass is

$$A \approx -a_0/\omega^2.$$

This effect is even smaller than that of a bass note because ω is larger than ω_0. The minus sign tells us that the glass vibrates exactly opposite to the way it is being pushed by the sound waves.

But if the impinging sound waves have precisely the frequency ω_0 then something startling happens. According to Eq. (21.3), no matter how small a_0 is, as ω approaches ω_0, the resulting amplitude of the system becomes arbitrarily large, blows up, and so does the system. We have resonance.

In real life, a glass is seldom shattered by the voice of a singer. Instead an audio generator, a device which produces pure tones, tuned precisely to the natural frequency of the glass is needed. And the natural frequency of the glass must be determined by a microphone held close to the glass which detects at what frequency of the audio generator the glass vibrates the most. The difficulty in breaking glasses is useful, because sound of every possible frequency, although at low intensities, is always in the air. There wouldn't be a glass left in the world if they really broke easily.

Our analysis, however, does not explain why it is actually difficult to break a wine glass. The reason is that we've oversimplified the situation to expose the essential phenomenon. In describing the forces acting on the harmonic oscillator, we ignored friction, viscosity, and so on. These forces are always present for any oscillator in the real world, and they are often proportional to the velocity dx/dt. In this case the differential equation of motion takes the form

$$\frac{d^2x}{dt^2} + a\frac{dx}{dt} + \omega_0^2 x = a_0 \sin \omega t, \tag{21.4}$$

where a is a positive constant due to friction or viscosity. A detailed analysis of this equation (which is not difficult but somewhat lengthy and so will be omitted) reveals that if $a^2 < 4\omega_0^2$ the solution $x(t)$ consists of two parts, a purely sinusoidal term plus a damped sinusoidal term. The damped term has a damping factor $e^{-at/2}$ which decreases to zero very rapidly as $t \to \infty$; the only visible part of the motion is the purely sinusoidal part, which can be written as

$$x(t) = A \sin(\omega t - \alpha), \tag{21.5}$$

where $\alpha = \arctan[a\omega/(\omega_0^2 - \omega^2)]$ is the phase angle and A is the amplitude, given by

$$A = \frac{a_0}{\sqrt{(\omega_0^2 - \omega^2)^2 + (a\omega)^2}}.$$

If there is no resistive force, the friction coefficient a is 0 and the formula for the amplitude becomes

$$A = a_0/|\omega_0^2 - \omega^2|,$$

in agreement with the result derived earlier in Eq. (21.3). If the friction coefficient a is small but nonzero, then the amplitude A exhibits a resonance peak of finite height nearly equal to $a_0/(a\omega_0)$ if ω is near to ω_0. The larger the friction coefficient a becomes, the lower the peak height, as illustrated in Fig. 21.3.

Example 2

Why do frictional forces prevent the amplitude of a driven oscillator from becoming infinite at resonance?

Let's first recall what we know about simple harmonic oscillators when friction is present, without any driving force. In Chapter 20 we argued that the effect of friction was to cause the amplitude (but not the frequency) of the oscillations to diminish. The reason is that the oscillator loses energy in the form of heat. Now if the oscillator is forced to oscillate, friction not only causes it to lose energy with each oscillation but also to lag behind the driving force. Recall that we found that for a frequency much greater than the natural frequency, the amplitude is negative, indicating that the oscillator vibrates in a way opposite to the driving force. A similar effect is caused by friction and depends upon the amount of friction. As a result, the oscillator can never be completely in step with the driving force and moreover it keeps on losing energy. Consequently, the amplitude can never become infinite, only very large if friction is small.

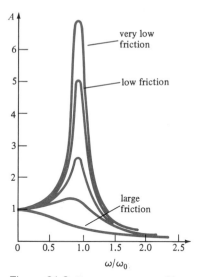

Figure 21.3 Resonance curves (A vs. ω/ω_0) for different amounts of frictional force acting on an oscillator.

In the idealized situation where there is no friction or viscosity, so $a = 0$ in Eq. (21.4), the damping factor $e^{-at/2}$ which we neglected in Eq. (21.5) is not there to carry the second term to 0. In this case, $\alpha = 0$ and the solution consists of two purely sinusoidal terms:

$$x(t) = C \sin(\omega_0 t - \beta) + \frac{a_0}{\omega_0^2 - \omega^2} \sin \omega t \tag{21.6}$$

provided that $\omega^2 \neq \omega_0^2$. Again, the amplitude of the second term is large if ω is near to the natural frequency ω_0. But if the friction coefficient $a = 0$ *and* $\omega^2 = \omega_0^2$ the solution takes yet a different form. In this case it becomes

$$x(t) = C \sin(\omega_0 t - \beta) - \frac{a_0}{2\omega_0} t \cos \omega_0 t. \tag{21.7}$$

The first term is purely sinusoidal, as above, but the second term oscillates with increasing amplitude as t increases because of the presence of the factor t multiplying the cosine. In this type of resonance the amplitude of the oscillation becomes arbitrarily large as t increases. An example of a solution $x(t)$ with $C = 0$ is shown in Fig. 21.4.

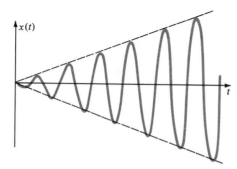

Figure 21.4 Resonant displacement of an oscillator in the absence of friction.

Breaking wine glasses is just one minor albeit dramatic example of the phenomenon of resonance. Many other things in everyday life exhibit resonance. Most cars have something that starts to rattle at certain motor speeds. This means that there is a natural harmonic oscillator somewhere (usually you can never quite pinpoint it) whose oscillation has the frequency of that motor speed. When the motor speed reaches that particular value, vibrations set in and the object begins to rattle. Another example is the rattling of windows when an airplane flies overhead. The windows have a natural frequency that is excited by the sound of the airplane engines.

Similarly, the sounding board of a piano, which is a piece of wood with many natural frequencies, resonates when a vibrating string is attached to it. Resonance also occurs in the cavity of a violin; and air inside can have large oscillations for certain frequencies.

Sometimes the effects of resonance can be ominous. In an earthquake, seismic waves are sent out from the epicenter in a range of frequencies, mostly at low frequencies compared to audible sound, known as infrasound. What happens if a structure has a resonance at one of those frequencies? Buildings between 5 and 40 stories high are typically resonant at earthquake frequencies. In an earthquake, these structures can literally come apart as a result of resonance. Architects try to minimize the resonant response of buildings by increasing friction in the joints.

Questions

4. Find the resonant frequencies for the following systems:

(a) $k = 2000$ N/m,   **(b)** $L = 0.35$ m,
     $m = 2.5$ kg; $m = 0.20$ kg.

(a) $k = 2000$ N/m   (b) $L = 0.35$ m
     $m = 2.5$ kg $m = 0.20$ kg

5. If pendulum E is given a slight push, which pendulum(s) will also begin to oscillate? Explain why.

6. Every child knows that by blowing across an empty Coke bottle a sound can be produced; the frequency of the sound is the natural frequency of the air inside the bottle. Explain what happens to the frequency produced when the bottle is partially filled with liquid.

7. A cabinet next to a refrigerator contains pots and pans which make a vibrational noise when the refrigerator motor runs. What is the source of this sound? How can it be eliminated?

8. A pendulum is forced to oscillate at a frequency $\omega = \frac{1}{4}\omega_0$, that is, at one-quarter of its natural frequency. Compare the amplitude of the oscillations to those which occur at $\omega = \frac{1}{2}\omega_0$.

9. In 1831 a bridge collapsed near Manchester, England, when soldiers marched across it in step. Ever since then, soldiers break step while marching across a bridge. Why?

10. Use the solution $x(t) = A \sin \omega t$, where A is given by Eq. (21.3), to find the total energy of a forced harmonic oscillator. What happens to the energy at resonance?

11. When the base of a vibrating tuning fork is touched to a table, the sound is amplified. Explain why.

21.3 SWINGING AND SINGING WIRES IN THE WIND

One fascinating example of resonance is provided by telephone wires singing in the wind. Imagine a taut wire suspended in the wind. The air flow around a cross section of the wire is illustrated in Fig. 21.5a. This smooth flow of air around the wire becomes unstable if the wind speed is great enough. The wind tries to move around the wire and prevent a vacuum from forming. If the speed is too high, the wind can't achieve this in a smooth flow, and instead it forms an eddy on both sides, as in Fig. 21.5b. The eddies start to occur behind the wire, and after a short time these vortices begin peeling off on alternate sides and running downstream in the wake of the wire, as shown in Fig. 21.5c. This complicated, yet very stable, flow pattern was first explained by the aerodynamicist Theodore von Kármán. The full pattern with a line of vortices in opposite directions is called the von Kármán Vortex Street; Fig. 21.6 is an actual picture of the pattern.

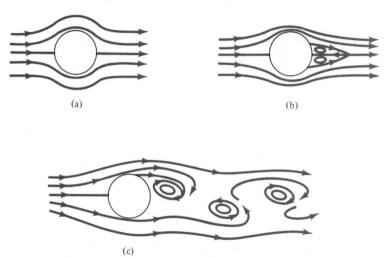

Figure 21.5 Formation of vortices which lead to singing wires.

Figure 21.6 Picture of a von Kármán vortex street in a fluid. [From Prandtl, L., and Tietjens, O. G. Applied Hydro- and Aerodynamics. McGraw-Hill Book Co., New York (1934). By permission of the publisher.]

Every time a vortex peels off the wire, it imparts a weak impulsive force. The reason is that the vortex has some momentum, and the vortex and wire conserve momentum; so every one of the vortices peeling off gives a push to the wire. At some wind speed, the vortices will start peeling off at the resonant frequency of the wire and thereby set it into motion. This resonant effect is why wires in the wind at the right wind speed begin to sing. The ancient Greeks noticed the eerie sound produced by harps as a result of this effect; we call the effect an aeolian harp.

Question

12. Suppose you were to simulate wind whistling wires by waving a fork with long tines. Which way would you wave it, in the plane of the tines or perpendicular to it? After deciding which way will create a sound, check your prediction.

21.4 A FINAL WORD

On July 1, 1940, a new bridge was opened up at the narrowest point in Puget Sound, connecting Tacoma to the Olympic peninsula. At the time it was the third-longest suspension bridge in the world. Right from the beginning, even before construction was completed, the bridge behaved in a peculiar way. Whenever there was a light breeze, ripples would run along the bridge. After a while local people began calling the bridge affectionately by the name Galloping Gertie. Driving across the bridge on a windy day became a favorite local pastime because it was like riding a roller coaster, although it was disconcerting to people driving across the bridge to see the car in front of them disappear over the crest of a wave.

On November 7, 1940, four months after the bridge was opened, a new mode of oscillations showed up in the bridge in a prevailing southwesterly wind of about 42 mph. Instead of rippling motions down the bridge, twisting motions set in. The peculiarities of the bridge were being studied by a hydrodynamicist from the University of Washington, Bert Farquharson. He rushed down to take pictures of the new mode of oscillations. At 11 o'clock in the morning that day, the Tacoma Narrows bridge collapsed. An inquest into the collapse determined that the bridge had been built according to the best engineering standards of the day. No one was guilty of wrongdoing, but also no one could figure out why the bridge collapsed.

A national commission investigating the collapse included aerodynamicist Theodore von Kármán of Caltech. He explained that vortices were pouring off the top and bottom of the bridge, driving the bridge at its resonant frequency, which eventually led to its collapse. His explanation was confirmed by experiments conducted in wind tunnels with structural models both at the University of Washington and at Caltech. In spite of the confirmation, the bridge building community was very reluctant to accept the explanation. Why? Bridge architects were concerned with static forces. They built in brute strength to confront maximum load, water flow, wind, etc. They didn't consider dynamic forces. Von Kármán said that the shape the roadway presented to the wind acted like an airplane wing. The displaced air formed vortices whose action induced vibrations in the deck. Since that disastrous event, models for all major bridges have been tested in wind tunnels, and bridge engineers have been forced to consider the aerodynamics of their designs.

Figure 21.7 depicts stages in the collapse of the Tacoma Narrows Bridge on that fateful day. In the twisting mode of Fig. 21.7b, the center line hardly moves at all – the

Figure 21.7 Collapse of the Tacoma Narrows Bridge. (Photo by Farquharson. Historical Photography Collection, University of Washington Libraries.)

Questions

1. A common rule of thumb for finding the distance to a lightning flash is to begin counting when the flash is seen and stop when the clap of thunder is heard. The number of seconds counted is then divided by 5 to get the distance in miles. Why is this justified? What speed of sound (in mi/h) does this method imply? What do you divide by to get the speed in km/h?

2. For Newton's measurements placing the speed of sound between 984 and 1109 ft/s, determine the pendulum lengths he must have used.

22.2 COUPLED OSCILLATORS

In Chapter 20 we studied the harmonic motion of a particular oscillator, the simple pendulum. The restoring force for this motion is the component of the pendulum's weight along the circular arc it follows. As a result of this restoring force, if we disturb a pendulum when it is at rest, in stable equilibrium, it oscillates back and forth about its equilibrium position with a definite frequency.

Now suppose we have two pendulums connected by a weak spring as shown in Fig. 22.1a. Such a system is called coupled pendulums or *coupled oscillators*. Can we predict what will happen if we disturb this system?

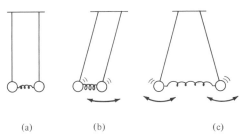

(a)   (b)   (c)

Figure 22.1 (a) Two pendulums of equal length coupled by a weak spring. (b) One mode of oscillation in which the spring is always unstretched. (c) Oscillations of the system for which the spring is alternately stretched and compressed.

If the coupled pendulums are at rest and then disturbed, oscillations will occur. But the exact nature of the oscillations depends upon how the pendulums are disturbed. Suppose you push both pendulums in the same direction by the same amount. Because they have identical lengths, they swing back and forth in unison, always staying the same distance apart. In this mode of oscillation for the system, known as a *normal mode*, the spring continually remains unstretched, as in Fig. 22.1b, and might as well not be there. The period of these oscillations is precisely the same as either pendulum would have alone, specified by $T = 2\pi \sqrt{L/g}$.

Another possibility is to start the pendulums swinging in opposite directions, as shown in Fig. 22.1c. This motion is more complicated because the motion of one pendulum is affected by the other pendulum. The oscillators are coupled together by the spring, and the two differential equations coming from Newton's second law, which describe the two motions, will also be coupled together in the sense that the solution of each will depend on the solution of the other. The net result is that the spring force on each pendulum is in step with gravity, creating a new normal mode. The frequency of this normal mode is higher than that of a single pendulum.

Example 1

Sketch the amplitude versus time graph for each of the two pendulums connected by a weak spring if the pendulums are pulled an equal distance apart from the equilibrium position and released.

If we take positive amplitude to be to the right as shown, then the pendulums initially have opposite amplitudes.

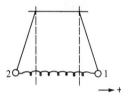

Once released, the pendulums will oscillate about their equilibrium positions at the same frequency. Therefore, we can easily sketch each pendulum's amplitude as a function of time because each is a cosine curve:

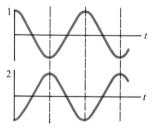

To make the pendulums swing in one of the normal modes we've described, we have to start them out in particular ways. In other words, the particular normal mode in which the system oscillates depends on the initial conditions. What if we had different initial conditions from those described previously? Suppose we start one pendulum swinging while the other is at rest. On each swing of the first pendulum the second gets a small, but resonant, push from the weak spring. The oscillations of this second pendulum grow larger while those of the first pendulum diminish. Eventually the second pendulum is swinging while the first is standing still. Then the process starts over in reverse. This exchange of energy between the two pendulums would continue forever, but of course there is friction and viscosity, so the oscillations eventually stop.

Example 2

A popular demonstration consists of five balls suspended by long strings. When one ball is pulled aside and released, it strikes the row of remaining balls and one ball at the end of the row swings away. Is this motion related to coupled pendulums?

The difference between this device and coupled pendulums is that as the first ball strikes the row, all its kinetic energy is almost *immediately* transferred to the end ball, which proceeds to swing away. The transfer of energy doesn't occur gradually as in the case of pendulums connected by a weak spring. We say that the pendulums in this device are strongly coupled – a complete exchange of motion occurs in each swing.

Questions

3. An astronaut using an air hammer to break up a boulder on the moon need not wear earplugs. Why?

4. Shown below is a graph of the amplitude of oscillations as a function of time for one of two pendulums connected by a weak spring. Initially one pendulum was at rest while the other was set into oscillations. Roughly sketch the amplitude versus time graph of the other pendulum.

5. Sketch the amplitude versus time curve for the two balls of Example 2, which are · strongly coupled.

22.3 WATER WAVES AND WAVE CHARACTERISTICS

Let's now begin to apply the idea of coupled oscillators to a familiar system – water waves. Imagine a quiet pond. Gravity tends to make the surface of the water smooth and horizontal. If you disturb the water, say, by dropping a stone into it, the surface of the water is forced down, leaving a hole where the stone fell. The water forced out of this hole at first piles up and then begins to flow back into the hole, setting up oscillations. But something else happens: a ripple spreads across the surface. Because of the inertia of the water, this action doesn't cease when the hole has been filled up, but continues until a depression is formed where previously the water was heaped up. As the process repeats, ripples spread out radially. But what moves across the water surface are not individual molecules of water; the individual molecules move in small circles about their equilibrium position. What travels across the water is a disturbance.

If there were a cork floating in the pond, you would see it bob up and down as ripples pass it. It acquires kinetic energy from the disturbance. But it does not move along with the spreading disturbance. We define a *wave* as a disturbance which transfers energy without net motion of a medium.

The transfer of energy in the spreading water wave is much like oscillations passing from one pendulum to another when they're coupled. If we had many coupled pendulums and started one in motion, we would see that disturbance passed down along the row, eventually reaching the last pendulum. We now have a new way to think about what a wave is: *the water behaves like many oscillators strongly coupled together*. When you disturb an oscillator, the disturbance passes on to the others, going along at some rate of speed. The water wave is the propagation of that disturbance from one oscillator to another, as Fig. 22.2 illustrates.

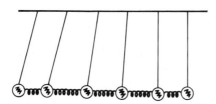

Figure 22.2 Coupled oscillators as a model for water waves.

Before we can use our model to determine, for example, the speed of waves, we need to characterize waves by observing their properties. If you watch one isolated spot of the pond as the water wave passes by, you will observe that the water bobs up and down. Figure 22.3 shows a wave at successive times, illustrating how one point of the water surface moves up and down like a harmonic oscillator. These oscillations occur at a definite frequency. So one characteristic of a wave is its *frequency*, which we define as the frequency of the oscillating matter of the medium.

If you look along the surface of the pond, you notice that the surface resembles a sine function, as indicated in Fig. 22.4. The terms *crest* and *trough* refer to the highest and lowest parts of the curve or wave it represents. The distance from one crest to another is called the *wavelength*, λ, because that's what it is – the length of the wave.

As the wave moves along, the time for one point to move down from a crest then bob back up again is exactly the time it takes for one ripple (from crest to trough) to pass that point: those are just two different ways of describing the phenomenon. The

deep water, as in the middle of the ocean, doesn't disturb the water all the way down to the bottom, so the depth can't be important.

For balls and springs we reasoned that the speed is roughly the frequency of an oscillator times the distance between oscillators. How can we apply this to water waves? Imagine a section of water oscillating as in Fig. 22.9. The length of that part of water is one wavelength, being the distance from one trough to another. This suggests that the distance we're searching for is the wavelength λ. The only other distance in evidence is the amplitude, and we don't expect the speed to depend on amplitude. Thus the speed

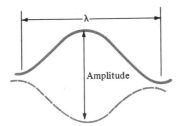

Figure 22.9 A water wave in deep water.

of the wave should be proportional to $\sqrt{g\lambda}$. A careful, elaborate analysis based on Newton's laws reveals that this is a good guess. The actual speed is

$$v \text{ (deep water)} = \sqrt{\frac{g\lambda}{2\pi}}. \tag{22.5}$$

No dimensional argument would have led us to the factor $1/\sqrt{2\pi}$ that we were missing.

Equation (22.5) tells us that water waves in the deep ocean have speeds that depend on their wavelengths. The long-wavelength, low-frequency waves travel faster than the short-wavelength, high-frequency waves. The low-frequency waves pass underneath the high-frequency ones.

In shallow water, the situation is different because the depth becomes important. The reason for this is that in water waves, as Fig. 22.6 illustrated, the motion of the water is circular. In deep water the size of the circle is roughly the size of the wavelength. But in shallow water it must be less than the depth. At the bottom the water hardly moves at all on account of the friction between the bottom and the nearby water. As we discussed in Chapter 11, this friction is the cause of viscosity. The result is that in shallow water, the factor λ in the expression for the wave speed is replaced by the depth of the water h. A detailed analysis yields a wave speed

$$v \text{ (shallow water)} = \sqrt{gh}. \tag{22.6}$$

This result is reasonable because it means the bottom slows down the wave as the water gets shallower. The difference between shallow and deep water is a matter of comparison between depth and wavelength. If $h \ll \lambda$, the water is shallow, and if $h \gg \lambda$ the water is deep.

When the wave speed is not constant, but depends on the wavelength of the disturbance, the medium is said to be dispersive. In this type of medium, waves of different wavelengths starting out together tend to disperse with time because they travel at different velocities. A glass prism is an important example of a medium which disperses light waves into a rainbow.

Example 3

An intricate and careful application of Newton's laws to water waves yields a wave speed for all wavelengths λ and depths h given by

$$v = \sqrt{\frac{g\lambda}{2\pi} \tanh\left(\frac{2\pi h}{\lambda}\right)},$$

where the hyperbolic tangent, $\tanh x$, is given by

$$\tanh x = \frac{e^x - e^{-x}}{e^x + e^{-x}}.$$

Show that this expression reduces to Eqs. (22.5) and (22.6) in the appropriate limit.

Since $x = 2\pi h/\lambda$, we need to examine the behavior of $\tanh x$ for large x ($h \gg \lambda$) and for small x ($h \ll \lambda$). In the first case, if we imagine x becoming very large, then e^x also becomes very large whereas e^{-x} becomes very small. But

$$\tanh x = \frac{e^x(1 - e^{-2x})}{e^x(1 + e^{-2x})} = \frac{1 - e^{-2x}}{1 + e^{-2x}},$$

so as x gets large the fraction on the right gets closer and closer to 1. In other words

$$\tanh x \approx 1 \quad \text{for large } x.$$

Using this limiting value, we see that the wave speed becomes

$$v \text{ (deep water)} = \sqrt{\frac{g\lambda}{2\pi}},$$

which is Eq. (22.5).

For the case of shallow water, we can imagine x becoming very small. To see how $\tanh x$ behaves for small values of x we first obtain an approximation to e^x for small x. From the second fundamental theorem of calculus we have

$$e^x - 1 = \int_0^x e^t \, dt.$$

If x is small the integrand e^t is nearly 1 and therefore the integral is nearly $\int_0^x 1 \, dt = x$. Hence we have the approximate formula $e^x - 1 \approx x$, or

$$e^x \approx 1 + x \quad \text{for small } x.$$

Replacing x by $-x$ we also have $e^{-x} \approx 1 - x$ for small x, hence

$$\tanh x = \frac{e^x - e^{-x}}{e^x + e^{-x}} \approx \frac{(1 + x) - (1 - x)}{(1 + x) + (1 - x)} = \frac{2x}{2} = x.$$

Therefore, for shallow water the speed is

$$v \text{ (shallow water)} = \sqrt{\frac{g\lambda}{2\pi}\frac{2\pi h}{\lambda}} = \sqrt{gh},$$

which is Eq. (22.6). Shallow waves, unlike deep-water waves, all have the same speed, regardless of the frequency and wavelength. This feature makes these waves more like longitudinal waves in our balls-and-springs model. It also makes them more like another important example we'll discuss later – sound.

From our understanding of the speed of water waves we can explain a common observation. As waves approach the beach from the deep ocean, they gradually change from deep water to shallow water waves. Then something else happens: the waves break.

Near the beach, the depth becomes so small that it is not much greater than the amplitude of the wave, as illustrated in Fig. 22.10. From Eq. (22.6), $v_w = \sqrt{gh}$, the speed is slower at h_2 than at the larger h_1. Therefore the thick part of the wave begins to catch up with the thin part, spoiling the sinusoidal shape as Fig. 22.11 illustrates. And that's why breakers form as waves wash up on a beach.

Figure 22.10 Near the beach the water depth is comparable to the amplitude of a wave.

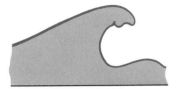

Figure 22.11 Breakers form as waves wash up on a beach on account of the dependence of wave speed on water depth.

Questions

9. Based upon the balls-and-springs model, would you expect a wave to travel faster in a liquid, where the molecules can move relatively easily, or in a solid, where they are more tightly bound to one another?

10. Determine the speed of a 5.0-m-long water wave in the open ocean at a water depth of 20 m. Compare your answers with the exact expression given in Example 3.

11. Do you expect waves to travel faster in shallow water, or in the same depth of molasses? What else would you guess about the behavior of waves in molasses?

12. Calculate the ratio of h_1 to h_2 for the large part of a wave in Fig. 22.10 to travel twice as fast as the thinner part.

13. What type of beach would you expect produces larger breakers, one with a gentle slope or one with a steep slope?

22.5 SOUND

The most common longitudinal wave is a sound wave in air. Sound is generally initiated by something vibrating in air, like vocal chords or the cardboard cone of a loudspeaker. The vibrating object sets the air next to it into motion, compressing it, then expanding it with each vibration, as Fig. 22.12 illustrates. That layer of air passes the motion on to the next layer, and so on. As a result a disturbance propagates away in the same

Figure 22.12 Production of sound waves by a tuning fork.

direction as the air molecules are vibrating. The sound wave has the same frequency as the source that caused it, and a wavelength given by Eq. (22.1),

$$\lambda f = v_{\mathrm{w}}, \tag{22.1}$$

where v_{w} is the speed of sound.

Isaac Newton, in a bold application of his dynamics, was the first to figure out the speed of sound. Instead of trudging through Newton's analysis, which is based on his second law and presented in the *Principia*, we will "guess" the answer in much the same way as we did for water waves. To accomplish this, we must know how air acts as a harmonic oscillator, and what provides the restoring force.

Let's begin with a simple piston-type device, shown in Fig. 22.13. The cylinder contains air and has an airtight sliding seal attached to the piston. If you push the piston in, thereby compressing the air inside, and then release it, the piston pops back out. That's the secret to the propagation of sound – the compression and expansion of air. Newton thought he understood perfectly how sound propagates: He thought that compressing air makes its pressure (the force per unit area) rise in exact proportion to the amount its volume decreases. He knew this from the experiments of Boyle which indicated

Figure 22.13 A sealed cylinder of air with a movable piston.

$$PV = \text{const},$$

where P is the pressure of the gas and V its volume. According to Boyle's law, pushing the piston in decreases V, so P increases. It is the higher pressure that makes the piston move back when released.

Now that we have an idea of what causes the air to oscillate, let's reason as we did for water waves. Remember that we argued that the mass of the water should not enter into the frequency. What about the mass of air? If we consider the cylinder of air, it has a definite mass m; the amount of mass, of course, depends on the size of the cylinder. If the gas is not compressed, it has the same *density* as the air outside. The ratio of mass to volume is the density ρ:

$$\rho = m/V.$$

Using this relation, we can rewrite Boyle's law as

$$P/\rho = \text{const},$$

where the constant is different from the previous one because it has absorbed the mass m. It is also a different kind of constant, because it doesn't depend on the size of the cylinder. This is important, because air near the surface of the earth has no particular value of volume associated with it, but it does have approximately the same pressure and density everywhere. When a sound wave travels through air, both the pressure and density of the air oscillate up and down, slightly higher and lower than the average values of the air.

When Newton analyzed the problem of sound in air, he found the speed to be

$$v_{\text{w}} \text{ (Newton)} = \sqrt{P/\rho}. \tag{22.7}$$

Let's understand why. Already we know that the restoring force on the air molecules is proportional to the pressure. So P is like the spring constant k, and ρ is proportional to mass. Therefore, P/ρ plays the same role as k/m. If we examine units, we have the following: Pressure has SI units of N/m^2, and those of density are kg/m^3. The units of P/ρ turn out to be

$$\frac{P}{\rho} = \frac{\text{N}}{\text{m}^2} \frac{1}{\text{kg/m}^3} = \frac{\text{kg m}}{\text{s}^2 \text{m}^2} \frac{\text{m}^3}{\text{kg}} = \frac{\text{m}^2}{\text{s}^2}.$$

We see that $\sqrt{P/\rho}$ has the same units as speed.

Isaac Newton knew the pressure (10^5 N/m^2) and density (1.29 kg/m^3) of air and arrived at his theoretical prediction of sound, 979 English feet per second. That's why he was experimenting with pendulums and echoes in the corridor at Trinity College; he was testing his theory. As it turns out, experiment disagrees with theory in this case. The real speed of sound is a little bigger than $\sqrt{P/\rho}$. We'll return to this point shortly.

Questions

14. When asked how deep a well is, a farmer drops a stone into the well and upon hearing the splash, answers that the well is 13 seconds deep.

(a) Determine the depth of the well on the assumption that the time for the sound of the splash to reach the farmer is negligible.

(b) Calculate the depth of the well accounting for the finite speed of sound, 340 m/s.

15. A coiled Slinky with mass m and spring constant k is stretched to a length L. Present an argument that the speed of longitudinal waves on it should be given by $v_s = L\sqrt{k/m}$.

22.6 SOUND INTENSITIES

Sound engineers measure the intensity I of sound in units of decibels (abbreviated dB), defined as

$$I = 20 \log_{10}(P/P_{ref}), \tag{22.8}$$

where P_{ref}, the reference pressure, is 2×10^{-10} times normal atmospheric pressure. This particular reference pressure is used because it corresponds to the threshold of hearing – the weakest pressure signal the human ear can detect. Because of the enormous range of intensities to which the ear is sensitive and because the sensation of loudness varies with intensity not directly, but more nearly logarithmically, a logarithmic scale is used.

Table 22.1 lists the sound levels of many common sounds. The threshold of pain, which is about the loudest sound a person can stand, has an intensity of 120 dB, corresponding to pressure signals a million times stronger than the weakest the ear can detect.

Example 4

A barking dog creates sound at a level of 68 dB. How many times greater than the reference pressure are the pressure variations the dog makes?

From Eq. (22.8), we know that

$$I = 20 \log_{10}(P/P_{ref}),$$

and that here $I = 68$ dB. From this information, we have

$$\log_{10}(P/P_{ref}) = 68/20.$$

Taking the antilogarithm (base 10), which is simply raising 10 to the power 68/20, on a calculator, we find $P/P_{ref} = 2500$.

Questions

16. If the pressure oscillations of a sound wave are doubled, by how many decibels does the sound intensity change?

Table 22.1 Intensity Levels for Common Sounds

Source	P/P_{ref}	dB	Description
	$10^0 = 1$	0	Hearing threshold
Normal breathing	10^1	20	Barely audible
Library	10^2	40	Quiet
Conversation	10^3	60	
Factory	10^4	80	
Subway train	10^5	100	Danger to hearing
Rock concert	10^6	120	Pain threshold
Jet takeoff	10^7	140	
Rocket launch	10^9	180	

17. Suppose there are five dogs, each barking at the same time and at the same intensity as the one in Example 4 above. How many times greater than the threshold pressure will their collective pressure oscillations be? What will be the intensity?

18. If 26 persons at a cocktail party all babble simultaneously, each producing an intensity of 72 dB, what is the total sound intensity?

22.7 SOUND AND HEAT

There is still one subtlety about sound we haven't taken into account. If a gas is compressed, not only does its density increase, but the gas also heats up. If you've ever pumped up a bicycle tire with a hand pump, you probably have noticed this heating effect. If the heat can escape keeping the temperature constant, then the heat makes no difference, and the pressure and density increase together, just as Newton thought. In this case the ratio P/ρ is constant. But this is not the case with sound waves; they travel too fast for the heat to escape. Consequently, compression increases the temperature of the gas and causes the gas to bounce back from the disturbance more strongly than it would otherwise. The net result of the heating is to make sound travel a little faster than what we have so far been led to expect. In the words of Chapter 16, air in a sound wave is heated (and cooled) by adiabatic compression (and expansion).

The phenomenon of heat causing sound to travel faster has some dramatic effects. A familiar example occurs when a jet airplane moves through air faster than the speed of sound. When it does so, it creates a bow wave because the air must move out of the way. A similar effect occurs on a lake when a boat travels faster than the speed of waves it creates on the water. Figure 22.14 illustrates the bow wave for a supersonic jet airplane.

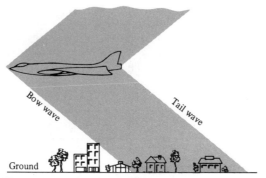

Figure 22.14 Shock wave created by a jet flying faster than the speed of sound.

The wave front is where the disturbance moving outward from the nose of the airplane has arrived moving at the fastest speed it can have, the speed of sound.

That fastest speed is the speed of sound in the undisturbed medium, which doesn't yet know that the jet is approaching. Behind the wave front, the air has been heated by compression, and the speed of sound is therefore faster than the speed of the front itself. As a result, the disturbance caused by the rest of the body of the craft catches up with the wave front and the change in pressure and density at the front builds up. In a way, it is similar to what happens to a water wave when it washes up on a beach, except that the sound disturbance doesn't fall over and "break." Instead, a shock wave is formed. You may know it better as a sonic boom.

The interplay between heat and the speed of sound is not only a problem for designers of supersonic aircraft, it was also unfortunate for Isaac Newton. His misfortune was that he was entirely unaware of it. Consequently, he found the wrong speed for sound. The actual speed is not simply $\sqrt{P/\rho} = 979$ ft/s, but something larger (it is actually $\sqrt{\gamma P/\rho}$, where γ is $\frac{7}{5}$ for air, as in Chapter 16). The best measurement at Newton's time, made by William Derham, was 1143 ft/s.

Question

19. How well does Derham's measurement compare to the correct prediction, $v_w = \sqrt{\gamma P/\rho}$?

22.8 A FINAL WORD

Newton had to make a value judgment about his calculation of the speed of sound. The value he calculated differed from the measured value by about 20%. He had to decide whether his value was good or bad.

Looking back on his calculation today, we know that it is stunningly good, an intellectual accomplishment of vast proportions. Before Newton, no one had the foggiest notion of why the speed of sound ought to be related to anything else in the world. Newton had exactly the right idea of what sound is, and found not just the right order of magnitude for its speed, but a result only 20% off from being exactly correct. The remaining cause of error, the heating effect, is so subtle that it would not be understood for a century after Newton. But Newton decided that his prediction must be brought into line with experiment. Why?

Newton had proposed the universal law of gravity, which describes the force bodies like the earth and moon exert on each other without touching. One hundred and fifty years earlier, Copernicus had initiated the scientific revolution, and one of the characteristics of this revolution was to sweep the occult out of physics. There had to be a natural, mechanical explanation for everything. To Newton's contemporaries, as well as to Newton himself, the action-at-a-distance nature of gravity seemed occult. Newton had to defend himself from charges, led by followers of Descartes, most notable of whom was Leibniz, that his law of gravity reintroduced occult qualities into physics. Newton's defense was that his theory worked so well. He found that he could perform calculations and always obtain precise results which agreed with observations. But, having decided that precise agreement between theory and experiment was the criterion for scientific success, he had to have agreement between the measured and predicted speeds of sound.

Impassioned by the struggle with Cartesians and in the midst of his plagiarism campaign against Leibniz, Newton set out to find corrections to his calculations for the second edition of the *Principia*. His argument went like this: The basic speed of sound in air is 979 ft/s, but in the derivation it was assumed that air molecules are point masses. But in fact, air molecules have some finite volume, and because the molecules themselves are rigid, sound travels through the molecules instantaneously. Consequently, he argued, sound actually travels an extra distance in one second due to the distance it travels through the molecules themselves. He estimated this effect by saying that the ratio of the density of air to the density of water is 1/870; this indicates what fraction of the volume of a gas is occupied by the hard molecules. The fraction of a linear distance is the cube root of the ratio. Therefore, sound travels an extra $\frac{1}{9}$ of the distance in a second, which is an extra $\frac{1}{9}$(979 ft) or 109 ft. Adding this to his calculated speed of sound gives 1088 ft/s, a value still less than Derham's.

Since this correction was insufficient to reconcile his calculation with the measured value, Newton argued that a further correction must be made for water vapor. Claiming that approximately 10% of the apparent density of air is due to water vapor, he said that the speed should be increased by the factor $(1/0.9)^{1/2}$, which is nearly equal to 21/20. Therefore, the actual speed should be (21/20)(1088 ft/s) or 1142 ft/s. The calculated value of 1142 ft/s is stated in the *Principia*. It agrees precisely with Derham's value, which itself was only an average, to within one part in 1000. Such was Newton's way of having science triumph over the occult.

CHAPTER

23

ANGULAR MOMENTUM

It is amazing! Although I had as yet no clear idea of the order in which the perfect solids had to be arranged, I nevertheless succeeded . . . in arranging them so happily, that later on, when I checked the matter over, I had nothing to alter. Now I no longer regretted the lost time; I no longer tired of my works; I shied from no computation, however difficult. Day and night I spent with calculations to see whether the proposition that I had formulated tallied with the Copernican orbits or whether my joy would be carried away by the winds . . . Within a few days everything fell into its place. I saw one symmetrical solid after the other fit in so precisely between the appropriate orbits, that if a peasant were to ask you on what kind of hook the heavens are fastened so that they don't fall down, it would be easy for thee to answer him. Farewell!

Johannes Kepler, Preface to *Mysterium Cosmographicum* (1596)

23.1 THE SEARCH FOR ORDER

In the sixteenth and seventeenth centuries, that brilliant awakening known as the Renaissance was brought to an end as Europe turned its attention to a spirited debate on the finer points of Christian theology. This period, known as the Reformation and Counter-Reformation, culminated in the bloody Thirty Years' War (1618–48).

At that time, three individuals devoted to physics and astronomy laid the foundations for the enormous scientific achievements of Isaac Newton. They were Galileo Galilei, Johannes Kepler, and Tycho Brahe. Each of these men searched for order in the heavens, but none perhaps had a life filled with more turmoil than Kepler.

Kepler was born in 1571 and died in 1630. He was a small, frail, nearsighted man, whose life was punctuated by illness of all sorts; he described himself as a mangy dog. As a child with an avid interest in astronomy, he was considered an intolerable egghead by his classmates. But as he grew older, he transformed his inner torments into creative achievements. Although he suffered periods of profound depression, he also showed soaring elation and bubbling enthusiasm for his discoveries, not all of which were more than brilliant speculation. But in everything he did he was rigorously honest. When he wrote about his scientific discoveries, his books revealed everything – not just what he had discovered, but also all of his wrong turns, as well as how he felt. No scientist writes that way any more; they disguise what they actually did and simply write down their final results. But in the case of Kepler, we know everything he did and exactly how he did it. There is hardly a page in Kepler's writings (in some 20 volumes) which is not full of vitality.

In 1594 Kepler assumed the post of mathematician in Graz, Austria. He thought himself a poor teacher because whenever he got excited, which was most of the time, he burst into speech without taking time to consider whether or not he was saying the right thing. His eagerness was harmful because it continually led him into digressions, new subjects, or new ways of proving his point. No wonder that in his first year he had only a handful of students in his class, and in his second, barely any at all.

Another duty Kepler had and secretly enjoyed during his four years in Graz was the publication of an annual calendar of astrological forecasts. This was a traditional obligation which brought a much needed supplement to the miserable salary of the official mathematician. With his first calendar Kepler was exceedingly lucky; he correctly predicted a bitter cold winter, peasant uprisings, and an invasion by the Turks – all of which came to pass. Although Kepler pursued astrology, which he called the foolish little daughter of respectable astronomy, his reaction to it was mixed. He had a profound feeling for harmony in the universe, which included a belief in a concord between the cosmos and the individual.

During his first year in Graz, Kepler stumbled upon an idea which remained an inspiration throughout his life. As he was drawing a figure on the blackboard for his class, he was struck with what he felt was the key to the universe. The idea was that the universe is built around certain symmetrical solids which form an invisible skeleton. Kepler realized that there existed only five regular solids, each with identical faces, and there were precisely five intervals between the planets. These regular solids are known as the Platonic solids. As shown in Fig. 23.1, they are (1) the tetrahedron (pyramid), (2) the cube, (3) the octahedron (8 equilateral triangles), (4) the dodecahedron (12 pentagons), and (5) the icosahedron (20 equilateral triangles). Being symmetrical, each of these solids can be inscribed in a sphere so that all corners are on the surface of the sphere. A sphere can also be surrounded by each solid so that each face of the solid touches the sphere.

Kepler fervently believed that it was impossible that the number of solids and intervals between the planets were equal only by chance. To him, it provided the complete answer

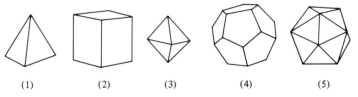

Figure 23.1 The five Platonic solids (1) tetrahedron, (2) cube, (3) octahedron, (4) dodecahedron, (5) icosahedron.

to why there were just six planets and not twenty or a hundred. It also explained why the distances between the orbits were as they were: they had to be spaced so that the five solids could be exactly fitted into the intervals, forming an invisible frame. So into the sphere of Saturn's orbit, Kepler inscribed a cube, then into the cube he inscribed the sphere of Jupiter's orbit, into that a tetrahedron, then the sphere of Mars, as shown in Fig. 23.2. Between the orbit of Mars and the earth came a dodecahedron, and between the earth and Venus was an icosahedron, then finally an octahedron between the spheres of Venus and Mercury.

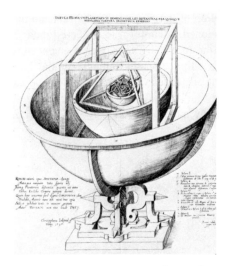

Figure 23.2 Kepler's model of the universe based on the Platonic solids. (Courtesy of the Archives, California Institute of Technology.)

At the age of 25, the ardent, restless Kepler wrote the *Mysterium Cosmographicum*, in which he elaborated on his model of the universe and enthusiastically supported the Copernican system. It was the first public support of that system by a professional astronomer in the 50 years since the death of Copernicus. This work thrust him into the front rank of astronomers, for he asked questions about the motion of planets which had never been asked before.

In the *Mysterium Cosmographicum*, Kepler concerns himself with finding reasons for the number and distribution of the planets. Satisfied that the five Platonic solids provided all the answers (which they did not) and citing discrepancies between his model and data as due to Copernicus's inaccurate measurements, Kepler turned to a more promising problem. He began an immensely fertile search for a mathematical relation between a planet's distance from the sun and the time needed to complete one revolution.

The periods of the five planets' orbits were well known; the greater the planet's distance from the sun, the longer its period. A precise mathematical ratio, however, was lacking. Saturn, for example, is twice as far as Jupiter from the sun, but its 30-year period is not twice the 12-year period of Jupiter. No simple ratio relates distances and periods of the other planets as well. The orbital period of a planet increases the farther the planet is from the sun, but not in the same proportion as the distance. Nobody before Kepler had asked why this should be so.

Kepler theorized that there must be a force emanating from the sun which drives the planets in their orbits around the sun. The outer planets move more slowly because this driving force diminishes with distance. Kepler's proposal had revolutionary significance. For the first time an attempt was made not only to describe heavenly motion in geometrical terms, but to assign a physical cause to it. After a divorce of 2000 years, astronomy and physics met again. This reunion led to Kepler's three laws, the pillars on which Newton built his universe.

After years of false starts and drudgery searching for a more accurate description in the Copernican system of the motion of the planets, Kepler realized that the ancient idea of circular orbits had to be abandoned to conform to observed data. Through obstinate perseverance, he discovered on the basis of Tycho Brahe's new observations that the actual orbits correspond more to elliptical paths rather than the circular paths of the Copernican system. The sun is at one focus of the ellipse, and the earth, for example, moves around it as shown in Fig. 23.3. This is *Kepler's first law: the planets orbit the sun along elliptical paths*. We'll say more about this law in Chapter 25.

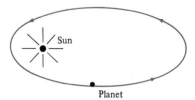

Figure 23.3 Greatly exaggerated illustration of Kepler's first law: the orbits of planets are ellipses with the sun at one focus.

Since antiquity, astronomers had known that the planets do not move along their orbits at constant speed. Each planet moves faster when it is close to the sun than when it is far away. This means that if we draw the angles swept out by the earth moving around the sun in equal intervals of time at two different parts of the orbit, one angle is smaller than the other, as illustrated in Fig. 23.4. In searching for some regularity in this orbit, Kepler discovered a law of amazing simplicity. As a planet moves in its orbit, *the radius vector from the sun to the planet sweeps out equal areas in equal times*; this is *Kepler's second law*. The solution to the problem we saw him struggling with earlier is his *third law*, which states that *the square of a planet's period* (i.e., year) *is proportional to the cube of its mean distance from the sun.*

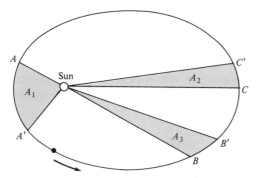

Figure 23.4 An illustration of Kepler's second law: an imaginary line from the sun to a planet sweeps out equal areas in equal times.

Questions

1. If a sphere has a radius R, what must be the length of the edge of a cube which is inscribed in it, as Kepler proposed for the skeleton inside the sphere of Saturn? What would be the radius of a sphere, like that of Jupiter, inscribed in this cube?

2. Between the spheres of the orbits of Jupiter and Mars, Kepler proposed to inscribe a tetrahedron. If the radius of the outer sphere is R, what must be the length of an edge of one of the four equilateral triangles forming the faces of the tetrahedron?

23.2 THE LAW OF EQUAL AREAS

Nearly 100 years after Kepler, Isaac Newton formulated his laws of mechanics and demonstrated that Kepler's laws can be deduced from his mechanics. We'll now demonstrate precisely how Kepler's second law of equal areas arises from Newton's dynamics. But first we need an accurate mathematical description of the law of equal areas.

Imagine a radius vector at some time t, $\mathbf{r}(t)$, drawn from the sun to a planet as in Fig. 23.5. If the planet moves through a displacement $\Delta\mathbf{r}$ in a short time Δt, then its new

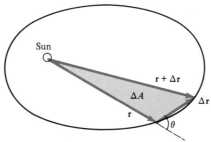

Figure 23.5 Position vector of a planet sweeping out an area ΔA in time Δt.

position is $\mathbf{r}(t + \Delta t) = \mathbf{r} + \Delta\mathbf{r}$. As illustrated in Fig. 23.5, the three vectors $\mathbf{r}(t)$, $\Delta\mathbf{r}$, and $\mathbf{r}(t + \Delta t)$ form a triangle. The area ΔA of this triangle is one-half the base r times the height, $\Delta r \sin \theta$; thus

$$\Delta A = \tfrac{1}{2}r(\Delta r \sin \theta),$$

where θ is the angle between $\mathbf{r}$ and $\Delta\mathbf{r}$.

Writing the area this way suggests using the vector cross product to represent the area. Recall that in Chapter 5 we found that the magnitude of the cross product between two vectors is equal to the area of the parallelogram formed by the vectors, as Fig. 23.6 shows. And the area of the triangle is half the area of the parallelogram. In vector notation, we can write the area as

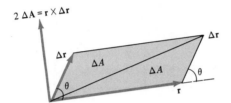

Figure 23.6 Area vector $\Delta\mathbf{A}$ as a vector cross product.

$$\Delta A = \tfrac{1}{2}|\mathbf{r} \times \Delta\mathbf{r}|.$$

The direction of the cross product, given by the right-hand rule, is perpendicular to the plane of the vectors. So we can introduce the vector

$$\Delta\mathbf{A} = \tfrac{1}{2}\mathbf{r} \times \Delta\mathbf{r}$$

and call it the area vector $\Delta\mathbf{A}$. Its length is the area of the triangle formed by $\mathbf{r}$ and $\Delta\mathbf{r}$.

The rate of change of this area vector, $d\mathbf{A}/dt$, is the limit of $\Delta\mathbf{A}/\Delta t$ as the time interval shrinks to zero:

$$\frac{d\mathbf{A}}{dt} = \lim_{\Delta t \to 0} \frac{\Delta\mathbf{A}}{\Delta t} = \lim_{\Delta t \to 0} \frac{\tfrac{1}{2}\mathbf{r} \times \Delta\mathbf{r}}{\Delta t} = \tfrac{1}{2}\mathbf{r} \times \lim_{\Delta t \to 0} \frac{\Delta\mathbf{r}}{\Delta t}.$$

But the last limit is simply the velocity $\mathbf{v}$ of the planet. Therefore we find that the rate of change of the area vector is

$$\frac{d\mathbf{A}}{dt} = \tfrac{1}{2}\,\mathbf{r} \times \mathbf{v}, \tag{23.1}$$

and hence

$$\left|\frac{d\mathbf{A}}{dt}\right| = \tfrac{1}{2}\,|\mathbf{r} \times \mathbf{v}|.$$

Now we can express the law of equal areas in a precise mathematical form. It says that the vector $d\mathbf{A}/dt$ has constant length:

$$\left|\frac{d\mathbf{A}}{dt}\right| = \text{const.}$$

But we can invoke Kepler's first law to show that $d\mathbf{A}/dt$ also has constant direction. The first law states that each orbit is an ellipse, so the orbit lies in a plane which contains both the position vector $\mathbf{r}$ and the velocity vector $\mathbf{v}$. Therefore the vector $d\mathbf{A}/dt$, being half the cross product of $\mathbf{r}$ and $\mathbf{v}$, must be perpendicular to the orbital plane. In other words, $d\mathbf{A}/dt$ has constant direction as well as constant magnitude. This means that

$$\frac{d\mathbf{A}}{dt} \quad \text{is a constant vector.}$$

This mathematical statement contains Kepler's second law of equal areas and also part of the first law, which states that each orbit lies in a plane.

Example 1

At its closest approach (perihelion), the earth is 1.47×10^8 km from the sun and its speed is 30.2 km/s. What is the speed at its farthest (aphelion) point from the sun, a distance of 1.52×10^8 km?

As illustrated with the ellipse below, when a planet is either at its closest or farthest point from the sun, its velocity is perpendicular to the radius vector.

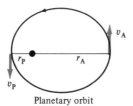

Planetary orbit

For these points Eq. (23.1) implies that

$$\tfrac{1}{2}\, r_P v_P = \tfrac{1}{2} r_A v_A.$$

Solving for v_A, the speed of the planet at its farthest distance, we have

$$v_A = v_P(r_P/r_A) = (30.2 \text{ km/s})(1.47/1.52) = 29.2 \text{ km/s}.$$

Questions

3. Referring to the elliptical orbit shown in Fig. 23.4, where is the velocity of the planet maximum? Where is the velocity minimum?

4. If the speed of the earth is 29.8 km/s when it is 1.49×10^8 km from the sun, what possible angle does the velocity vector make with its position vector? (Use data in Example 1.)

5. What does Kepler's second law imply about the speed of a planet moving in a circular orbit?

6. In what type of orbit, elliptical or circular of the same average radius and speed, would you expect a planet to have greater total energy?

23.3 ANGULAR MOMENTUM

Let's now begin to analyze planetary motion in terms of Newtonian mechanics. According to Newton, the force of gravity, given by

$$\mathbf{F} = -G\frac{M_s m_p}{r^2}\,\hat{\mathbf{r}}, \tag{8.1}$$

is responsible for the motion of a planet of mass m_p. This force produces an acceleration, according to the second law:

$$\mathbf{F} = m_p\frac{d\mathbf{v}}{dt}. \tag{6.1}$$

Looking at Eq. (23.1), we see that we really want to know something about the quantity $\mathbf{r} \times \mathbf{v}$, where $\mathbf{r}$ is the vector from the planet to the sun and $\mathbf{v}$ is its velocity vector. To do this, let's form the cross product of $\mathbf{r}$ with both sides of Eq. (6.1):

$$\mathbf{r} \times \mathbf{F} = m_p\mathbf{r} \times \frac{d\mathbf{v}}{dt}.$$

The force of gravity always acts along the line connecting the sun and planet; in other words it is parallel to $\mathbf{r}$, as Eq. (8.1) indicates. But the cross product of any vector with another vector parallel to it is always zero. Therefore we have

$$\mathbf{r} \times \mathbf{F} = \mathbf{0},$$

which implies that

$$m_p\mathbf{r} \times \frac{d\mathbf{v}}{dt} = \mathbf{0}. \tag{23.2}$$

This result is obvious for uniform circular motion. In Chapter 9 we found that the acceleration of a body moving uniformly on a circle always points radially inward. Consequently, on a circle the cross product of the radius and acceleration is zero. However, what we've just found holds not only for uniform circular motion, but for any motion in which the acceleration is radial, in particular, for planetary motion.

Equation (23.2) is a differential equation which describes the motion of any planet. By solving it for the unknown function $\mathbf{r}(t)$ we could describe explicitly the motion of any planet. Yet, without solving it, we can gain considerable insight into planetary motion.

Note that $\mathbf{r} \times \mathbf{v}$ enters into Eq. (23.1) and $\mathbf{r} \times d\mathbf{v}/dt$ appears in (23.2). The two cross products $\mathbf{r} \times \mathbf{v}$ and $\mathbf{r} \times d\mathbf{v}/dt$ are related. To find this relation, consider

$$\frac{d}{dt}(\mathbf{r} \times \mathbf{v}) = \frac{d\mathbf{r}}{dt} \times \mathbf{v} + \mathbf{r} \times \frac{d\mathbf{v}}{dt}.$$

Realizing that dr/dt is the velocity $\mathbf{v}$ and that the cross product of any vector with itself is zero, that is, $\mathbf{v} \times \mathbf{v} = \mathbf{0}$, we have the relation

$$\frac{d}{dt}(\mathbf{r} \times \mathbf{v}) = \mathbf{r} \times \frac{d\mathbf{v}}{dt}.$$

Using this to substitute for $\mathbf{r} \times d\mathbf{v}/dt$ in Eq. (23.2) and bringing the mass m_p inside the derivative (because it's constant), we arrive at

$$\frac{d}{dt}(m_p\mathbf{r} \times \mathbf{v}) = \mathbf{0}. \tag{23.3}$$

We now have a differential equation which we can integrate; when the derivative of something is zero, that something must be a constant. Therefore the quantity inside the parentheses of Eq. (23.3) must be equal to a constant vector which we'll call $\mathbf{L}_p$,

$$\mathbf{L}_p = m_p\mathbf{r} \times \mathbf{v}. \tag{23.4}$$

The quantity $m\mathbf{r} \times \mathbf{v}$ is called the *angular momentum* of any object of mass m. The SI unit of angular momentum is the kilogram-meter per second (kg m/s). In our derivation, we naturally used the sun as the origin for calculating angular momentum; $\mathbf{r}$ starts from the sun. In the general definition of angular momentum, the vector $\mathbf{r}$ starts from some reference point 0 to the position of the particle of mass m, and the angular momentum is said to be about 0.

Equation (23.3) implies that as a planet orbits the sun its angular momentum about the sun is constant. Because

$$\frac{d\mathbf{L}}{dt} = \mathbf{0},$$

we say that the angular momentum $\mathbf{L}$ of the planet is conserved.

Now we can complete our derivation. From Eq. (23.1) we see that $d\mathbf{A}/dt$ is simply a scalar multiple of the angular momentum:

$$\frac{d\mathbf{A}}{dt} = \tfrac{1}{2}\mathbf{r} \times \mathbf{v} = \frac{\mathbf{L}_p}{2m_p} = \text{const.} \tag{23.5}$$

This shows that Kepler's empirical law of equal areas,

$$\frac{d\mathbf{A}}{dt} = \text{const,}$$

is a consequence of the conservation law $d\mathbf{L}/dt = \mathbf{0}$, which, in turn, follows from Newton's gravitational law and his second law of motion.

Note. If we use Newton's second law but *not* the gravitational law, the foregoing analysis shows that

$$\mathbf{r} \times \mathbf{F} = \frac{d}{dt}(\mathbf{r} \times m\mathbf{v}) = \frac{d\mathbf{L}}{dt}$$

for any motion caused by any force $\mathbf{F}$. The angular momentum is conserved if $d\mathbf{L}/dt = \mathbf{0}$. This happens if $\mathbf{r} \times \mathbf{F} = \mathbf{0}$ or if $\mathbf{r} \times m\mathbf{v}$ is constant.

Example 2

A particle of mass m moves with constant velocity $\mathbf{v}$ along a straight line which is a distance b from the origin of a coordinate system.

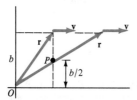

(a) Find the angular momentum of the particle at any instant about point 0.
(b) Show explicitly that the angular momentum is conserved.
(c) Calculate the angular momentum about point P.

(a) We know that the angular momentum is given by

$$\mathbf{L} = m\mathbf{r} \times \mathbf{v}. \tag{23.4}$$

The magnitude of the angular momentum is $L = mr_\perp v$, where $r_\perp$ is the component of the position vector $\mathbf{r}$ perpendicular to the velocity $\mathbf{v}$. From the diagram we see that this distance is b. Therefore

$$L = mr_\perp v = mbv.$$

By the right-hand rule, the direction of the angular momentum is into the plane of the page.

(b) From our expression for L, we see that as the particle continues to move along the straight line, the component of $\mathbf{r}$ perpendicular to $\mathbf{v}$ remains equal to b. Therefore the angular momentum is conserved. The reason for the conservation of angular momentum here is that no net force (by Newton's first law) acts on the particle. Looking over the steps leading to Eq. (23.4), we see $\mathbf{F} = \mathbf{0}$ is a special case leading to conservation of angular momentum.

(c) The angular momentum about point P is simply $L = mr_\perp v = m(\frac{1}{2}b)\, v$, which remains constant. Although the value of the angular momentum depends on the reference point, if the angular momentum is conserved, it is conserved for *any* reference point.

Example 3

An object of mass m moves in a circle of radius R with a constant speed v. What is its angular momentum?

At any instant, the velocity $\mathbf{v}$ of the object is perpendicular to the radius vector $\mathbf{R}$. Therefore the magnitude of the angular momentum is

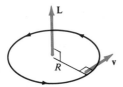

$$L = mvR \sin 90° = mvR.$$

The direction of the angular momentum is perpendicular to the plane of the circle, as shown in the figure.

Questions

7. Suppose an object were acted upon by the force $\mathbf{F} = -k\mathbf{r}$, where $\mathbf{r}$ is the radius vector from some origin. Would the angular momentum about the origin of the particle be conserved? Explain why.

8. Two identical masses, each of mass m, move along straight lines with velocity $\mathbf{v}$, as shown in the diagram. Calculate the angular momentum of the system with respect to the origin 0.

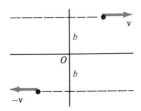

9. Two identical masses, each of mass 3.0 kg, are connected by a light rod and move in a circle as shown. The speed of each is 1.5 m/s.

(a) What is the total angular momentum of the objects about the center?
(b) If the rod contracts to one-third of its original length, will the speed of the objects change? If so, how?

10. In a certain coordinate system, a 0.5-kg particle is at a position $\mathbf{r} = (3\hat{\mathbf{i}} - 2\hat{\mathbf{j}} + \hat{\mathbf{k}})$ m moving with a velocity $\mathbf{v} = (-5\hat{\mathbf{i}} + 3\hat{\mathbf{j}} - 2\hat{\mathbf{k}})$ m/s.

(a) Find the angular momentum of this particle with respect to the origin of the coordinate system.

(b) Calculate the angular momentum relative to the point $(1,1,1)$.

11. A mass m is twirled in a circle of radius r_1 with a constant speed v_1. If the string is pulled so that the mass moves in a circle of radius r_2 as shown, find

(a) the speed of the particle,
(b) the angular speed of the particle.

12. A toy train is initially at rest on a track fastened to a bicycle wheel with its axis vertical. When the train moves clockwise, the wheel also rotates. What happens when the train stops? Explain why.

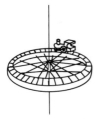

23.4 TORQUE AND ANGULAR MOMENTUM

Let's summarize what we've found so far. Considering an object of mass m and velocity $\mathbf{v}$ acted upon by a force $\mathbf{F}$, we formed $\mathbf{r} \times \mathbf{F}$, where $\mathbf{r}$ is a vector from some arbitrary origin to the object. We found that in general this quantity is related to the time rate of change of something we identified as the angular momentum:

$$\mathbf{r} \times \mathbf{F} = \frac{d}{dt}(\mathbf{r} \times m\mathbf{v}) = \frac{d\mathbf{L}}{dt}.$$

The quantity $\mathbf{p} = m\mathbf{v}$ is called the *linear momentum* of the object. What we've found is a "twisted" version of Newton's second law. We call the quantity $\mathbf{r} \times \mathbf{F}$ a *torque* and use the symbol $\boldsymbol{\tau}$ (the Greek letter tau) to denote it:

$$\boldsymbol{\tau} = \mathbf{r} \times \mathbf{F}. \tag{23.6}$$

The SI unit of torque is the newton-meter (N m).

Torque plays much the same role in rotation, as force does in linear motion. The relation between torque and angular momentum is as follows: *the torque acting on an object is equal to the rate of change of angular momentum,*

$$\boldsymbol{\tau} = \frac{d\mathbf{L}}{dt}. \qquad (23.7)$$

For linear motion, *force* is equal to the rate of change of *linear momentum*. And when no net external force acts on a system, linear momentum is conserved.

In particular, for any system on which no torques act, that is, for which

$$\sum \boldsymbol{\tau} = \mathbf{0},$$

then by Eq. (23.7), it follows that $d\mathbf{L}/dt = \mathbf{0}$ and the angular momentum is conserved. So a necessary condition for the angular momentum of a system to be conserved is that no net external torques act on it.

Example 4

An ice skater holds two heavy masses, each of mass 8.0 kg. Originally the skater is rotating at an angular speed of 2 rad/s with the masses 0.4 m from the axis of rotation. What will be the skater's angular speed if the masses are pulled in to a distance of 0.3 m?

Since the ice has very little friction on skates, there are essentially no torques acting on the skater. Therefore the angular momentum of the system is conserved. Each mass moves in a circle about the axis of rotation, so the total angular momentum is the sum of the individual angular momenta. Since the angular momentum of both masses is in the same direction, the magnitude of the total angular momentum is

$$L_1 = mvr_1 + mvr_1 = 2mvr_1,$$

where r_1 is the distance from the axis and v is the speed of each mass. We can write this in terms of angular speed by recalling that $v = r_1\omega_1$:

$$L_1 = 2mr_1^2\omega_1.$$

When the masses are pulled in to a new distance r_2, the angular momentum will be

$$L_2 = 2mr_2^2\omega_2.$$

Conservation of angular momentum implies that $L_1 = L_2$. Setting the two angular momenta equal and solving for ω_2, we get

$$\omega_2 = \omega_1(r_1^2/r_2^2).$$

Substituting $\omega_1 = 2$ rad/s, $r_1 = 0.4$ m, and $r_2 = 0.3$ m, we find $\omega_2 = 3.6$ rad/s.

Example 5

A force of 15.0 N acts on the particle shown. Compute the torque acting on the particle with respect to (a) point A, (b) point B.

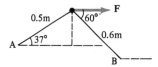

By definition, the torque is the cross product of the vector drawn from the reference point to the particle and the force vector. The magnitude of the torque can be expressed as

$$\tau = r_\perp F,$$

where $r_\perp$ is the component of **r** perpendicular to **F**; this distance is also known as the *lever arm* of the force. Using this, we easily see from the diagram that the torque about point A is

$$\tau_A = (0.50 \text{ m})(\sin 37°)(15.0 \text{ N}) = 4.5 \text{ N m}.$$

By the right-hand rule, the direction of the torque is into the plane of the page. Similarly, the torque about point B is

$$\tau_B = (0.60 \text{ m})(\sin 60°)(15.0 \text{ N}) = 7.8 \text{ N m}.$$

The direction of this torque is also into the plane of the page.

Questions

13. Of the four forces shown acting on the wheel, which produces the greatest torque about the axis?

14. A worker finds it very tough to twist a stubborn bolt with a wrench, so he attaches a rope to the wrench as shown, and pulls just as hard on the rope. How does the torque applied with the rope compare to that without the rope?

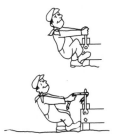

15. A particle of mass m is at the end of a rod of negligible mass and length L. If at any instant the rod makes an angle θ with the horizontal as it falls, calculate the torque about 0 acting on the mass as it falls.

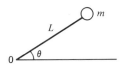

16. A 1.5-kg particle has position and velocity as shown with $r = 0.4$ m and $v = 3.0$ m/s. The magnitude of the force acting on it is 4.0 N.

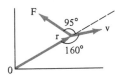

Relative to the origin find

(a) the angular momentum of the particle,
(b) the torque acting on it.

17. A particle at position $\mathbf{r} = (3\hat{\mathbf{i}} + 6\hat{\mathbf{j}} - 2\hat{\mathbf{k}})$ m is acted upon by a force $\mathbf{F} = (-4\hat{\mathbf{i}} - 3\hat{\mathbf{j}} + 5\hat{\mathbf{k}})$ N. Find the torque about the origin acting on the particle.

23.5 VORTICES AND FIRESTORMS

We've just found that the law of conservation of angular momentum explains why planets orbiting the sun sweep out equal areas in equal times – an empirical law which Kepler discovered by examining voluminous data. The conservation of angular momentum applies not only to planets moving silently around the sun, but to any moving body for which $\mathbf{r} \times m\mathbf{v}$ is constant. One interesting application of this law is to vortices.

Suppose you have a huge bowl (a bathtub will suffice) which you fill with water through a hose, as shown in Fig. 23.7. As water pours into the bowl, it moves in some type of circulatory motion. If you turn off the water, that rotary motion will die out very slowly; it may not completely disappear for hours or even days. What happens if you pull the plug on the bottom of the bowl while the water is still rotating? Of course, the water will run out. When you first pull the plug, the water flows straight out through the hole, but after a while, the water goes into a new kind of flow state. Instead of flowing straight out, it forms a whirlpool. Why?

Aside from a small amount of viscosity from the walls of the bowl, there are no torques acting on the water. Consequently, the water conserves angular momentum as it runs out. The angular momentum of a small portion of water at any instant as it spirals toward the center of the bowl is $\mathbf{L} = \mathbf{r} \times m\mathbf{v}$, about an axis through the center of the bowl. Since $\mathbf{r}$ and $\mathbf{v}$ are perpendicular, the magnitude of the angular momentum of this chunk is

Figure 23.7 Water flowing out of a tank through a hole in the bottom forms a whirlpool.

$$L = mvr = \text{const},$$

where we are ignoring viscosity. The precise value of the constant depends on the initial conditions of rotation.

As each small bit of water moves down toward the hole, it conserves angular momentum. So as the distance from the central axis becomes smaller, the velocity becomes increasingly larger, according to

$$v = (\text{const})/mr.$$

When the water is moving in very small circles, it is moving very rapidly. But an inward force must keep this bit of water moving in a circle. That force which provides the centripetal acceleration is the tensile strength of the water – the ability of the water to keep itself together.

As the distance r becomes very small, the velocity correspondingly becomes extremely large. A large force is required to hold the water together. But when this force exceeds the tensile strength of the water, the water can no longer keep itself moving in a circle, so the surface ruptures and forms a hole. The hole in the center of the whirlpool is called a vortex.

The same dynamics occurs in other instances on a much larger scale. Whenever there is a large fire, such as a forest fire, great destruction can occur. This happened in the Chicago fire of 1871, and in the bombing of Hamburg, Germany during the Second

World War. The cause of the destruction was not that of a simple fire, but something more devastating, a firestorm. Firestorms are vortices similar to those in a bathtub, and are a result of conservation of angular momentum.

The heat of the fire causes the air to swell upward (just as we had water flowing downward). The resulting low pressure region at the bottom draws in oxygen from the sides, so the fire burns faster. If there is any circulatory motion in the air, the air circulates faster and faster as it moves toward the center of the fire, where the upwelling occurs. Since the air cannot provide the centripetal force to keep itself moving so rapidly, a vortex forms. Thus the same phenomenon occurs as in a bowl of water, except upside down. This vortex is a long-lived and violent state – the firestorm.

Meteorologists investigating historical records for the Great Fire of London in 1666 discovered that the air in the vicinity at that time had no distinct circulatory motion. Consequently, they believe that a firestorm never developed. Their conjecture is borne out by a comparison in the tolls: in the London fire, only four people died directly as a result of the fire and only 436 acres were burnt in 87 hours; on the other hand, in the Chicago fire, 2124 acres were burnt and 250 people killed in two days.

The same phenomenon can occur in the atmosphere without a fire. If there is an upwelling of air, such as that above warm water, and if there is any circulatory motion, a vortex can form. In this case it is called a hurricane. Hurricanes can be long-lived (several days), persistent, and very destructive. The longest-lived hurricane we know of is on the planet Jupiter; it is the Great Red Spot of Jupiter, which has persisted for at least 300 years. Figure 23.8 is a photograph of the Red Spot taken by *Voyager 1* on its flight past the planet. Years before the Voyager fly-by, it was suggested that the Red Spot was a huge hurricane, three times the size of the earth. Data from Voyager support this idea.

Figure 23.8 Photograph from *Voyager 1* of the Great Red Spot on Jupiter. (Courtesy JPL/NASA.)

Questions

18. In what way would you expect the "width" of a whirlpool to depend on the tensile strength of the liquid?

19. What happens to the kinetic energy of a bit of fluid as it spirals toward the center of a vortex? If you agree that it increases, what is the source of this energy for a bathtub whirlpool?

23.6 ANGULAR MOMENTUM AND THE ARCHITECTURE OF THE HEAVENS

The conservation of angular momentum has consequences on all scales, from the atom to the Cosmos. An especially fascinating application is something that not even the imaginative Kepler would have envisioned – the contraction of colossal clouds of gas and dust which form galaxies and solar systems.

Suppose initally there is a spherical cloud of gas (mostly hydrogen) and dust in intergalactic space. By considering the energy and angular momentum of a particle on the edge of this cloud, we can follow the evolution of the cloud as a whole.

Suppose the mass of this particle is m. The particle is acted upon by the gravitational force due to all of the mass M in the sphere. The mutual gravitational attraction of all particles in the cloud causes it to contract. Since by assumption the cloud is spherical, we can treat it as a point mass located at the center. (We found this out in Chapter 8.) Therefore, we can write the potential energy of the particle (recalling Chapter 14) as

$$U(r) = -GMm/r, \tag{23.8}$$

where we've imagined bringing in the particle from an infinite distance to the edge of the cloud, a distance r from the center; that's why the potential energy is negative – because of our choice of the reference point.

Astrophysicists believe that these clouds have initial rotations. We will assume that the entire cloud is rotating about a single axis. Thus the particle, through its rotation with the cloud, has angular momentum **L**, of magnitude

$$L = mvr.$$

To illustrate the physics, we'll assume in addition that the particle is on the equator of the cloud.

Because gravity, directed toward the center of the cloud, applies no torque on the particle, the angular momentum of the particle must be conserved as the particle moves. This means that the speed of the particle depends on the distance from the center:

$$v = \frac{L}{mr},$$

where L is a constant which depends on the initial rotation of the cloud.

We can express the kinetic energy of the radial motion of the particle in terms of r by substituting for v from above:

$$K(r) = \tfrac{1}{2} mv^2 = \tfrac{1}{2} m \left(\frac{L}{mr} \right)^2 = \frac{L^2}{2mr^2}. \tag{23.9}$$

The total energy of the particle, therefore, is

$$E(r) = K + U = \frac{L^2}{2mr^2} - G\frac{Mm}{r}. \tag{23.10}$$

Figure 23.9 illustrates graphs of the kinetic, potential, and total energy of the particle.

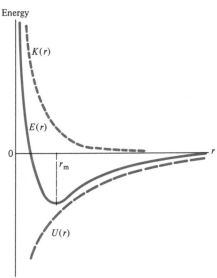

Figure 23.9 Kinetic, potential, and total energy of a particle moving under a gravitational force.

The particle's potential energy is always negative and its kinetic energy is always positive. The combination of the two energies produces a curve which has a minimum value, a point of stability. In Chapter 14 we dealt with the stability of a stationary body, and found that the minimum in the potential-energy curve is a point of stable equilibrium – one to which the object will eventually return if slightly displaced.

The curve we have now is not a potential-energy curve, but rather a graph of the total energy, which is a function of r; it is a kind of effective potential. Our interpretation of the graph is slightly different from that in Chapter 14 because now the object is moving.

In a condensing gas cloud, a particle spirals in, losing energy by collisions with other particles until it eventually reaches the lowest energy possible. Then, because it has no way to gain energy, the particle remains moving around at that certain distance from the center. In other words, the stable orbit is a circle. As Fig. 23.10 indicates, particles with different initial values of angular momentum will have different minimum values of energy corresponding to circular orbits of different radii.

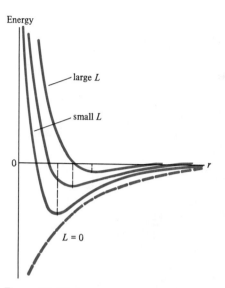

Figure 23.10 Stable orbit of a particle depends on the initial angular momentum.

We can mathematically determine the radius of the circular orbit which corresponds to minimum energy; it is the value of r for which the slope of the curve is zero. In other words, the minimum energy occurs where $dE/dr = 0$. Taking the derivative of Eq. (23.10),

$$\frac{dE}{dr} = -\frac{L^2}{mr^3} + G\frac{Mm}{r^2},$$

setting it equal to zero, and solving for r_m, the minimum distance, we find

$$r_m = \frac{L^2}{GMm^2}. \tag{23.11}$$

We now have a picture of the evolution of a rotating gas cloud. As Fig. 23.11 depicts, the cloud begins to contract toward a point under the mutual attraction of gravity. As it contracts in the equatorial direction, the cloud can't move all the way in because of conservation of angular momentum. Particles will settle into circular orbits whose radii are prescribed by Eq. (23.11). However, in the vertical direction, contraction can take place without any increase in rotational kinetic energy and therefore continues after the limit to contraction in the equatorial direction has been reached. Consequently, the cloud tends to flatten out, forming galaxies like those in Fig. 23.12.

Not only do galaxies often have this pancake shape, but so does our solar system, in which all the planets are in nearly the same plane. And planetary systems – the moons and rings of each planet – also have this same basic structure. Presumably all these phenomena are consequences of the conservation of angular momentum.

In addition to the pancakelike shape, some galaxies exhibit spiral arms. The simple explanation we found for the flattening is not sufficient to explain the structure of the spiral arms. Different phenomena are occurring. Likewise, when gas clouds condense to

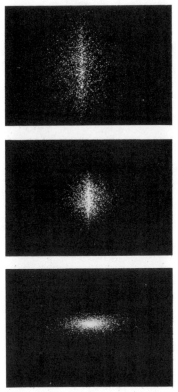

Figure 23.11 Successive stages in the contraction of a rotating sphere of gas.

form solar systems, planets are formed, rather than a uniformly flattened structure as we have in galaxies. The same thing happens in planetary systems; moons are formed out of the gas and dust condensing into planets. Sometimes rings are formed. Since at least three planets in our solar system have rings, we know that ring formation is not unusual.

In all these phenomena, there presumably is no underlying physics involved other than Newton's laws. If we could solve the differential equations for all sorts of conditions, then we should be able to predict all the spectacular phenomena, such as Saturn's braided rings, which were unexpected before the Voyager exploration. There are solutions to Newton's simple equations which nobody ever guessed existed. Commenting on this, the Nobel laureate Richard P. Feynman wrote in 1962:

> There are those who are going to be disappointed when no life is found on other planets. Not I. I want to be reminded and delighted and surprised once again through interplanetary exploration of the infinite variety and novelty of phenomena that can be generated from such simple principles. The test of science is its ability to predict. Had you never visited earth, could you predict the thunder storms, the volcanoes, the ocean waves, the auroras, and the colorful sunsets? A salutory lesson it will be when we learn of all that goes on, on each of those dead planets – those eight or ten balls – each agglomerated from the same dust cloud and each obeying exactly the same laws of physics.

Figure 23.12 Pictures of galaxies exhibiting pancakelike structure.
(Palomar Observatory Photograph.)

Questions

20. The planetary orbits in our solar system are very nearly circular. What does this say about the energy of the orbits?

21. Suppose two gas clouds initially have the same angular momentum but one is much more massive than the other. If these clouds eventually form galaxies, which will be flattened out more?

23.7 A FINAL WORD

We have now derived the third of the three great conservation laws in physics: conservation of energy, conservation of linear momentum, and conservation of angular momentum. All three laws came from Newton's law, $\mathbf{F} = m\mathbf{a}$. On the other hand, the conserved quantities involved are simple things which must be true *all* the time; they are useful bookkeeping devices for applying Newton's laws in complicated situations. For example, to have written Newton's laws for the water in the vortex bowl as it spirals down and out through the hole would have been exceedingly difficult. But the principle of why a hole develops in the middle is not difficult to understand, once we understand why angular momentum is conserved. Nevertheless, the conservation laws may be more significant than simply special ways of applying Newton's laws.

In the twentieth century the world witnessed a scientific revolution which is comparable in magnitude to the original scientific revolution of the period from Copernicus to Newton. We no longer believe that Newton's laws are a sufficient, adequate description of the way the world actually works. His laws are a special case, an approximation to deeper and more exact laws, known as the laws of quantum mechanics. Surprisingly, though, the three conservation laws of energy, linear momentum, and angular momentum exist in quantum mechanics also. Remember that they were derived from Newton's laws, which we no longer believe are universally correct. But in the most profound and correct theory that we have, the conservation laws themselves persist.

CHAPTER 24

GYROSCOPES

To those who study the progress of exact science, the common spinning top is a symbol of the labours and the perplexities of men who had successfully threaded the mazes of planetary motions. The mathematicians of the last age, searching through nature for problems worthy of their analysis, found in this toy of their youth ample occupation for their highest mathematical powers.

No illustration of astronomical precession can be devised more perfect than that presented by a properly balanced top, but yet the motion of rotation has intricacies far exceeding those of the theory of precession.

James Clerk Maxwell, "On a Dynamical Top" (1857)

24.1 AN ANCIENT QUESTION

In ancient times, people much more familiar with the night sky than we are helped themselves memorize its configurations by seeing heroes and creatures in clusters of stars. The constellations were patterns formed by stars fastened to a great sphere that surrounded the earth and formed the boundary of the universe. This celestial globe rotated on an axis through the earth, causing the stars to move along circular paths across the sky.

Figure 24.1 Time exposure photograph of the night sky when a camera is pointed at the North Star. (Lick Observatory Photograph.)

Likewise the life-giving sun was fixed to its own sphere whose rotation made the sun seem to travel across the sky each day, rising in the east and setting in the west. But unlike the stars, the sun gradually changed its path each day, rising and setting more northerly in the summer and more southerly in the winter. Ancient astronomers accounted for the yearly motion of the sun by an extra annual rotation of the sun's sphere about an axis tilted by $23\frac{1}{2}°$. They named the plane of the sun's orbit the ecliptic, and the zodiac was the circular zoo of constellations through which the sun moved on its yearly journey around the stationary earth. Figure 24.2 illustrates the ecliptic and zodiac.

Twice a year, once in the spring and once in the autumn, the orbit of the sun passes through the equatorial plane of the earth. On these dates, called the *vernal* and *autumnal equinoxes*, the sun rises due east and sets due west, and the lengths of day and night are equal. The points were marked by the position of the sun in the zodiac.

Through amazingly careful observations in the second century B.C. the Greek astronomer Hipparchus discovered that the position of the equinoxes in the zodiac slowly drifts westward. For this phenomenon, called the *precession of the equinoxes*, he reported a value of 36 seconds of arc per year. But to him the precession of the equinoxes merely represented an empirical fact, like so many astronomical facts, which had to be considered in the compilation of calendars for planting and harvesting.

Not until 1543 was an explanation proposed for the precession of the equinoxes. In his book *De Revolutionibus*, Copernicus conjectured that the earth orbits the sun and that the earth spins on an axis tilted by $23\frac{1}{2}°$ to the ecliptic. In his model of the universe, the precession of the equinoxes was due to the fact that the axis of the earth slowly traces out a circle, as shown in Fig. 24.3. As the axis shifts, the star it points at, the pole star, also seems to drift until it's replaced by another. Copernicus concluded that the precessional rate is about 52 seconds of arc per year, which corresponds to a precessional period of 26,000 years.

By the time of Newton, the descriptive account of the precession of the equinoxes was well documented. However, a physical cause remained a mystery until Newton himself solved this great astronomical problem, the key to which lies in the gyroscope.

Figure 24.2 Celestial sphere with ecliptic and zodiac indicated.

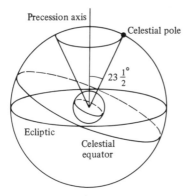

Figure 24.3 Precession of the equinoxes is caused by the rotation of the earth's axis.

24.2 THE GYROSCOPE

Before we can explain the precession of the equinoxes, we first need to understand the underlying physics of a type of motion called *gyroscopic precession*. The simplest gyroscope is a wheel which is free to spin on an axle, which we'll call the *spin axis*. A spinning top is another example of a gyroscope.

Figure 24.4a shows a nonspinning bicycle wheel of mass M at the center of a horizontal axle OP of negligible mass along the x axis. Initially the axle is supported at both endpoints O and P. If the support at P is removed, the torque due to the weight Mg causes the wheel to fall and, in so doing, to rotate counterclockwise about the y axis as shown in Fig. 24.4b, where the wheel is viewed from along the positive y axis.

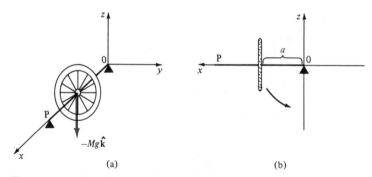

Figure 24.4 (a) Axle supported at both ends. (b) Support at P removed.

If the same wheel is set spinning rapidly on its axle, as indicated in Fig. 24.5, an extraordinary phenomenon occurs when the support is removed at P. The wheel does not fall as before but instead the axle remains almost horizontal and begins to revolve, or *precess*, about the x axis, as shown in Fig. 24.5. This apparently paradoxical motion, called *gyroscopic precession*, can be explained by changes in angular momentum.

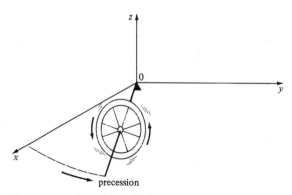

Figure 24.5 A rapidly spinning wheel does not fall but exhibits precession.

As the nonspinning wheel in Fig. 24.4 begins to topple we can calculate its torque about O as follows. Let $a\hat{\mathbf{i}}$ denote the vector from O to the center of the wheel. When the support at P is removed the torque has initial value

$$\boldsymbol{\tau} = (a\hat{\mathbf{i}}) \times (-Mg\hat{\mathbf{k}}) = aMg(\hat{\mathbf{k}} \times \hat{\mathbf{i}}) = aMg\hat{\mathbf{j}}.$$

When the axle has dropped through an angle θ from the x axis, as in Fig. 24.6, where $0 \leq \theta \leq \pi/2$, the corresponding torque about O is

$$\boldsymbol{\tau} = [a(\cos\theta)\hat{\mathbf{i}}] \times (-Mg\hat{\mathbf{k}}) = aMg(\cos\theta)\hat{\mathbf{j}}.$$

This torque is the rate of change of some angular momentum vector $\mathbf{L}_0$,

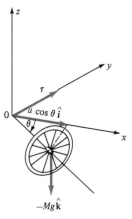

Figure 24.6 Torque about O when axle drops through angle θ.

$$\boldsymbol{\tau} = \frac{d\mathbf{L}_0}{dt} ,$$

where $\mathbf{L}_0$ has the same direction as $\boldsymbol{\tau}$ (parallel to the plane of the wheel) since $\mathbf{L}_0 = \mathbf{0}$ when $t = 0$.

Now suppose you start the wheel spinning with a large constant angular speed in the counterclockwise direction, as shown in Fig. 24.5, before the support at P is removed. This gives the wheel an amount of angular momentum about the center which we denote by $\mathbf{L}$, a vector of constant length. We call this the *spin angular momentum* of the wheel. The faster the wheel spins the longer the vector $\mathbf{L}$. Let's analyze what happens now if you remove the support at P.

The spin angular momentum $\mathbf{L}$ plus the angular momentum $\mathbf{L}_0$ due to the weight Mg have vector sum $\mathbf{L} + \mathbf{L}_0$. If the wheel spins rapidly enough, $\mathbf{L}_0$ will be very small compared to $\mathbf{L}$. For the moment we neglect $\mathbf{L}_0$ and assume all the angular momentum is represented by the spin vector $\mathbf{L}$. Gravity still exerts a torque $\boldsymbol{\tau}$ about the point O and this torque is parallel to the plane of the wheel. The torque is related to $\mathbf{L}$ by the equation

$$\boldsymbol{\tau} = \frac{d\mathbf{L}}{dt} \tag{23.7}$$

where now $\mathbf{L}$ is perpendicular to $\boldsymbol{\tau}$ because a vector of constant length is always perpendicular to its derivative. On the other hand, $\mathbf{L}$ is always directed along the axle and, since the torque $d\mathbf{L}/dt$ is not zero, $\mathbf{L}$ must change its direction, so $\mathbf{L}$ moves in a circle. As the axle moves about this circle in response to the torque, the direction of $\mathbf{L}$ changes and because of Eq. (23.7) the direction of $\boldsymbol{\tau}$ changes as well. This is described by saying that $\mathbf{L}$ tries to follow $\boldsymbol{\tau}$, which leads it around in a circle, as suggested in Fig. 24.7. The resulting motion is uniform gyroscopic precession.

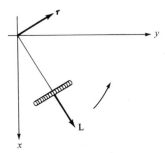

Figure 24.7 Angular momentum vector **L** tries to follow torque
vector **τ**.

Example 1

A gyroscope with angular momentum **L** is spinning in space, where it is acted on by two
equal but opposite forces **F**, shown in the diagram, equidistant from the center of mass
at O.

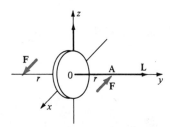

(a)   What is the torque acting on the gyroscope about point O?
(b)   For the instant shown in the diagram, in what direction is endpoint A moving as a
      result of precession?

   (a)   From $\boldsymbol{\tau} = \mathbf{r} \times \mathbf{F}$, we can find the direction of the torque created by each force
using the right-hand rule. Gravity creates no torque about O. Each force **F** creates a
torque $rF\hat{\mathbf{k}}$ in the z direction. Therefore the total torque is

$$\boldsymbol{\tau} = 2rF\hat{\mathbf{k}}.$$

   (b)   Since the angular momentum vector **L** tries to follow the torque, endpoint A
(along with **L**) actually moves *upward* as a result of the horizontal forces! This precessional
motion is such that **L** describes a circle in the yz plane; so A moves counterclockwise
when viewed from the positive x axis.

Note that although the net force acting on the gyroscope is zero, the net torque is nonzero. A pair of forces equal in magnitude and oppositely directed are known as a couple; they produce rotation without acceleration of the center of mass.

Now that we have some idea of the cause of gyroscopic motion, let's consider the angular momentum once more as well as energy associated with this motion.

In the foregoing description of gyroscopic precession we neglected $\mathbf{L}_0$ and assumed that all the angular momentum of the gyroscope is purely spin momentum lying in the horizontal xy plane. But when you remove the support at P of a real gyroscope as in Fig. 24.4a you observe that the axle actually tilts down a bit, so $\mathbf{L}$ now has a vertical component $-L_z$ as well as horizontal components. Since there is no torque to produce angular momentum in the z direction, any angular momentum in that direction must be conserved. How does this take place?

It is the gyroscopic precession itself that creates the angular momentum that balances the downward vertical component $-L_z$. The resulting rotation of the axle about the z axis is directed so that the angular momentum associated with it is in the positive z direction.

The energy of the gyroscope must also be conserved. The precession of the gyro has a kinetic energy associated with it. That kinetic energy must come from somewhere, and the source is a change in gravitational potential energy of the center of mass. When the center of mass drops slightly as the gyro initially falls, the gravitational potential energy of the center of mass decreases. That decrease in potential energy appears as kinetic energy of the precession of the gyroscope.

The simple gyroscope displays even more intricacies in its motion. Consider what happens to the center of mass. Originally, the center of mass lies in the xy plane of Fig. 24.5. Once the gyro is released, the center of mass falls a bit as precession begins. The new (lower) height of the center of mass is one of stable equilibrium. But the gyroscope didn't start off in a stable equilibrium position. We recall from Chapter 14 that a system slightly disturbed from its stable equilibrium position oscillates. That's what happens to the center of mass of the gyroscope: the center of mass oscillates about the stable equilibrium position. Figure 24.8 illustrates the potential-energy curve as a function of the angle of precession. This oscillatory motion which takes place in addition to the precession is called *nutation* (after the Latin word for *nodding*). Figure 24.9 illustrates the nutation we've just described, where the curved, cusplike path represents the motion of the center of mass. Eventually these oscillations are damped out by friction at the pivot point and air resistance and the gyro's motion turns into uniform precession.

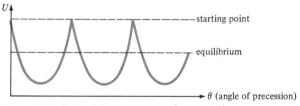

Figure 24.8 Potential-energy curve for a nutating gyroscope.

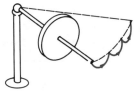

Figure 24.9 In nutation, the center of mass of a gyroscope traces a
path like the one shown.

Let's briefly recapitulate the motion of a gyroscope. There are three different kinds
of motion simultaneously occurring. First, there is rotation about the gyro's spin axis;
this *spin* is at some angular velocity which we'll call **ω**. Second, there is precession
about the point of suspension. The *precession* has a different angular velocity which we'll
call **Ω** (capital omega), the angular velocity of precession. The third kind of motion is
the oscillation about the direction of precession, and that's called *nutation*.

Example 2

Suppose that instead of releasing one end of a spinning horizontally oriented gyroscope,
you impart a slight horizontal velocity to the end as you release it. Describe the nutation
you would expect.

By imparting a slight velocity to the gyroscope when releasing it, you are adding
kinetic energy to it. At first, you might think that this added energy will become the
needed kinetic energy of precession. However, the gyroscope still must conserve angular
momentum in the vertical direction, and the only way it can do that is to drop down a
little. Therefore the gyroscope will nutate for reasons cited above. If the initial velocity
is in the same direction as the precession, then the nutation will consist of oscillations
about the stable equilibrium position which are "stretched out" more along the path of
the center of mass, as illustrated below:

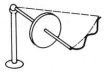

Questions

1. Suppose you had two gyroscopes of identical dimensions with one twice as massive
as the other. Would you expect the rate of precession of the more massive gyroscope to be greater than, less than, or equal to that of the lighter one? Why?

2. Consider the two gyroscopes shown in the diagram. They have equal masses, but the center of mass of gyroscope A is twice as far from the suspension point as that of B. How would their angular velocities of precession compare?

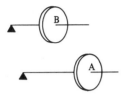

3. The angular momentum of the gyroscope shown below points along the negative y axis. If the support at B is removed, describe the precessional motion.

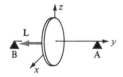

4. Imagine a gyroscope spinning in outer space. At one instant, its intrinsic angular momentum is directed along the negative x axis as shown in the diagram. Two equal but oppositely directed forces act at ends A and B. From the resulting precession, end A is moving in the xy plane in the positive y direction.

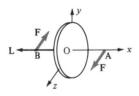

(a) In what direction is the torque about O acting on the gyroscope?
(b) In what direction are the forces at A and B acting?

5. Shown below is a photograph of the motion of the end of a nutating gyroscope taken with a stroboscope. As the end moves around the loops, it doubles back as

From Kleppner, D., and Kolenkow, R. J. *An Introduction to Mechanics*. McGraw-Hill Book Co., New York (1973). By permission of the publisher.

the dots show. This nutation was achieved by imparting a small velocity to the end of the gyroscope as it was released. Relative to the way the end is precessing, in what direction was the initial velocity? Explain your reasoning.

6. Everyone knows that to turn a bicycle to the right, you lean in that direction. In terms of torque and precession of the bicycle wheel explain how leaning makes the bicycle turn.

24.3 THE GYROCOMPASS

Another simple and initially surprising behavior of gyroscopes makes them useful in navigation. Suppose you have a toy gyroscope (you should try this experiment for yourself) and tie strings to the frame at points A and B on oppostie sides midway between the bearings, as shown in Fig. 24.10. With the gyroscope spinning, suppose that you hold the strings taut at arm's length with the spin axis horizontal. Now if you slowly pivot to your left arm so that the gyroscope moves in a circle with arm's length radius, what happens? Surprisingly, the gyroscope suddenly flips over and tilts out of the horizontal plane. After a few oscillations, which are damped out by friction, the spin axis of the gyroscope comes to rest with its axis vertical, parallel to your axis of rotation.

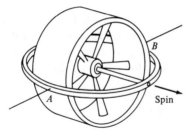

Figure 24.10 A gyroscope moved in a circle flips over.

The gyrocompass is based on this effect. At the heart of a gyrocompass is a gyroscope which has an axis of rotation which is itself free to rotate about the horizontal axis. Figure 24.11 illustrates a gyrocompass; the outer gimbal allows the inner gimbal and spin axis to rotate (taking the place of the strings in our experiment). We can understand the behavior of a gyrocompass by simple vector arguments.

When the outer frame is rotated, the pivots on which the inner gimbal is mounted provide a torque. Let's find the direction of this torque. Rotating the outer frame in a circle, that is, turning it about the z axis, produces horizontal forces at the gimbal mounts, as shown in Fig. 24.11. These forces create a torque τ in the z direction. Since torque is equal to the change in angular momentum,

$$\tau = \frac{d\mathbf{L}}{dt} ,$$

(23.7)

the spin angular momentum moves in the direction of the torque and swings toward the z direction. This motion is precession, with the torque now provided not by gravity but by the gimbal pivots.

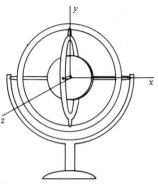

Figure 24.11 A gyrocompass consists of a gyroscope mounted on a gimbal mounted inside another gimbal which allows rotation of the gyroscope about a horizontal axis.

We can understand why the effect is so pronounced by considering angular momentum in the y direction of Fig. 24.12. An attempt to rotate the spin axis of the gyroscope in the horizontal plane would, by itself, introduce a component of angular momentum in the y direction. However, the pivots at A and B allow the gyroscope to rotate freely about the y axis, so angular momentum in that direction must be conserved. Since the angular momentum along the y direction is initially zero, it must remain zero. Now as the gyrocompass begins to rotate about the z axis, the spin angular momentum L begins to point slightly in the y direction. At the same time, the gyroscope and its mount begin to tilt and rotate about the y axis. The angular momentum arising from motion is in the negative y direction and exactly cancels out the component of L in the y direction; this

Figure 24.12 Torque creating a rotation of a gyrocompass about a vertical axis causes the spin axis to rotate about a horizontal axis.

way angular momentum in the y direction remains zero. When **L** finally comes to rest in the z direction, parallel to the axis of rotation, the motion of the frame no longer changes the direction of **L**, so the spin axis remains stationary in the z direction.

Example 3

Suppose a gyrocompass frame is slowly rotated in a horizontal plane. How does the rate at which the spin axis flips over depend on the magnitude of the spin angular momentum?

From the discussion in the text and Fig. 24.12, we see that rotating the gyrocompass causes the spin angular momentum to acquire a y component. Since angular momentum in the y direction is conserved and is initially zero, the flipping of the spin axis about the y axis creates the necessary angular momentum to cancel out the newly acquired y component of the spin angular momentum **L**. The greater the magnitude of **L**, the faster the gyroscope and its mount must rotate about the y axis. Therefore, the faster the gyrocompass is spinning, the faster it will flip over as it is rotated in a horizontal plane.

The rotation of the earth causes a torque like that described above on a gyroscope mounted and spinning in a horizontal plane. The gyroscope would, therefore, slowly flip until its axis is pointing parallel to true north as shown in Fig. 24.13. After the spin axis is pointed northward, it will continue to act as a compass indicating that direction. A gyrocompass is a nonmagnetic compass.

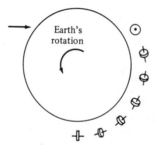

Figure 24.13 Precession of a gyroscope on the rotating surface of the earth.

There is another way in which gyroscopes can be used to aid navigation. It depends on the fact that a spinning gyroscope, carefully balanced and mounted to eliminate torques, maintains its spin axis in a fixed direction in absolute space. This application of gyroscopes is called *inertial guidance*.

One important application of inertial guidance is in aircraft navigation. A gyroscope can give a constant indication of a chosen direction. Another gyroscope which precesses about the vertical direction under the torque provided by gravity can be used to identify the true vertical and so define an artificial horizon against which the angles of banks and climbs can be measured. Inertial guidance is also indispensable in the navigation of submarines and spacecraft.

Questions

7. Suppose you had a toy gyroscope mounted on two strings as described in the text with the spin angular momentum pointing away from you. Next suppose you hold the strings taut and rotate your body to your right. How will the gyroscope flip?

8. If you repeat the experiment in Question 7 with the spin angular momentum pointing toward you, how will the gyroscope flip?

9. In the experiment with a toy gyroscope described in the text, the gyroscope quickly flips over as you turn it in a circle. But a similarly aligned gyroscope on the surface of the earth gradually flips over. Why does one motion occur rapidly and the other slowly?

10. Why is a spinning Frisbee more stable in flight than one which is not spinning? (For the same reason, bullets are given a spin as they leave the gun barrel.)

11. A typical prank of the fun-loving physicist is to load a suitcase with a heavy flywheel which is rotating. If a bellhop were to take such a suitcase as shown, with its spin angular momentum pointing along the x direction, and suddenly make a turn to the left what would happen to the suitcase?

12. According to Galileo and the principle of inertia, uniform motion in a constant direction cannot be detected. This means if you were sealed in a room with no windows you could not tell by any experiment done totally within that room whether you were at rest or moving with a constant velocity. But suppose you had a gyrocompass; then would you be able to detect whether you were moving with constant velocity or at rest?

24.4 ANGULAR VELOCITY OF PRECESSION

The basic component of any inertial guidance system is a gyroscope. Since such systems guide missiles and other weapons of war, details of the best gyroscopes are military secrets. But what factors make a good gyroscope? Once a gyroscope is pointed in a certain direction, it's never perfectly stable. Its direction always appears to drift because there is always some precession. Let's now find out what determines the velocity of precession.

Earlier we described precession as the rotation of the angular momentum vector **L**, as shown in Fig. 24.14. Because of the constant torque due to gravity, a gyroscope precesses with a certain precessional angular velocity $\mathbf{\Omega}$. Consider the angular momentum vector at times t and $t + \Delta t$, as shown in Fig. 24.14. The magnitude of the small change in the angular momentum vector, $|\Delta \mathbf{L}|$, is given by

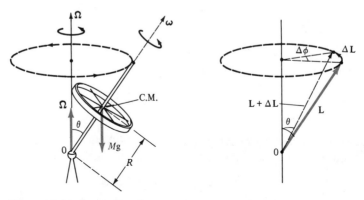

Figure 24.14 Precession of a gyroscope.

$$|\Delta \mathbf{L}| = (L \sin \theta)\Delta \phi,$$

where $\Delta \phi$ is the angle the tip of $\mathbf{L}$ moves through in time Δt, θ is the angle between $\mathbf{L}$ and the axis of precession, and $L \sin \theta$ is the radius of the circle traced out by the tip of $\mathbf{L}$. Therefore we have

$$\left| \frac{\Delta \mathbf{L}}{\Delta t} \right| = L \sin \theta \frac{\Delta \phi}{\Delta t}.$$

Taking the limits as $\Delta t \to 0$, we find the rate of change of the magnitude of the angular momentum vector to be

$$\left| \frac{d\mathbf{L}}{dt} \right| = \lim_{\Delta t \to 0} \left| \frac{\Delta \mathbf{L}}{\Delta t} \right| = \lim_{\Delta t \to 0} L \sin \theta \frac{\Delta \phi}{\Delta t} = L \sin \theta \frac{d\phi}{dt}. \tag{24.1}$$

The direction of this vector is perpendicular to $\mathbf{L}$ and at any instant is tangent to the circle that the tip of $\mathbf{L}$ traces out.

The quantity $d\phi/dt$ is the rate at which the angular momentum vector precesses and is the magnitude of what we have called the angular velocity of precession $\boldsymbol{\Omega}$. The direction of this vector is taken to be along the axis of rotation in such a way that if you curl the fingers of your right hand around in the direction the angular momentum vector is precessing, your thumb points in the direction of $\boldsymbol{\Omega}$.

With the direction of $\boldsymbol{\Omega}$ as specified above, we see that the angle between $\boldsymbol{\Omega}$ and $\mathbf{L}$ is θ. Moreover, Eq. (24.1) shows that $|d\mathbf{L}/dt| = |\boldsymbol{\Omega} \times \mathbf{L}|$. This suggests that we can write $d\mathbf{L}/dt$ as a vector cross product:

$$\frac{d\mathbf{L}}{dt} = \boldsymbol{\Omega} \times \mathbf{L}. \tag{24.2}$$

This is easily verified because dL/dt is perpendicular to both Ω and L and is at any instant tangent to the circle that the tip of L traces out. Therefore the vectors dL/dt and $\Omega \times L$ have the same length and the same direction.

Since torque is equal to the rate of change of angular momentum, the equation that describes the precession of a gyroscope is

$$\tau = \frac{dL}{dt} = \Omega \times L. \tag{24.2}$$

This is sometimes called *the gyroscope equation*.

A good gyroscope is one which drifts as little as possible, or in other words, precesses as slowly as possible. Therefore, we want Ω to be as small as possible. From Fig. 24.14, we see that the torque due to gravity about the pivot is

$$\tau = R \times Mg,$$

where R is the vector from the pivot to the center of mass and M is the mass of the wheel. The magnitude of this torque is

$$\tau = MgR \sin \theta.$$

As we saw earlier, the magnitude of dL/dt is

$$\left| \frac{dL}{dt} \right| = \Omega L \sin \theta.$$

Since these two expressions are equal, we can solve for the magnitude of the precessional angular velocity:

$$\Omega = \frac{MgR}{L} . \tag{24.3}$$

We can express the angular momentum of the bicycle wheel gyro in terms of its mass, radius, and angular velocity. If we imagine the wheel to be a rim of mass M rotating about its center, each little bit of mass is moving around in a circle of radius r, and in doing so makes a contribution of $r \times p$ to the angular momentum. For each piece of the wheel, $r \times p$ is vector pointed along the axis, as Fig. 24.15 illustrates. Therefore, each little bit of angular momentum adds up the same way, and the result is that the

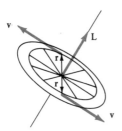

Figure 24.15 Angular momentum of a rotating wheel has magnitude $Mr^2\omega$.

angular momentum of the wheel is the total mass of the wheel times the velocity of rim times the radius r:

$$L = Mvr.$$

Since we can write the velocity of any point on the wheel as

$$v = r\omega, \tag{9.5}$$

where ω is the angular speed of the rotating wheel, the angular momentum can be written as

$$L = Mr^2\omega.$$

Therefore we can express the rate of precession of the idealized gyroscope as

$$\Omega = \frac{gR}{\omega r^2}. \tag{24.4}$$

If we want to minimize Ω, we can make ω as large as possible, that is, have the gyroscope spin extremely fast. Or we could make r very large by concentrating the mass as far as possible from the axis of rotation. One other way to minimize Ω is to make the distance from the pivot to the center of mass, R, as small as possible. In principle, gyroscopes, like that in Fig. 24.11, are suspended about the center of mass so that the gravitational torque won't make it precess. In practice, perfect balance can never be achieved, and so the gyroscope will drift.

How good are the best gyroscopes? The answer is probably a military secret, but it is estimated that the best inertial guidance gyroscopes precess at about 100 seconds of arc (about 0.03°) per day.

Questions

13. Consider a point on a sphere of radius r which is rotating with an angular velocity ω. Show that

$$\frac{d\mathbf{r}}{dt} = \omega \times \mathbf{r},$$

where $\mathbf{r}$ is the position vector of the particle. This relation is analogous to Eq. (24.2).

14. A simple gyroscope consisting of a 3.0-kg wheel having a 0.4-m radius spins at 100 rad/s. If the distance from the point of suspension to the center of mass is 0.6 m, what is the magnitude of the angular velocity of precession?

15. A toy gyroscope precesses faster as friction acting on the axle causes its spinning to slow down. Explain why.

16. A gyroscope consisting of a wheel spinning at 400 rad/s has a mass of 10.0 kg and a radius of 0.5 m. What is the distance from the point of suspension to the center of mass if it precesses at a rate of 100 seconds of arc per day?

24.5 THE EARTH AS A GYROSCOPE

Among the numerous and perplexing mysteries Isaac Newton removed from the pages of history was the precession of the equinoxes. In the *Principia*, Newton gives an explanation of observed precession based on his dynamics. His stunning insight was to realize that the earth itself acts like a gyroscope.

As a good approximation, we usually think of the earth as being spherical. If it were spherical, the sun (or moon) could not exert a torque on it because the force of the sun (or moon) acts on the center of mass at the center of the earth (just as gravity acts on the center of mass of the gyroscope). To this approximation, the angular momentum of the spinning earth is always in the same direction.

However, if we analyze the earth more closely, we find it is not a perfect sphere, on account of its rotation, which causes a bulge at the equator, as exaggerated in Fig. 24.16. The width of each bulge at the equator is about 1/300 the polar radius of the earth; this corresponds to a deviation of 13 mi (21 km) at the equator from the radius of a perfect sphere.

Figure 24.16 Forces from the sun on equatorial bulges in (a) the summer and (b) the winter. The torque produced is in the same direction in both cases.

In addition, the earth's spin axis makes an angle of $23\frac{1}{2}°$ with respect to the plane of its orbit. Consequently, the bulge is oriented asymmetrically as shown in Fig. 24.16. Although the sun (or moon) creates no torque on the spherical distribution of mass, it does exert a torque about the center on the bulges. The bulge nearer to the sun is more strongly attracted than the bulge that is farther away. This results in a torque τ pointed up from the plane of Fig. 24.16a. Since the precession occurs very slowly – taking 26,000 years to complete one circle – by the time the sun gets to the other side of the earth, the bulges have hardly moved at all. So when the sun is on the opposite side of the earth six months later as shown in Fig. 24.16b, the force on the bulge nearer the sun is greater and creates a torque in the *same* direction. (You can easily verify this with the right-hand rule.)

Since the earth rotates from west to east, its spin angular momentum is directed toward the North Pole. The torque acting on the spinning earth causes a precession in which the tip of the angular momentum vector traces out a circular path from east to west, as shown in Fig. 24.17, observed as the precession of the equinoxes and the change in pole star. In the *Principia*, Newton calculated the rate of precession to be 51 seconds of arc per year, knowing that the earth is a gyroscope.

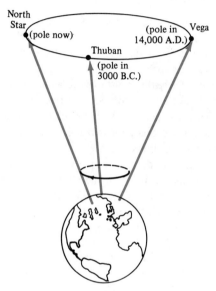

Figure 24.17 Precession of the earth.

Question

17. If the angular momentum of the earth is 7.100×10^{33} kg m²/s, and the observed rate of precession of the equinoxes is 46.74 seconds of arc per year, what torque is exerted on the earth?

24.6 A FINAL WORD

The gyroscope is a strange, amusing, and important device. What is perhaps even stranger and more beguiling is that Isaac Newton himself understood how it worked and managed to explain the precession of the equinoxes. For most of us, that little detail is lost in the magnitude of Newton's grander accomplishments. But if some other person who did nothing else in his lifetime had managed to do that, we would remember him today as an important scientist.

One day in 1684, Newton casually told a young friend named Edmund Halley that he had discovered that a $1/r^2$ force law leads to orbits in the form of conic sections. It's difficult to imagine what Halley felt at that moment. Having Newton as a friend cannot have been very relaxing at best, but Newton had just told him that he had discovered the key to the universe.

Halley prevailed upon his friend to publish the result, which Newton proceeded to do. And indeed it *was* the key to the universe, although the known universe was smaller then than it is today. In Chapter 25, our job is to begin to turn that key.

25

KEPLER'S LAWS AND THE CONIC SECTIONS

I was almost driven to madness in considering and calculating the matter. I could not find out why the planet [Mars] would rather go on an elliptical orbit. Oh ridiculous me! As if the libration on the diameter could not also be the way to the ellipse. So this notion brought me up short, that the ellipse exists because of the libration. With reasoning derived from physical principles agreeing with experience, there is no figure left for the orbit of the planet except for a perfect ellipse. . . .

Why should I mince my words? The truth of Nature, which I had rejected and chased away, returned by stealth through the back door, disguising itself to be accepted. That is to say, I laid [the original equation] aside, and fell back on ellipses, believing that this was a quite different hypothesis, whereas the two, as I shall prove in the next chapter, are one and the same. . . I thought and searched, until I went nearly mad, for a reason why the planet preferred an elliptical orbit. . . Ah, what a foolish bird I have been!

Johannes Kepler, *Astronomia Nova* (1609)

25.1 THE QUEST FOR PRECISION

Not long after Copernicus published his revolutionary book, Tycho Brahe (1546–1601) provided a multitude of new observations that, despite his own intentions, provided crucial support for the Copernican hypothesis. At that time the furious debate between the Copernican and Ptolemaic systems was no longer waged in words alone; observations and exact measurements carried a new significance. Tycho realized the need for more precise astronomical observations and the instruments to make them. He built giant measuring devices on the island at Hveen, Denmark. Before Tycho's time, the locations of the heavenly bodies were known with a precision of about ten minutes (10′) of arc. Tycho's painstakingly careful measurements reduced the uncertainty to about 2′ of arc. His contribution as an astronomer was based on a method of observation that would become obsolete just ten years after his death, when the telescope was invented, transforming astronomy forever.

Tycho was a foul-tempered Danish lord who tongue-lashed kings, tormented peasants, sported a silver nose (his own having been lost in a youthful duel over mathematics), kept a clairvoyant dwarf as his court jester and a tame elk that got drunk one night, fell down stairs, broke a leg, and died. Yet Tycho was a measuring maniac, a fussily precise man who opened a new age of observation in science.

The princes of Europe vied to acquire the services of Tycho as court astronomer and astrologer. King Frederick II of Denmark provided Tycho with the island of Hveen off the coast of his kingdom. There Tycho built Uraniborg, his hilltop observatory, which housed a windmill, a paper mill, fishponds, and a prison for unruly peasants, as well as workshops for artisans who built his magnificent devices. His instruments were constructed on a huge scale. One of his quadrants, for example, was 38 ft in diameter and contained a life-size portrait of Tycho seated within its arc.

At Uraniborg Tycho held court like a lord rather than a scholar. He ate and drank excessively, and at odd moments would pop off his gleaming metal nose to rub ointment on what lay beneath. When his feudal tenants complained of mistreatment, he tossed them into jail. And when the young King Christian IV reduced his benefits, Tycho, indignant, left the country. But at Uraniborg, he had so precisely determined the position of 777 stars and refined the measurements of Mars that today's measurements have only fractionally refined them.

Figure 25.1 Tycho seated at his giant quadrant. (Courtesy of the Archives, California Institute of Technology.)

In Prague in 1597, Rudolph II, King of Bohemia and Emperor of Germany, welcomed Tycho to his court as Imperial Mathematician with the fattest salary in the realm. On January 1, 1600, the impoverished Protestant Kepler left the Roman Catholic town of Graz for Prague to join the great Tycho. Tycho and Kepler knew that they needed each other. Kepler, in order to do his theoretical cosmology, the work that was his mission in life, needed Tycho's superb astronomical data. He was insolent and resentful when Tycho would only provide him with faint hints of key observations. And Tycho, in order to get his data organized into a useful form, needed Kepler's mathematical genius; he sensed that his hope for lasting fame lay in Kepler's penetrating intelligence. This tenuous relationship lasted for 18 months, but Tycho kept his secrets to the end.

At a banquet, Tycho overdrank and then held back his water beyond the demands of courtesy. An acute urinary infection set in, and within 11 days Tycho died. His last words to Kepler were, "Let me not seem to have lived in vain."

25.2 KEPLER'S LAWS

During the titanic period when the Aristotelian world would be replaced by the Copernican universe, Tycho believed in neither. He had his own model of the universe – the Tychonic theory, illustrated in Fig. 25.2. Tycho's theory was that the earth is stationary at the center of the universe and that the sun orbits the earth, with all of the planets orbiting the sun. Tycho fervently believed in his model and hoped that Kepler would build the universe on it. Kepler was to do just the opposite, to use the Tycho's data to establish the Copernican universe.

Figure 25.2 The Tychonic model of the universe. (Courtesy of the Archives, California Institute of Technology.)

Immediately after Tycho's death, Kepler stealthily took Tycho's prized data for fear that they would be lost in the settlement of his estate. Rudolph appointed Kepler to the vacant post of Imperial Mathematician, and Kepler at last could settle down and work with the data he so much needed. Kepler eventually published in 1627, a full set of tables generated from Tycho's data, the *Rudolphine tables*.

Kepler devoted the six years following Tycho's death to a battle with the planet Mars. Because the observed irregularities of its motion were greater than those of any other planet, the motion of Mars could not easily be described in terms of the Platonic ideal of uniform circular motion. Kepler, fully accepting a heliocentric universe, sought the smooth, continuous curve in which the planets orbited the sun. The problem he faced was that he had to find this curve on the basis of observations from a moving platform, the earth, which itself orbits the sun in some nonuniform way.

In his book *Astronomia Nova* (*New Astronomy*), Kepler discusses the ingenious method by which he determined first the path of the earth itself. Kepler knew that the length of the Martian year (the time needed to complete one orbit) is 687 days. He used this information to identify the dates on which Mars would return to a given point in its orbit. The particular point chosen was point M in Fig. 25.3, when the earth at point E_0 was on a straight line between the sun and Mars. In the 687 days it takes Mars to return to point M, the earth moves 687/365 = 1.88 revolutions, or through an angle of 677°. In other words, the earth moves 43° less than two complete revolutions. Thus, when Mars is again at point M, the earth is at point E_1. One Martian year later, Mars will again be at point M, but the earth will be at point E_2, which is 43° from E_1. By locating the earth's position on successive Martian years, Kepler succeeded in constructing a plot of the earth's orbit. He found this plot to be indistinguishable from a circle, except that the sun was slightly displaced from the center.

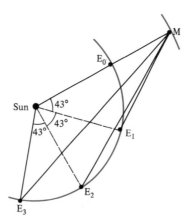

Figure 25.3 Kepler's determination of the earth's orbit from knowledge of the position of Mars on successive Martian years.

Kepler's plot of the earth's orbit also revealed that the earth moves fastest when it is nearest the sun. From an analysis of the speed of the earth at various points of its orbit, Kepler deduced what became known as his *second law*:

> A line directed from the sun to a planet
> sweeps out equal areas in equal
> times.

As we've seen in Chapter 23, this law can be understood as a consequence of the conservation of angular momentum.

Knowing the orbit and timetable of the earth, Kepler then reversed his analysis to find the shape of Mars' orbit as seen from the sun. Again he used observations of the position of Mars separated by one Martian year. By using the position of the earth at the same stage of successive Martian years, Kepler could triangulate to find a point on the orbit of Mars, as depicted in Fig. 25.4, where point M was fixed by sighting along the lines E_0M and E_1M. Then he could choose a second point, for example M', the position of Mars the next time the earth – now at E_0' – was on a straight line between the sun and Mars (Fig. 25.4). When Mars returned to M' after one further Martian year, the earth would be at E_1' as shown in Fig. 25.4, allowing a second triangulation to fix M'. Tycho's data amassed over a quarter of a century allowed Kepler to fix 12 points on the orbit of Mars in this manner. He could not force the points to fit a circular orbit, but rather he found the orbit of Mars to be an oval, as shown in Fig. 25.5. The disagreement between the data and the best circular path was about 8' of arc. Here the improvement from the 10' uncertainty in the data available to Copernicus to the 2' uncertainty in Tycho's data proved crucial. Had Copernicus not tried to use epicycles but attempted instead to fit the orbit of Mars to a circle in this way, he would have succeeded in matching the best observations available to him. But Kepler, working with Tycho's improved data, could not. Faced with the choice of giving up the Platonic ideal of circular motion or

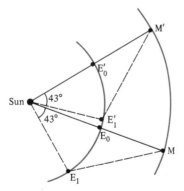

Figure 25.4 Kepler's triangulation to determine the orbit of Mars.

violating Tycho's magnificent observations, he chose to believe the observations. After months of agonizing calculations, Kepler realized that the orbit of Mars could be fitted by an elegant curve whose special properties had been known for hundreds of years. The orbit of Mars, he realized, is an ellipse, as is, indeed, that of every planet. Thus he formulated what is now called his *first law*:

> Each planet orbits the sun along an elliptical path
> with the sun at one focus.

What made the orbit of Mars resist description as a circle is its large *eccentricity*. The eccentricity is a measure of how distant the sun is from the center of the ellipse. As depicted in Fig. 25.5, if the length of the semimajor axis (half the longer dimension of the ellipse) is a, then the sun is located at a point (called a focus) at a distance ea from the center, where e is the eccentricity. If $e \neq 0$ there are two foci equidistant from the center. If $e = 0$ the foci coincide with the center of the curve, which then becomes a

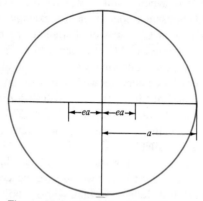

Figure 25.5 Elliptical orbit of Mars.

circle. As indicated in Table 25.1, Mars has the greatest eccentricity of those planets which Kepler could observe.

Kepler discovered many laws; three of them turn out to be correct. Buried within the pages of his book *Harmony of the World*, amid a long list of speculations, is what we call his *third law*:

> The square of the period of a planet is proportional
> to the cube of its semimajor axis.

In other words, $T^2 = ka^3$, where T is the period (the time to go once around the ellipse) and k, a constant of proportionality, is the same for all planets. Our task now is to formulate a mathematical description of Kepler's second law. In Chapter 26 we will deduce it from Newton's laws of motion and gravity.

25.3 CONIC SECTIONS

The ellipse is one of a family of curves that can be formed by the intersection of a cone with a plane. The curves obtained by slicing a cone with a plane not passing through the vertex are called *conic sections*, or simply *conics*. If the cutting plane is parallel to the

We know that any point of an ellipse satisfies $r + r' = 2a$, so let's apply this to the point directly above the center at distance b as shown. In this case $r = r' = a$. For the right triangle shown we have

$$a^2 = b^2 + (ea)^2,$$

which implies that

$$b = a\sqrt{1 - e^2}.$$

Example 2

Show that the aphelion distance r_a and the perihelion distance r_p are, respectively, the maximum and minimum values of r in the polar equation (25.3), and that their ratio is

$$\frac{r_a}{r_b} = \frac{1 + e}{1 - e}.$$

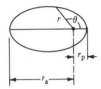

Because r is largest when $\cos\theta$ is smallest ($\theta = \pi$) we get

$$r_a = \frac{a(1 - e^2)}{1 - e} = a(1 + e),$$

and because r is smallest when $\cos\theta$ is largest ($\theta = 0$) we find

$$r_p = \frac{a(1 - e^2)}{1 + e} = a(1 - e),$$

hence

$$\frac{r_a}{r_p} = \frac{1 + e}{1 - e}.$$

The polar equation (25.3) is, for our purposes, the most useful equation for an ellipse. It shows that the shape of an ellipse is determined entirely by its eccentricity; the factor a governs the size of its major axis. Figure 25.9 shows several ellipses with the same rightmost focus and the same value of a, but with various eccentricities. From Eq. (25.3) we see that $r \rightarrow a$ as $e \rightarrow 0$, which tells us that the smaller the eccentricity the rounder the ellipse. This is illustrated in Fig. 25.9.

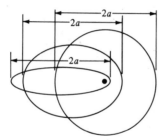

Figure 25.9 The shape of an ellipse is determined by its eccentricity.

As the eccentricity approaches zero, the distance between foci approaches zero and the ellipse approaches a circle of radius a. On the other hand, the ellipse becomes flatter as the eccentricity approaches 1 because $b = a\sqrt{1 - e^2}$ approaches 0 as $e \to 1$.

One other property, which we state without proof, is that the area A of the region enclosed by an ellipse is

$$A = \pi ab. \tag{25.5}$$

When $a = b$ this becomes the familiar formula for the area of a circle.

Example 3

Show that the focal distance r' and the angle ϕ shown in the ellipse below of eccentricity e satisfy the equation

$$r' = \frac{a(1 - e^2)}{1 - e \cos \phi}. \tag{25.6}$$

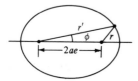

Arguing as we did to derive Eq. (25.3) from Fig. 25.8 we have

$$r^2 = (r' \cos \phi - 2ae)^2 + (r' \sin \phi)^2$$
$$= r'^2 - 4aer' \cos \phi + 4a^2e^2.$$

But, $r = 2a - r'$, hence we also have

$$r^2 = 4a^2 - 4ar' + r'^2$$

which, when subtracted from the foregoing equation for r^2, leads to Eq. (25.6). Note the similarity to Eq. (25.3). The denominator contains a minus sign when we use the leftmost focus and a plus sign when we use the rightmost focus.

Questions

3. Using the data in Table 25.1, find the ratio of the perihelion distance to aphelion distance for (a) the earth and (b) Mars.

4. On a certain ellipse it is 5.0 cm from one focus to the farthest point, and 2.0 cm from the same focus to the nearest point. Find (a) the eccentricity, (b) the length of the semimajor axis, and (c) the length of the semiminor axis.

5. What is the area of the ellipse in Question 4?

6. Knowing that the perihelion distance of Mercury is 45.8×10^6 km and its eccentricity is 0.206, calculate the aphelion distance.

7. A satellite placed into elliptical orbit about the earth is described by the polar equation

$$r = \frac{(8000 \text{ km})}{1 + 0.4 \cos \theta}.$$

Find (a) the eccentricity, (b) the length of the semimajor axis, and (c) the length of the semiminor axis.

8. Using Eq. (25.3) and conservation of angular momentum, derive an expression for the ratio of the speed of a planet at aphelion to that at perihelion.

9. For what eccentricity is the ratio in Question 8 a minimum?

10. A satellite in elliptical orbit around the earth is 7500 km from the center of the earth at perihelion and has a speed of 8000 m/s there. Its aphelion distance is 12,000 km. Find (a) the eccentricity of the orbit, (b) the length of the semimajor axis, and (c) the speed at aphelion.

11. Use the equation $r + r' = 2a$ together with Eqs. (25.3) and (25.6) to show that the two angles θ and ϕ are related to the eccentricity e as follows:

$$\frac{1}{1 + e \cos \theta} + \frac{1}{1 - e \cos \phi} = \frac{1}{1 + e} + \frac{1}{1 - e}.$$

25.5 THE CONICS AND ECCENTRICITY

The concept of eccentricity can be used to give a unified treatment of all the conics. A conic section can be defined as a curve traced out by a point moving in a plane in such a way that the ratio of its distance from a fixed point (a focus) and a fixed line (a directrix) is constant. This constant ratio is called the eccentricity and is denoted by e. If $0 < e < 1$, the conic is called an *ellipse*; if $e = 1$, it is called a *parabola*; and if $e > 1$, it is called a *hyperbola*.

In Fig. 25.10, F denotes the focus, the vertical line is the directrix, P is any point on the conic, Q is the nearest point to P on the directrix, and the eccentricity is the ratio $e = \text{FP/QP}$.

From this definition we can easily find a polar equation for any conic with a vertical directrix. Let d be the distance from the focus to the directrix, and introduce r and θ as shown in Fig. 25.10. If both P and F are to the left of the directrix, as shown, we have

$$FP = r$$

and

$$QP = d - r \cos \theta,$$

so the relation FP $= e$QP becomes

$$r = e(d - r \cos \theta).$$

Solving for r we obtain

$$r = \frac{ed}{1 + e \cos \theta}. \qquad (25.7)$$

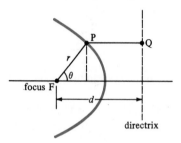

Figure 25.10 Geometry for obtaining a polar equation for conics.

Note that Eq. (25.7) resembles the one we obtained for the ellipse in Eq. (25.3), except that ed appears in the numerator instead of $a(1 - e^2)$. Comparison of Eq. (25.7) with (25.3) shows that for an ellipse we have $ed = a(1 - e^2)$, so the length of the semimajor axis is

$$a = \frac{ed}{1 - e^2}.$$

The three types of conics are shown in Fig. 25.11. Because $\cos(-\theta) = \cos \theta$, all three conics are symmetric about the horizontal axis. On an ellipse, because $0 < e < 1$, the distance from the focus to a point on the curve is always less than the distance from the directrix to that point. In this case the curve crosses the horizontal axis at two points, when $\theta = 0$ and when $\theta = \pi$. Because of the symmetry, the curve is closed as shown in Fig. 25.11a.

What happens if $e = 1$? The curve is a parabola, and to visualize it we examine its polar equation, which now becomes

$$r = \frac{d}{1 + \cos \theta}.$$

When $\theta = 0$ the parabola cuts the axis at $r = \frac{1}{2}d$, the point equidistant from the focus and directrix. But it never again crosses the axis. In fact, as θ increases toward π, the denominator $1 + \cos \theta$ approaches 0 and the distance r becomes infinite. In other words, the curve spreads out at arbitrarily large distances from the axis as θ increases toward

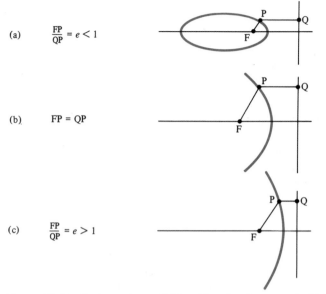

(a) $\dfrac{FP}{QP} = e < 1$

(b) $FP = QP$

(c) $\dfrac{FP}{QP} = e > 1$

Figure 25.11 Conic sections. (a) The ellipse has $0 < e < 1$. (b) The parabola has $e = 1$. (c) The hyperbola has $e > 1$.

π, as shown in Fig. 25.11b. The symmetric lower half corresponds to values of θ in the interval $-\pi < \theta < 0$.

When $e > 1$ the polar equation (25.7) shows that when $\theta = 0$ the hyperbola crosses the axis at $r = ed/(1 + e)$. It, too, never crosses the axis again, as we can see by writing the polar equation in the form

$$r = \frac{d}{1/e + \cos\theta},\tag{25.8}$$

where we have divided top and bottom of the right-hand side of (25.7) by e. The term $1/e$ in the denominator is now less than 1, so there is a specific angle α between $\pi/2$ and π whose cosine is $-1/e$. Hence Eq. (25.8) can be written as

$$r = \frac{d}{\cos\theta - \cos\alpha},$$

where $\cos\alpha = -1/e$. As θ approaches α, the denominator approaches 0 and again r becomes infinite. The hyperbola is also open ended but it spreads out in a different manner from the parabola. As r becomes larger the angle θ increases but never reaches the value α.

Questions

Each of Questions 12 through 16 gives the equation for a conic section with focus F at the origin and a vertical directrix lying to the right of F. In each case, determine the eccentricity e, the type of conic, and the distance d of F from the directrix.

12. $r = \dfrac{2}{1 + \cos \theta}$

13. $r = \dfrac{6}{3 + \frac{1}{2} \cos \theta}$

14. $r = \dfrac{4}{\frac{1}{2} + \cos \theta}$

15. $r = \dfrac{1}{1 + 3 \cos \theta}$

16. $r = \dfrac{4}{6 + 3 \cos \theta}$

25.6 CARTESIAN EQUATIONS FOR CONIC SECTIONS

In our treatment of Kepler's laws it is convenient to use polar equations for analyzing the conic sections. But in discussing trajectories of projectiles, it is more natural to use rectangular coordinates to discuss the Galilean parabola. This section describes briefly how all conics can be described in rectangular coordinates. The material in this section is not needed for the later chapters and may be considered optional.

After the advent of analytic geometry in the seventeenth century, the conic sections were studied by algebraic methods. It was shown that no matter how the coordinate axes are chosen, the rectangular coordinates (x, y) of every point on a conic section satisfy a quadratic equation of the form

$$Ax^2 + Bxy + Cy^2 + Dx + Ey + G = 0,$$

where A, B, C, D, E, and G are constants. We will illustrate this in Example 4 for the ellipse and in Example 5 for all three types of conics through the origin. If the conic has eccentricity e it can be shown that the quantity $4AC - B^2$ has the same algebraic sign as $1 - e$, so the type of conic can be recognized from the equation. It is an ellipse, parabola, or hyperbola, depending on whether $4AC - B^2$ is positive, zero, or negative.

Example 4
Take the x axis through the foci of an ellipse and the origin at its center and show that the Cartesian coordinates (x, y) of each point on the ellipse satisfy

$$\frac{x^2}{a^2} + \frac{y^2}{b^2} = 1, \tag{25.9}$$

where a and b are the lengths of the semiaxes.

Referring to Fig. 25.8, we see that by the theorem of Pythagoras

$$r^2 = (x - ae)^2 + y^2$$

and

$$r'^2 = (x + ae)^2 + y^2.$$

Adding and subtracting these two equations we find

$$r^2 + r'^2 = 2(x^2 + y^2 + a^2e^2) \qquad (25.10)$$

and

$$r^2 - r'^2 = (x - ae)^2 - (x + ae)^2 = -4axe. \qquad (25.11)$$

But $r^2 - r'^2 = (r + r')(r - r') = 2a(r - r')$ since

$$r + r' = 2a, \qquad (25.1)$$

hence (25.11) implies $2a(r - r') = -4axe$, or

$$r - r' = -2xe. \qquad (25.12)$$

Adding and subtracting (25.1) and (25.12) we find

$$r = a - xe \qquad \text{and} \qquad r' = a + xe.$$

Putting these values in (25.10) and rearranging terms we obtain

$$x^2(1 - e^2) + y^2 = a^2(1 - e^2),$$

which gives (25.9) after we divide by $b^2 = a^2(1 - e^2)$.

Example 5

A conic of eccentricity e with its focus on the x axis has a vertical directrix and passes through the origin. Show that the rectangular coordinates (x, y) of each point on the conic satisfy an equation of the form

$$y^2 = (e^2 - 1)x^2 - cx, \qquad (25.13)$$

where c is a constant depending on the conic.

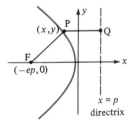

If we let the line $x = p$ be the directrix, then the focus will be at $(-ep, 0)$ because the origin lies on the curve. If the directrix is to the right of the origin, as shown here, p is positive. Squaring the fundamental defining relation $PF = ePQ$ we have

$$(PF)^2 = e^2(PQ)^2.$$

From the figure, we see that

$$(PF)^2 = (x + ep)^2 + y^2 \qquad \text{and} \qquad (PQ)^2 = (x - p)^2$$

so the defining relation gives us

$$(x + ep)^2 + y^2 = e^2(x - p)^2.$$

Solving for y^2 and simplifying we find

$$y^2 = (e^2 - 1)x^2 - 2ep(1 + e)x.$$

This is Eq. (25.13) with $c = 2ep(1 + e)$.

Questions

17. Use the method of Example 5 to show that a parabola has a Cartesian equation of the form

$$y = ax^2$$

if the focus is on the y axis, the directrix is parallel to the x axis, and the curve passes through the origin. Make a sketch of the curve and indicate the geometric meaning of a.

18. For an ellipse through the origin with directrix $x = p$, show, using Eq. (25.13), that the curve also crosses the x axis when $x = 2ep/(e - 1)$.

19. Modify the argument in Example 4 as needed to show that each point (x, y) on a hyperbola of eccentricity e satisfies the equation

$$x^2(1 - e^2) + y^2 = a^2(1 - e^2),$$

where ae is the distance from the origin to the focus. This can also be written

$$\frac{x^2}{a^2} - \frac{y^2}{b^2} = 1,$$

where $b^2 = (e^2 - 1)a^2$. Since the equation remains unchanged when (x, y) is replaced by $(-x, -y)$, the hyperbola is symmetric about the origin. Make a sketch showing the symmetry and indicate the geometric meanings of a and b on your sketch.

20. The point $(2, 2)$ is a focus and the line $x + y = 2$ is a directrix of a hyperbola with eccentricity $e = \sqrt{2}$. Proceed directly from the definition in terms of eccentricity to show that each point (x, y) on the hyperbola satisfies the equation

$$xy = 2.$$

25.7   A FINAL WORD

The English language contains words and constructions similar in form and definition to the conic sections. The parabola corresponds to the word *parable*, the ellipse corresponds to the word *ellipsis*, and the hyperbola corresponds to the word *hyperbole*. This is no accident, because as any dictionary will reveal, these words are actually derived from the same words as the respective conic sections.

Ellipsis is a construction in which a grammatically necessary element in a sentence is omitted because it can be understood. For example, *Physics is more fun than chemistry*. This is an ellipsis, because a correct statement would be *Physics is more fun than chemistry is*.

A parable is a short story that has some hidden meaning, such as a moral lesson. An example would be the legend of Newton and the apple, or Adam and Eve and the other apple. Mathematically, parabolas correspond to $e = 1$ and are formed by the intersection of a cone with a plane parallel to the generator of the cone. The parable represents a parallel to life.

Hyperbole is an extravagant exaggeration or overstatement. The hyperbola corresponds to $e > 1$. An example of a hyperbole is *Physics is the greatest course in the world*.

In any case, it is a breathtaking fact that planets flying around in space follow particular mathematical curves with special properties. This fact brings us face to face with the great mystery that has awed everyone from Galileo down to Albert Einstein: mathematical relationships describe the laws of nature.

CHAPTER 26

SOLVING THE KEPLER PROBLEM

From your remarks on the moon, I infer that your telescope is of such an inferior effectiveness, that perhaps it is not suitable for observing the planets. Since July 5, I have seen and noted these planets in the east with Jupiter in the morning. . . .

Therefore, let it lie concealed in hell and likewise let us make nothing of the insults of the entire crowd. For not even the Giants, much less the pygmies, stood against Jupiter. Let Jupiter stand in the heavens, and let the slanderers bark away as much as they wish. . . .

What must be done? Must we stand with Democritus and Heraclitus? I wish, Kepler, that we could laugh at the extraordinary foolishness of the public. What do you say about the foremost philosophers of this university, who filled with the stubborness of vipers have never wished to see the planets, the moon or the telescope, although I have willingly offered a thousand times.

But as a man stops up his ears, so those men have stopped up their eyes against the light of truth. . . .

Why am I not able to laugh with you long since? What a laugh you would have Kepler, if you could hear what things have been put forward against me in the presence of the Grand Duke at Pisa by a distinguished philosopher of this university, while he tried with logical arguments, as though with magical incantations, to tear away and remove from the sky the nine planets.

Letter of August 19, 1610, from Galileo to Kepler

26.1 SETTING THE STAGE

In 1543 Copernicus wrote his famous book; a generation later Kepler formulated his three laws; and then 150 years after Copernicus's book, Isaac Newton took Kepler's third law and used it to deduce the law of universal gravitation. From the law of gravitation and his dynamics, Newton was able to deduce Kepler's other two laws.

The task of deducing Kepler's laws from Newton's laws is called the *Kepler problem*. Its solution is one of the crowning achievements of Western thought. It is part of our cultural heritage just as Beethoven's symphonies or Shakespeare's plays or the ceiling of the Sistine Chapel are part of our cultural heritage.

But it differs from a symphony or a play or a painting in an important way. It is a living idea. It is not something to be executed or performed by others and merely admired by us. We can absorb it, penetrate it, master it, and it becomes our very own, to take with us forever. For the same reason it is not necessary to try to do it in the same way Newton did it for himself, and we shall not. To make the task easier, we'll use ideas and techniques Newton didn't have: energy and vectors. We have been carefully preparing ourselves for this, and now we are on the threshold of the great discovery.

What we will do is to show that the differential equation we get from Newton's second law and the law of gravitation is satisfied only by the equation of the conic sections – the ellipse, the parabola, or the hyperbola. The orbits of planets turn out to be ellipses, but other heavenly bodies such as meteors or comets can travel along paths which are ellipses, hyperbolas, or even parabolas. The solution of the differential equation does not, by itself, reveal which type of conic the orbit will be. Energy considerations discussed in Chapter 27 will help us discover which orbits are ellipses and which are hyperbolas or parabolas.

Our task is not as formidable as it may seem because we already solved part of the problem when we derived the law of equal areas in Chapter 23. Recall that the rate of change of the area vector of any planet in its orbit is

$$\frac{d\mathbf{A}}{dt} = \tfrac{1}{2}\mathbf{r} \times \mathbf{v} = \frac{\mathbf{L}}{2M} = \text{const}, \tag{23.5}$$

where $\mathbf{A}$ is the area vector, $\mathbf{v}$ the velocity, M the mass of the planet, and $\mathbf{L}$ its angular momentum. The vector $\mathbf{L}/(2M)$ is always constant, because the sun acting on the earth by a central force can apply no torque to it, and therefore the angular momentum of the object (planet, meteor, comet) is conserved, as is its mass.

Since $d\mathbf{A}/dt$ is a constant vector it has constant magnitude and constant direction. Constant magnitude implies Kepler's second law, and constant direction tells us that the orbit lies in a plane perpendicular to that direction.

26.2 POLAR COORDINATES AND THE UNIT VECTORS $\hat{\mathbf{r}}$ AND $\hat{\boldsymbol{\theta}}$

In order to relate Eq. (23.5) to the polar equation for an ellipse, we must write it in terms of polar coordinates, which are a more natural way to describe planetary orbits. In Chapter 9 we introduced the polar coordinates r and θ, which are related to the rectangular coordinates (x, y) by the equations

$$x = r \cos \theta, \qquad y = r \sin \theta.$$

For any plane curve, the position vector $\mathbf{r} = x\hat{\mathbf{i}} + y\hat{\mathbf{j}}$, shown in Fig. 26.1, is given by

$$\mathbf{r} = r(\cos \theta)\hat{\mathbf{i}} + r(\sin \theta)\hat{\mathbf{j}} = r[(\cos \theta)\hat{\mathbf{i}} + (\sin \theta)\hat{\mathbf{j}}], \tag{9.4}$$

where $r = |\mathbf{r}|$. We will consider both r and θ as functions of time.

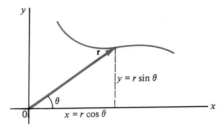

Figure 26.1 Position vector expressed in terms of polar coordinates r

The vector $(\cos\theta)\hat{\mathbf{i}} + (\sin\theta)\hat{\mathbf{j}}$ is a vector of unit length having the same direction as $\mathbf{r}$. We will denote this unit vector by our customary notation $\hat{\mathbf{r}}$, so we have

$$\mathbf{r} = r\hat{\mathbf{r}},$$

where

$$\hat{\mathbf{r}} = (\cos\theta)\hat{\mathbf{i}} + (\sin\theta)\hat{\mathbf{j}}. \tag{26.1}$$

We also introduce a unit vector $\hat{\boldsymbol{\theta}}$ which is perpendicular to $\hat{\mathbf{r}}$ and which is defined by

$$\hat{\boldsymbol{\theta}} = \frac{d\hat{\mathbf{r}}}{d\theta} = (-\sin\theta)\hat{\mathbf{i}} + (\cos\theta)\hat{\mathbf{j}}. \tag{26.2}$$

You can easily verify that

$$\hat{\mathbf{r}} \cdot \hat{\boldsymbol{\theta}} = 0,$$

which implies that $\hat{\mathbf{r}}$ and $\hat{\boldsymbol{\theta}}$ are perpendicular, and that

$$\frac{d\hat{\boldsymbol{\theta}}}{d\theta} = (-\cos\theta)\hat{\mathbf{i}} - (\sin\theta)\hat{\mathbf{j}} = -\hat{\mathbf{r}}.$$

The unit vectors $\hat{\mathbf{r}}$ and $\hat{\boldsymbol{\theta}}$ play the same role in polar coordinates as $\hat{\mathbf{i}}$ and $\hat{\mathbf{j}}$ play in Cartesian coordinates. However, $\hat{\mathbf{r}}$ and $\hat{\boldsymbol{\theta}}$ are not constant in direction; they change direction, as illustrated in Fig. 26.2.

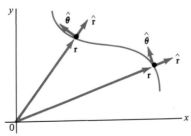

Figure 26.2 Unit vectors $\hat{\mathbf{r}}$ and $\hat{\boldsymbol{\theta}}$ attached to a curve at two points.

Observe that, by the chain rule, we have

$$\frac{d\hat{\boldsymbol{\theta}}}{dt} = \frac{d\hat{\boldsymbol{\theta}}}{d\theta}\frac{d\theta}{dt} = -\hat{\mathbf{r}}\frac{d\theta}{dt},$$

so we get the relation

$$-\hat{\mathbf{r}} = \frac{1}{d\theta/dt}\frac{d\hat{\boldsymbol{\theta}}}{dt}, \tag{26.3}$$

which we will use in solving the Kepler problem.

The vectors $\hat{\mathbf{r}}$ and $\hat{\boldsymbol{\theta}}$ share another property with $\hat{\mathbf{i}}$ and $\hat{\mathbf{j}}$. We recall that $\hat{\mathbf{i}} \times \hat{\mathbf{j}} = \hat{\mathbf{k}}$. Similarly, we have

$$\hat{\mathbf{r}} \times \hat{\boldsymbol{\theta}} = \hat{\mathbf{k}}, \tag{26.4}$$

which is easily seen from the formula

$$\hat{\mathbf{r}} \times \hat{\boldsymbol{\theta}} = \begin{vmatrix} \hat{\mathbf{i}} & \hat{\mathbf{j}} & \hat{\mathbf{k}} \\ \cos\theta & \sin\theta & 0 \\ -\sin\theta & \cos\theta & 0 \end{vmatrix} = (\cos^2\theta + \sin^2\theta)\hat{\mathbf{k}}.$$

We will now express the velocity vector of any plane motion as a combination of the unit vectors $\hat{\mathbf{r}}$ and $\hat{\boldsymbol{\theta}}$. The velocity is given by

$$\mathbf{v} = \frac{d\mathbf{r}}{dt} = \frac{d}{dt}(r\hat{\mathbf{r}}).$$

Since both r and $\hat{\mathbf{r}}$ may change with time we must be careful when we take the derivative. Applying the product rule, we have

$$\mathbf{v} = \frac{dr}{dt}\hat{\mathbf{r}} + r\frac{d\hat{\mathbf{r}}}{dt}.$$

We can write the derivative $d\hat{\mathbf{r}}/dt$ in terms of $\hat{\boldsymbol{\theta}}$ by applying the chain rule, which gives us

$$\frac{d\hat{\mathbf{r}}}{dt} = \frac{d\hat{\mathbf{r}}}{d\theta}\frac{d\theta}{dt} = \frac{d\theta}{dt}\hat{\boldsymbol{\theta}},$$

where we used Eq. (26.2) for $d\hat{\mathbf{r}}/d\theta$. Substituting this into our expression for the velocity, we obtain the formula we were seeking:

$$\mathbf{v} = \frac{dr}{dt}\hat{\mathbf{r}} + r\frac{d\theta}{dt}\hat{\boldsymbol{\theta}}. \tag{26.5}$$

The scalar factors dr/dt and $r\,d\theta/dt$ multiplying $\hat{\mathbf{r}}$ and $\hat{\boldsymbol{\theta}}$ are called, respectively, the *radial* and *transverse* components of velocity.

Since $\hat{\mathbf{r}}$ and $\hat{\boldsymbol{\theta}}$ are perpendicular unit vectors, we can easily determine the *speed*:

$$v = \sqrt{\mathbf{v}\cdot\mathbf{v}} = \sqrt{\left(\frac{dr}{dt}\right)^2 + \left(r\frac{d\theta}{dt}\right)^2}. \tag{26.6}$$

Note that the terms are not second derivatives but the squares of first derivatives.

Example 1

The polar coordinates of a particle are given by $r(t) = v_0 t$ and $\theta(t) = \omega_0 t$, where v_0 and ω_0 are constants. Determine the velocity and speed of the particle at any time t.

Using Eq. (26.5), we have

$$\mathbf{v} = \frac{dr}{dt}\hat{\mathbf{r}} + r\frac{d\theta}{dt}\hat{\boldsymbol{\theta}},$$

where $dr/dt = v_0$ and $d\theta/dt = \omega_0$. Therefore the velocity is given by

$$\mathbf{v} = v_0\hat{\mathbf{r}} + v_0\omega_0 t\hat{\boldsymbol{\theta}}.$$

From Eq. (26.5) we find the speed to be

$$v = \sqrt{v_0^2 + (v_0\omega_0 t)^2}.$$

Example 2

Show that for any plane motion we have

$$\mathbf{r} \times \mathbf{v} = r^2\frac{d\theta}{dt}\hat{\mathbf{k}}. \tag{26.7}$$

Taking the cross product of $\mathbf{r} = r\hat{\mathbf{r}}$ with $\mathbf{v}$ as given by (26.5) we find

$$\mathbf{r} \times \mathbf{v} = r\frac{dr}{dt}\hat{\mathbf{r}} \times \hat{\mathbf{r}} + r^2\frac{d\theta}{dt}\hat{\mathbf{r}} \times \hat{\boldsymbol{\theta}}.$$

But $\hat{\mathbf{r}} \times \hat{\mathbf{r}} = 0$, and in (26.4) we showed that $\hat{\mathbf{r}} \times \hat{\boldsymbol{\theta}} = \hat{\mathbf{k}}$, so this proves (26.7).

In Chapter 23 we introduced the angular momentum vector

$$\mathbf{L} = M\mathbf{r} \times \mathbf{v} \tag{23.4}$$

for any moving point mass of mass M. We can now obtain a simple formula for $\mathbf{L}$ in polar coordinates when the motion is in a plane. Using Eq. (26.7) we have

$$\mathbf{L} = Mr^2\frac{d\theta}{dt}\hat{\mathbf{k}}. \tag{26.8}$$

This tells us that for any plane motion the angular momentum vector has a fixed direction perpendicular to the plane of the motion and that its length is $Mr^2\omega$, where $\omega = |d\theta/dt|$ is the angular speed.

In Chapter 23 we showed that for any plane motion the vector quantity $\frac{1}{2}\mathbf{r} \times \mathbf{v}$ is the rate of change of the area vector,

$$\frac{1}{2}\mathbf{r} \times \mathbf{v} = \frac{d\mathbf{A}}{dt}. \tag{23.1}$$

Equation (26.7) now tells us that

$$\frac{d\mathbf{A}}{dt} = \frac{1}{2}r^2\frac{d\theta}{dt}\hat{\mathbf{k}},$$

(26.9)

so Kepler's law of equal areas is equivalent to the statement that

$$r^2\frac{d\theta}{dt} = \text{const},$$

in which case, by (26.8) the constant is L/M,

$$r^2\frac{d\theta}{dt} = \frac{L}{M}.$$

(26.10)

This property will be used to help solve the Kepler problem.

Questions

1. A particle moving in a plane traces out a spiral given by

$$r(t) = r_0e^t, \qquad \theta(t) = \omega_0t,$$

where r_0 and ω_0 are constants. Determine the velocity and speed of the particle.

2. How does the angular speed of a planet vary with its distance from the sun?

3. If an object has constant mass and angular momentum in a circular orbit, what must be true of $d\theta/dt$?

4. For circular motion in polar coordinates the vector

$$\boldsymbol{\omega} = \frac{1}{r^2}\mathbf{r} \times \mathbf{v}$$

is called the angular velocity vector. Show that

(a) $\boldsymbol{\omega} = \dfrac{d\theta}{dt}\hat{\mathbf{k}},$

(b) $\mathbf{v} = \boldsymbol{\omega} \times \mathbf{r},$

(c) $\mathbf{v} \times \boldsymbol{\omega} = \dfrac{v^2}{r^2}\mathbf{r},$

(d) $\boldsymbol{\omega} \times (\boldsymbol{\omega} \times \mathbf{r}) = -\omega^2\mathbf{r}$, where $\omega = \left|\dfrac{d\theta}{dt}\right|.$

26.3 SOLUTION OF THE KEPLER PROBLEM

We now have all the machinery in place to solve the Kepler problem. Assume we have a fixed sun of mass M_0 and a moving body of mass M attracted to the sun by a gravitational force $\mathbf{F}$. Newton's law of gravity states that

$$F = -G\frac{MM_0}{r^2}\hat{\mathbf{r}},$$

where $\mathbf{r} = r\hat{\mathbf{r}}$ is the radius vector from the sun to the body. Newton's second law of motion describes the acceleration due to that force,

$$F = M\frac{d\mathbf{v}}{dt}.$$

Equating the two expressions for F and canceling M, we find

$$\frac{d\mathbf{v}}{dt} = -G\frac{M_0}{r^2}\hat{\mathbf{r}}.$$

Using (26.3) and then Eq. (26.10) we can write this as

$$\frac{d\mathbf{v}}{dt} = \frac{GM_0}{r^2(d\theta/dt)}\frac{d\hat{\boldsymbol{\theta}}}{dt} = \frac{GMM_0}{L}\frac{d\hat{\boldsymbol{\theta}}}{dt} = \frac{D}{L}\frac{d\hat{\boldsymbol{\theta}}}{dt},$$

or as

$$\frac{L}{D}\frac{d\mathbf{v}}{dt} = \frac{d\hat{\boldsymbol{\theta}}}{dt},$$

where L/D is a constant and $D = GMM_0$. This last differential equation can be integrated at once to give

$$\frac{L}{D}\mathbf{v} = \hat{\boldsymbol{\theta}} + \mathbf{C},$$

where $\mathbf{C}$ is a constant vector which depends on the initial conditions. Let's measure t so that at time $t = 0$ the body is closest to the sun. Then $dr/dt = 0$ at $t = 0$ because r has a minimum there. Hence $\mathbf{v}(0)$ has the same direction as $\hat{\boldsymbol{\theta}}(0)$ since the radial component of $\mathbf{v}(0)$ is zero. If we measure θ so that it increases with t, then $\hat{\boldsymbol{\theta}}(0) = \hat{\mathbf{j}}$, as indicated in the diagram.

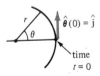

Hence $\mathbf{C}$ is a scalar times $\hat{\mathbf{j}}$. We call this scalar e and write $\mathbf{C} = e\hat{\mathbf{j}}$. (There is a good reason why we are using e; it will turn out to be the eccentricity.) The basic equation for $\mathbf{v}$ becomes

$$\frac{L}{D}\mathbf{v} = \hat{\boldsymbol{\theta}} + e\hat{\mathbf{j}}.$$

Taking the dot product with $\hat{\boldsymbol{\theta}}$ we have

$$\frac{L}{D}\mathbf{v}\cdot\hat{\boldsymbol{\theta}} = \hat{\boldsymbol{\theta}}\cdot\hat{\boldsymbol{\theta}} + e\hat{\mathbf{j}}\cdot\hat{\boldsymbol{\theta}} = 1 + e\cos\theta. \tag{26.11}$$

But from (26.5) and (26.10) we have

$$\mathbf{v} \cdot \hat{\mathbf{\theta}} = r\frac{d\theta}{dt} = \frac{1}{r}\left(r^2\frac{d\theta}{dt}\right) = \frac{1}{r}\frac{L}{M}$$

hence (26.11) becomes

$$\frac{L^2}{DM}\frac{1}{r} = 1 + e \cos\theta. \tag{26.12}$$

This equation implies that e is positive. To see this, let θ take the values 0 and π. Then $L^2/(DMr)$ takes the values $1 + e$ and $1 - e$, respectively. Since r has a minimum when $\theta = 0$, this implies $1 + e > 1 - e$, so $e > 0$ as asserted. When we solve (26.12) for r we get

$$r = \frac{L^2}{DM(1 + e \cos\theta)}, \tag{26.13}$$

the polar equation for a conic with eccentricity e.

For an ellipse the polar equation was shown in Chapter 25 to be

$$r = \frac{a(1 - e^2)}{1 + e \cos\theta}, \tag{25.3}$$

where a is the length of the semimajor axis. Comparing this with Eq. (26.13) we see that for an elliptical orbit we have

$$\frac{L^2}{DM} = a(1 - e^2). \tag{26.14}$$

It should be realized that the foregoing solution of the Kepler problem is based on simplifying assumptions which are not exactly true in the real solar system. We have assumed that the sun is fixed (which it is not) and that the only force acting on the body is the gravitational attraction of the sun. In reality, all the planets and other objects in the solar system also exert gravitational forces on the body but these are negligible compared to the massive attraction of the sun. In a solar system such as ours with one huge sun and a small number of little planets (called a *Keplerian system*) these simplifications seem reasonable because the predicted orbits agree with the actual observed orbits to a remarkable degree of accuracy.

We have now deduced Kepler's first and second laws from Newton's laws. Deducing Kepler's third law is a relatively simple matter, which we defer to Chapter 29.

26.4 A FINAL WORD

Johannes Kepler lived at the same time as William Shakespeare, Queen Elizabeth I of England, and, of course, Galileo Galilei. Although Kepler and Galileo were contemporaries, they never met, but they did exchange correspondence, the nature of which gives insight into the characters of these two giants.

In 1610 Galileo turned his newly invented telescope to the heavens and gazed upon new worlds. Kepler eagerly beseeched Galileo to send him a telescope so that he too could see mountains on the moon and the moons of Jupiter. But his letters went unanswered by Galileo except for cryptic messages sent to him through the Tuscan Ambassador in Prague. One communication was "SMAISMRMILMEPOETALEUMIBUNENUGTTAURIAS," an anagram in Latin. Galileo, wishing to maintain his priority on observations, frequently used anagrams as a safeguard. Kepler, after struggling to decipher it, thought it meant "Hail, burning twins, offspring of Mars" – that Galileo had discovered two moons around Mars. Only when Emperor Rudolph II expressed interest in the puzzle did Galileo disclose the anagram's meaning – that he had observed two moons around Saturn. The moons Galileo observed were really the rings of Saturn, which appeared as featureless blobs of light in his low-power telescope.

Kepler decoded another anagram to mean that Galileo had observed a rotating red spot on Jupiter, but the intended meaning, supplied by Galileo through similar means, was that he had discovered that Venus, like the moon, shows phases, which is proof that Venus orbits the sun.

Although Kepler's solutions to these anagrams were incorrect, what he thought they meant turned out to be correct: Mars has two moons, and Jupiter a red spot. But these were discovered after the deaths of both Kepler and Galileo.

In 1610 Galileo published his observations as *The Starry Messenger*, a concise book that could be read by anyone in an hour and that broke with the flowery style of scholarly writing of the times. But it was not the style of his writing for which he was attacked but for the general content of the book. People didn't believe him, and most philosophers chose to ignore him.

Galileo appealed to Kepler through intermediaries. Finally, in a letter, lines from which opened this chapter, he asked Kepler to support his case. Kepler immediately and warmly did that, saying that Galileo had made discoveries that would go down through the ages – of course he was correct about that. Yet Galileo completely ignored Kepler's discoveries. He was put off by Kepler's enthusiasm for uncovering hidden harmonies and geometric relations in the universe. In a sense, Kepler was the last of the ancient Pythagoreans, while Galileo was the first modern scientist. But the ironic consequence was that Galileo defended to the end of his life the circles and epicycles of Copernicus as the only conceivable form of heavenly motion.

CHAPTER 27

ENERGY AND ECCENTRICITY

Therefore, during the whole time of their appearance, comets fall within the sphere of activity of the circumsolar force, and hence are acted upon by its impulse and therefore (by Corollary 1, Proposition XII) describe conic sections that have their foci in the center of the sun, and by radii drawn from the sun describe areas proportional to the times. For that force propagated to an immense distance, will govern the motions of the bodies far beyond the orbit of Saturn.

Isaac Newton, *Principia* (1687)

27.1 CELESTIAL OMENS: COMETS

Ancient astronomers earned their keep and kept their heads by making accurate predictions of the arrival of the seasons and of such troubling celestial events as solar and lunar eclipses. As their technical expertise improved, astronomers learned to predict even the wandering motion of the planets. Yet at times, interlopers, such as meteors, appeared in the night sky. Although meteors seemed as unpredictable as the weather (and were thought to be related to it, which is why their name shares the same Greek root with the science of weather – meteorology), even meteor showers were observed to occur regularly. For example, the most spectacular meteor showers occur every year in mid-August.

Nevertheless, objects occasionally and mysteriously appeared in the heavens which were not planets or meteors. Trailing plumes of cold fire, these objects were named *comets*, from the Greek word meaning *thing with hair*. Because their appearance was unpredictable, comets were interpreted as omens of impending disaster. The Bayeux Tapestry, which tells the story of the Norman conquest of England, shows a comet that appeared in 1066. Later in that year, Harold, the pretender to the throne, was defeated by William the Conqueror at the Battle of Hastings. Shakespeare places a comet in the sky the night before the murder of Julius Caesar. And according to legend, Montezuma, the Aztec leader, fell into such a depression at the foreboding appearance of a comet that he could not lead his people against the invading conquistadors.

Figure 27.1 Part of the Bayeux Tapestry depicting the comet of 1066, known today as Halley's comet. (Courtesy of Science Graphics.)

For ages comets remained a superstitious and perplexing anomaly of the heavens. Neither the system of Ptolemy, nor the heliocentric theory of Copernicus, nor even the ellipses of Kepler made the appearance of comets understandable and predictable. In 1682 a spectacular comet blazed across the sky, and among the astronomers who charted its position was Isaac Newton. He saw in the orbit of the comet the same force and dynamics at work which govern the motion of the planets. Some comets, Newton realized, could swing past the sun in open curves – parabolas and hyperbolas – and so would never return. But other comets should move along elliptical paths like the planets' except on much longer leashes. Newton's penetrating insight revealed that comets are members of the solar system and thus he cast off the superstition that enshrouded them.

27.2 ENERGY IN SPACE

We have shown that a planet, comet, meteor, or any heavenly body that orbits the sun must move along a conic section with polar equation

$$r = \frac{L^2}{DM(1 + e \cos \theta)} , \tag{26.13}$$

where L is the angular momentum of the body, M is its mass, and $D = GMM_0$, where M_0 is the mass of the sun. To determine whether the orbit is an ellipse, parabola, or hyperbola we need to relate the eccentricity e to the energy E of the moving body.

The energy E consists of two parts,

$$E = K + U,$$

where $K = \frac{1}{2}Mv^2$ is the kinetic energy and $U = -D/r$ is the potential energy associated with the gravitational force. For plane motion we can express the kinetic energy in polar coordinates by using Eq. (26.6) for the speed and we obtain

$$K = \tfrac{1}{2}Mv^2 = \tfrac{1}{2}M\left[\left(\frac{dr}{dt}\right)^2 + r^2\left(\frac{d\theta}{dt}\right)^2\right]. \tag{27.1}$$

Therefore the total energy is

$$E = \frac{1}{2}M\left[\left(\frac{dr}{dt}\right)^2 + r^2\left(\frac{d\theta}{dt}\right)^2\right] - \frac{D}{r} . \tag{27.2}$$

The energy is a constant of the motion. At all points at all times, the energy has the same value.

Example 1

A satellite is fired from the surface of a spherical, nonrotating planet of mass M_0, radius R, which has no atmosphere, with a speed v_0 at an angle of 30° from the radial direction. In its subsequent orbit, the satellite reaches a maximum distance of $\frac{5}{2}R$ from the center of the planet. Using conservation of energy and angular momentum, find v_0 in terms of M_0, R, and G.

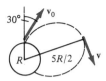

Let's take the mass of the satellite to be m, and write down its energy at point A, where it is fired off, and at point B in its orbit:

$$E_A = \frac{1}{2}mv_0^2 - G\frac{mM_0}{R} ,$$

$$E_B = \frac{1}{2}mv^2 - G\frac{mM_0}{\frac{5}{2}R} .$$

We know that $E_A = E_B$, but we have two unknowns, namely, v_0 and v. The additional information needed to solve the problem is provided by the equation expressing conservation of angular momentum. Recalling that $L = mvr$, where v is the component of velocity perpendicular to the radius, we have

$$L_A = mRv_0 \sin 30°,$$

$$L_B = mv(\tfrac{5}{2}R).$$

Since $L_A = L_B$, we can solve for v in terms of v_0:

$$v = \tfrac{1}{5}v_0.$$

Substituting this into E_B and setting E_A equal to E_B, we get

$$\frac{1}{2}mv_0^2 - G\frac{mM_0}{R} = \frac{1}{2}m\left(\frac{v_0}{5}\right)^2 - G\frac{2mM_0}{5R}.$$

After some algebra, we find that

$$v = \sqrt{\frac{5GM_0}{4R}}.$$

Questions

1. Can the total energy of a planet be negative? If so, what does this mean? Can kinetic energy ever be negative?

2. A satellite of mass m has a circular orbit around a planet of mass M_0. The angular momentum of the satellite is L. Find the total energy of the satellite in terms of m, M_0, and L.

3. A spherical, nonrotating planet with no atmosphere has mass M_0 and radius R. A spacecraft is fired from the surface with a speed $v_0 = \tfrac{3}{4}\sqrt{2GM_0/R}$. Considering conservation of energy and angular momentum, calculate the farthest distance it reaches from the center of the planet if it is fired off (a) radially and (b) tangentially.

4. Repeat Question 3 for the case of $v_0 = \sqrt{2GM_0/R}$.

5. A satellite of mass m is traveling at a speed v_0 in a circular orbit of radius r_0 about a planet. Show that the total energy of the satellite is $-\tfrac{1}{2}mv_0^2$.

6. A rocket is fired from Cape Canaveral with an initial speed v_0 at an angle θ from the horizon as shown.

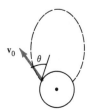

Neglecting air resistance and the earth's rotation, calculate the maximum distance from the center of the earth that the rocket reaches in terms of the mass and radius of the earth, M and R, v_0, θ, and G.

27.3 ENERGY AND ECCENTRICITY

Let's now combine conservation of angular momentum, given by Eq. (26.10),

$$r^2 \frac{d\theta}{dt} = \frac{L}{M}, \tag{26.10}$$

with our expression for the energy and orbit of an object moving about the sun. In doing so, we will find a connection between energy and the eccentricity of the object's path.

From Eq. (26.10) we see that

$$r^2 \left(\frac{d\theta}{dt}\right)^2 = \frac{L^2}{M^2 r^2}.$$

Substituting this into the expression for the kinetic energy, Eq. (27.1),

$$K = \frac{1}{2} M \left[\left(\frac{dr}{dt}\right)^2 + r^2 \left(\frac{d\theta}{dt}\right)^2\right], \tag{27.1}$$

we get

$$K = \frac{1}{2} M \left(\frac{dr}{dt}\right)^2 + \frac{1}{2} \frac{L}{Mr^2}.$$

Therefore the total energy is

$$E = \frac{1}{2} M \left(\frac{dr}{dt}\right)^2 + \frac{1}{2} \frac{L^2}{Mr^2} - \frac{D}{r}. \tag{27.3}$$

Next, we feed the equation of the orbit, Eq. (26.13), into this relation using the reciprocal form

$$\frac{1}{r} = c(1 + e \cos \theta), \tag{27.4}$$

where

$$c = \frac{DM}{L^2}. \tag{27.5}$$

Differentiating both sides of Eq. (27.4) with respect to time, we get

$$-\frac{1}{r^2} \frac{dr}{dt} = -ce \sin \theta \frac{d\theta}{dt},$$

which tells us that

$$\frac{dr}{dt} = ce \sin \theta \left(r^2 \frac{d\theta}{dt}\right).$$

Using Eq. (26.10), we eliminate the factor $r^2 \, d\theta/dt$ and get

$$\frac{dr}{dt} = \frac{L}{M} ce \sin \theta.$$

Substituting this expression into our expression for the energy, Eq. (27.3), we find

$$E = \frac{1}{2} \frac{L^2 c^2}{M} e^2 \sin^2 \theta + \frac{1}{2} \frac{L^2 c^2}{M} (1 + e \cos \theta)^2 - Dc(1 + e \cos \theta).$$

Since $c = DM/L^2$ we have

$$c^2 = \frac{MDc}{L^2} \quad \text{and} \quad \frac{L^2 c^2}{M} = Dc.$$

Therefore the last equation for E becomes

$$E = \tfrac{1}{2}Dce^2 \sin \theta + \tfrac{1}{2}Dc(1 + e \cos \theta)^2 - \tfrac{1}{2}Dc(2 + 2e \cos \theta),$$

$$E = \tfrac{1}{2}Dc(e^2 \sin^2 \theta + 1 + e^2 \cos^2 \theta + 2e \cos \theta - 2 - 2e \cos \theta).$$

Since $\sin^2 \theta + \cos^2 \theta = 1$, we get

$$E = \tfrac{1}{2}Dc(e^2 - 1).$$

Substituting for c from Eq. (27.5), we have

$$E = \frac{D^2 M}{2L^2} (e^2 - 1). \tag{27.6}$$

The formula for the energy is suddenly much simpler! Both r and θ have been eliminated and the total energy is constant, as it should be. We've expressed the energy in terms of the masses M and M_0 (since $D = GMM_0$), the angular momentum L, and the eccentricity of the orbit. This formula for the energy will give us important insight into the shapes of the orbits that we couldn't get from the equation of the conics alone.

For an elliptical orbit we can obtain an even simpler result. Equation (26.14) tells us

$$\frac{L^2}{DM} = a(1 - e^2), \tag{26.14}$$

so the energy in this case is simply

$$E = \frac{-D}{2a} = \frac{-GMM_0}{2a}. \tag{27.7}$$

This result is useful because it expresses the total energy of an elliptical orbit entirely in terms of the mass of the planet, the mass of the sun, and the length of the major axis of the ellipse.

Questions

7. Knowing that the aphelion and perihelion distances for the earth are 1.47×10^{11} m and 1.53×10^{11} m, respectively, calculate the energy associated with the earth's orbit. (Use data in Appendix D.)

8. Show that the ratio of the kinetic energy at aphelion to that at perihelion is given by

$$\frac{K_a}{K_p} = \frac{r_p^2}{r_a^2},$$

where r_p and r_a are the perihelion and aphelion distances, respectively.

9. The elliptical orbit of a 2500-kg satellite about the earth is described by

$$r = \frac{(8600 \text{ km})}{1 + 0.4 \cos \theta}.$$

(a) What is the eccentricity of the orbit?
(b) What is the energy of the orbit?
(c) What is the angular momentum of the orbit?

(Use data in Appendix D for the mass of the earth.)

27.4 ORBITS AND ECCENTRICITY

Let's briefly recapitulate what we've found thus far. A planet, or satellite, starting out in some orbit has a definite value of D, which depends on the masses; an initial energy E; and an initial angular momentum L. These factors should completely determine the type of orbit, which is measured by the eccentricity. Taking

$$E = \frac{D^2 M}{2L^2} (e^2 - 1) \tag{27.6}$$

and solving for the eccentricity we find

$$e = \sqrt{1 + \frac{2L^2 E}{D^2 M}}, \tag{27.8}$$

which indeed specifies the type of orbit.

Case I: $E < 0$.
First, let's consider the case of negative total energy, that is, $E < 0$. The total energy can only be negative if the negative gravitational potential energy is always greater in magnitude than the positive kinetic energy. According to Eq. (27.8), if $E < 0$, the quantity under the radical sign is less than one, so the eccentricity is less than one. In fact it is bounded between zero and one: $0 \leq e < 1$. But an eccentricity in this range describes an ellipse!

Since e is a real number, the quantity under the radical in Eq. (27.8) must be nonnegative. The minimum value of energy is that for which $e = 0$. From Eq. (27.6), we find that this minimum value of energy is given by

$$E_{min} = -\frac{D^2 M}{2L^2}.$$

If exactly this relationship holds between L^2 and E, the orbit will be a circle, a Platonic orbit. The orbits of most planets are very nearly circular.

Case II: *E = 0.*

For the case of zero energy, the potential energy is exactly equal to the kinetic energy. This is equivalent to saying that the planet starts out an infinite distance away from the sun and slowly begins to fall in toward it, with its decrease in potential energy appearing as kinetic energy. The eccentricity, according to Eq. (27.8), is equal to 1 when $E = 0$. The conic described by this case is a parabola. Parabolic orbits are theoretically possible but highly unlikely because it would require a perfect balance between the negative potential energy and the positive kinetic energy.

Case III: *E > 0.*

For positive energy, the object always has more kinetic energy than potential energy. In this case the object is in motion at infinite distance from the sun before it begins falling toward the sun. The orbit is not closed since it corresponds to an eccentricity greater than 1. The object will make one pass by the sun, along a hyperbola, and whip away to infinity, never to come back again.

The structure of the solar system, the motion of planets, comets, and satellites, has finally been revealed. We've seen that Newton's dynamics lead to Kepler's first two laws. Kepler's third law is derived in Chapter 29. In the rest of this chapter we discuss applications of orbital dynamics.

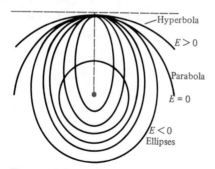

Figure 27.2 Possible orbits about the sun.

Example 2

Show that Galileo's parabolic trajectory is approximately a small segment of an ellipse with the earth's center at the more distant focus.

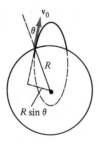

27.6 CALCULATING THE ORBIT FROM INITIAL CONDITIONS

Now that we understand the factors that determine the orbit of a planet, a comet, or a satellite, let's indicate just how the orbit can be found if some initial conditions are known. Suppose a satellite of mass m is launched, for example, from a space shuttle a distance r_0 from the center of the earth (of mass M), at an initial angle θ_0 from some reference line, as indicated in Fig. 27.4. As part of the initial conditions the initial speed v_0 as well as the angle of launch ϕ are also known. How do we find the size, shape, and orientation of the orbit?

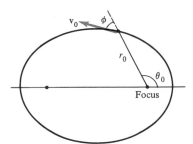

Figure 27.4 Initial conditions which determine the size, shape, and orientation of an orbit.

First, we know that the orbit will be closed only if the energy is negative. The energy, which remains constant, is specified by

$$E = \frac{1}{2}mv_0^2 - G\frac{Mm}{r_0}.$$

According to Eq. (27.7), this energy is also given by

$$E = -G\frac{Mm}{2a}, \tag{27.7}$$

where $2a$ is the length of the major axis, if the orbit is closed. Since E is known from the initial speed and distance, the length of the major axis is also specified. This tells us the size of the orbit.

The angular momentum likewise remains constant and equal to its initial value, which is given by

$$L = mv_0r_0 \sin \phi.$$

Since the energy and angular momentum are known, the eccentricity of the orbit can be determined from Eq. (27.8):

$$e = \sqrt{1 + \frac{2L^2E}{D^2m}}. \tag{27.8}$$

Once the eccentricity is known, the perigee and apogee can be found from

$$r_p = a(1 - e) \quad \text{and} \quad r_a = a(1 + e).$$

So far we've determined the size (given by a) and shape (given by e) of the orbit. The orientation, which we've denoted by θ_0, can be found from the equation of an ellipse, Eq. (26.13):

$$r = \frac{L^2}{Dm(1 + e \cos \theta)}, \tag{26.13}$$

when the values of r_0, e, m, and L are inserted.

Example 5

A 5×10^3-kg satellite is launched into space with an initial speed $v_0 = 4000$ m/s at a distance $r_0 = 6R = 3.6 \times 10^7$ m from the center of the earth, at an angle of 30° from the radial direction. Calculate (a) the length of the semimajor axis, (b) the angular momentum, (c) the eccentricity, (d) the orientation, and (e) the perigee and apogee distances of the orbit.

Using Eq. (27.7),

$$E = -G\frac{Mm}{2a}. \tag{27.7}$$

we can determine the length of the semimajor axis. We know that the initial energy is

$$E = \frac{1}{2}mv_0^2 - G\frac{Mm}{r_0} = -1.6 \times 10^{10} \text{ J},$$

when we insert the values of v_0, r_0, m, M, and G. Equating this to Eq. (27.7) and solving for a, we find that $a = 6.4 \times 10^7$ m, which is about 10.4 earth radii.

The angular momentum is simply $L = mv_0 r_0 \sin \theta_0$, which turns out to be equal to 3.6×10^{14} kg m^2/s. Once we know both E and L, we can use Eq. (27.8),

$$e = \sqrt{1 + \frac{2L^2 E}{D^2 m}}, \tag{27.8}$$

to calculate the eccentricity. Substituting values, we find that $e = 0.89$.

The orientation of the orbit comes from taking the equation of an ellipse, Eq. (26.13):

$$r = \frac{L^2}{Dm(1 + e \cos \theta)}, \tag{26.13}$$

and solving for $\cos \theta_0$:

$$\cos \theta_0 = \frac{1}{e}\left(\frac{L^2}{Dmr_0} - 1\right).$$

Here we find that $\theta_0 = 136°$.

Finally, the perigee and apogee distances are given by

$$r_p = a(1 - e) = 7.0 \times 10^6 \text{ m} = 1.2R,$$

$$r_a = a(1 + e) = 1.2 \times 10^8 \text{ m} = 20R.$$

The orbit is illustrated below:

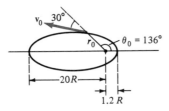

Questions

17. A satellite of mass m is launched with a speed $v_0 = (1.5GM/R)^{1/2}$ from a distance of $4R$ from a planet of mass M and radius R as shown in the sketch. In terms of G, m, M, and R find the following for the orbit: (a) the energy, (b) the type, and (c) the eccentricity.

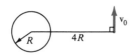

18. From a distance of $5R$ from the center of a planet of mass M and radius R, a satellite of mass m is launched with a speed $v_0 = (0.2GM/R)^{1/2}$ in the direction shown in the sketch. In terms of G, m, M, and R, determine the following quantities for the orbit: (a) the energy, (b) the angular momentum, and (c) the eccentricity.

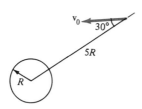

19. Using conservation of energy and angular momentum, calculate the speed of the satellite in Question 18 at both its perigee and apogee in terms of G, M, and R.

20. Knowing that the eccentricity of the orbit of Halley's comet is 0.967 and the length of its semimajor axis is 20 times that of the earth's orbit, use additional data in Appendix D to calculate the following:

(a) the perihelion and aphelion distances for the comet,

(b) its period.

21. Estimate the potential energy of the moon

    (a) with respect to the earth,

    (b) with respect to the sun.

27.7 A FINAL WORD

Aside from Newton himself, the first person entrusted with knowledge of the structure of the universe – really what we now call the solar system – was the British astronomer Edmund Halley. Halley realized that although the planets had nearly circular orbits, Newton's results implied that all bodies in space must have elliptical orbits in order to have been captured by the sun at all. Some of them, for example, comets that pass by infrequently, might have highly eccentric orbits.

In 1682 an awesome comet dominated the sky for months, and Halley made repeated measurements of its path. Later he applied Newton's method to find its orbit. Although his observations covered only a small segment of the complete orbit, Halley calculated that the comet had a highly elliptical orbit with a semimajor axis 20 times the earth's and a period of 76 years. Looking back into astronomical records, he found observations of comets in 1607 and 1531 so nearly identical that he concluded that they had to be the same object. The comet of the Bayeux Tapestry, it turns out, was also Halley's comet. Then Halley made the crucial test, which, of course, was to predict the comet's next return. He said that it would be seen at the end of 1758, a prediction which neither he nor Newton lived to see fulfilled.

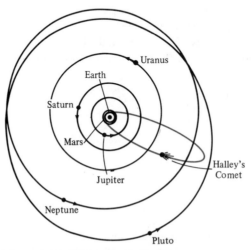

Figure 27.5 Orbit of Halley's comet.

Halley's comet was next seen at Christmas 1758. Of course, it has not stopped making its rounds. At the comet's nearest approach to the sun, it travels inside the orbit of Venus and shines brightly. At its most distant point, it goes beyond the orbit of Neptune as shown in Fig. 27.5. The most recent round trip brings it to this part of the solar system in 1910 and again, just as Halley predicted, in 1986.

> Of all the comets in the sky,
> There's none like comet Halley.
> We see it with the naked eye,
> And periodically.
>
>   – Anon.

must be fired to remove it from the transfer orbit and place it in the orbit of the target. This mode of travel requires the least amount of fuel, and that's much more important when you're traveling from Earth to Saturn than when you're driving across town.

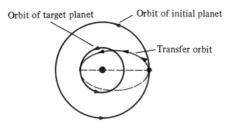

Figure 28.2 A transfer orbit is a semielliptical orbit with the sun at one focus which is tangent to the orbits of both the initial and target planets.

Transfer orbits put constraints on space travel. We can't send a probe to any planet at any time by this method. Instead the launch must take place when the earth and the target are in the correct relative positions, a happening known as a *launch opportunity*. The earth must be at one end of the major axis of the transfer orbit at launch, and the target planet must arrive at the other end simultaneously with the spacecraft. By this method, we can send spacecraft to Venus every 19 months, to Mars every 780 days, and to Jupiter every 13 months. Launch opportunities are a consequence of the different orbital periods of the planets. According to Kepler's third law, each planet has an orbital period which is related to the length of its semimajor axis by

$$T^2 = ka^3, \tag{28.1}$$

where k is a constant of proportionality which depends on the mass of the sun but is the same for all planets.

When a launch opportunity occurs, a spacecraft is initially placed in a temporary orbit about the earth known as a *parking orbit*. To enter a transfer orbit a spacecraft must somehow leave its parking orbit and escape the earth's gravity. Rocket thrusts supply a spacecraft with energy to escape, but when and how the thrusts are made depend on the destination.

Even during an opportunity the craft must be launched at the right place in its parking orbit. That place is called a *launch window*. If the spacecraft is headed for one of the inner planets (Mercury or Venus), the launch window occurs as the craft emerges into the sunlit side of the earth, as shown in Fig. 28.3a. To launch toward the outer planets, a craft must leave its parking orbit as it approaches the dark side of the earth, as shown in Fig. 28.3b. Here's why: The initial velocity of the craft includes a contribution due to the earth's orbital velocity in addition to the craft's orbital velocity about Earth. (The parking orbit around the earth is always in the same direction as the earth's rotation. Why?) When the craft is on the sunlit side of the earth, these contributions are in opposite directions (see Fig. 28.3). An additional rocket thrust in the direction of the craft's orbital motion about the earth allows the craft to escape from its Earth orbit. But since the velocity of the craft relative to the sun now is smaller than the earth's, the craft falls into an orbit closer to the sun – a transfer orbit. Although the craft takes off in a hyperbolic (escape) orbit relative to the earth, its orbit relative to the sun is elliptical.

To enter a transfer orbit to an outer planet, a spacecraft is launched from its parking orbit while approaching the dark side of the earth. When on the dark side the craft's orbital velocity about the earth is in the same direction as the earth's about the sun. Thus the two contributions will add, and a rocket blast causes the craft to escape from its Earth orbit and move into a larger orbit about the sun. (See Fig. 28.3b.)

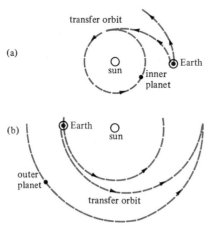

Figure 28.3 (a) Launch window for a journey to an inner planet. (b) Launch window for a journey to an outer planet.

Questions

1. A satellite is placed into orbit around the earth before voyaging to Jupiter. Explain how by firing its rockets both the energy and angular momentum of the satellite can be changed.

2. One method proposed for space travel is to use ion propulsion. By constantly burning fuel and ejecting ions, a spacecraft could gain speed smoothly rather than in spurts from the firing of a rocket. How does this method of propulsion, in comparison to rocket blasts, complicate calculating transfer orbits?

3. Using the idea of "falling" in the sun's gravitational field, explain how a spacecraft (a) gains speed traveling along a transfer orbit to an inner planet, and (b) loses speed traveling to an outer planet.

4. Show that the number of days between launch opportunities, T_L, for any planet is given by

$$T_L = \left| \frac{T_{earth} \, T_{planet}}{T_{earth} - T_{planet}} \right|,$$

where T_{earth} and T_{planet} are the orbital periods about the sun. Find the launch opportunity period for Saturn and the earth.

5. Launch opportunity is associated with a physical alignment of two planets and the sun. Imagine a planet on each of the two hands of a clock. If the correct alignment is that the two planets should be directly opposite each other, how many minutes will pass between successive favorable positions?

28.3 TRANSFER ORBITS

Let's now describe mathematically the steps involved in calculating a transfer orbit from, say, the earth to another planet. To simplify the calculations somewhat and emphasize the physical ideas, we'll asssume the orbits of the planets are circles. We'll also ignore the need to escape from the earth's gravity, and the speed of the spacecraft in its parking orbit around the earth. As discussed earlier, a transfer orbit is semielliptical with the two planets at perihelion and aphelion. In plotting a course for such an orbit, the first step is to find the length $2a$ of the major axis of the ellipse. As shown in Fig. 28.4, if r_1 and r_2 are the radii of the planetary orbits, then the length of the major axis of the transfer orbit is

$$2a = r_1 + r_2.$$

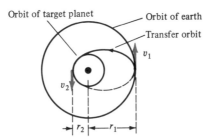

Figure 28.4 Quantities determining a transfer orbit.

Next we need to know the speed which the satellite must have to move into the transfer orbit. Recalling that the energy of any orbit about the sun is

$$E = -G\frac{mM_0}{2a}, \tag{27.7}$$

where M_0 is the mass of the sun and m is the mass of the spacecraft, we see that once $2a$ is known, the energy of the orbit is determined. The energy of the spacecraft is also given by

$$E = \tfrac{1}{2}mv_1^2 - G\frac{mM_0}{r_1},$$

so we can solve for v_1, the speed needed to begin the transfer orbit. From the orbital speed of the earth we can then calculate precisely the increase or decrease in speed necessary to send the spacecraft into the transfer orbit.

If the probe travels to an inner planet, it gains speed as it falls toward the sun. On the other hand, if it is traveling to an outer planet, it loses speed. But in both cases, the speed of the spacecraft must be changed to match that of the planet in its orbit about the sun. So when the spacecraft reaches the orbit of the target planet, we need to know its arrival speed. Since angular momentum is conserved along the transfer orbit, we can easily calculate this speed. If v_2 is its arrival speed, conservation of angular momentum applied to the two points of the transfer orbit indicated in Fig. 28.4 tells us that

$$mv_1r_1 = mv_2r_2$$

because the velocity is perpendicular to the radius vector from the sun at both points. From this equation we can solve for v_2.

From Kepler's third law, we can find the time for the spacecraft to travel along the transfer orbit; it's one-half of the period T. By using Kepler's third law $T^2 = ka^3$ twice, the constant k can be eliminated. Thus we have

$$\frac{T^2}{T_E^2} = \frac{a^3}{a_E^3},$$

where T_E is the period of the earth's orbit and a_E the length of its semimajor axis, and T and a are those for the transfer orbit. These are measured in years ($T_E = 1$ year) and in astronomical units (AU), where, by definition, $a_E = 1$ AU $= 1.5 \times 10^{11}$ m, the mean distance from the earth to the sun.

Finally, the launch opportunity can be found by deciding where the target planet should be relative to the earth at launch so that the spacecraft will arrive at the orbit of the target planet when the planet is also at the same point. Using the time of travel t and the orbital period of the planet T_p, the fraction $(360°)t/T_p$, measures the angle in degrees the planet will move through while the spacecraft is on its way. From this angle we can determine the opportunity for launch. The following examples illustrate these calculations.

Example 1

For a transfer orbit between Earth and Venus, assume that both orbits are circular and that the radius of Venus' orbit is 0.72 AU, and its orbital period and speed are 225 days and 35.0 km/s, respectively. Determine the following: (a) The length of the major axis, (b) the speed necessary to propel a spacecraft into the transfer orbit, (c) the change in speed required when the spacecraft reaches Venus, (d) the time of the trip, and (e) the relative positions of Venus and Earth at launch.

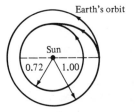

We assume that initially the spacecraft has the same speed as the earth in its orbit around the sun. The energy of a spacecraft moving in that orbit is

$$E_0 = -G \frac{mM_0}{2r_e},$$

where M_0 is the mass of the sun, m the mass of the spacecraft, and r_e the radius of the earth's orbit. Equating this energy to

$$E = \tfrac{1}{2} mv_0^2 - G \frac{mM_0}{r_e},$$

we find that the speed of the satellite as it moves with the earth is

$$v_0 = (GM_0/r_e)^{1/2} = 29.8 \text{ km/s}.$$

We'll express the other speeds in terms of this speed.

As indicated in the above diagram, the length of the major axis of the orbit is $2a = 0.72 \text{ AU} + 1.00 \text{ AU} = 1.72 \text{ AU} = 1.72r_e$. The energy of the transfer orbit is determined by $2a$, so we have

$$E = -G \frac{mM_0}{1.72r_e} = \tfrac{1}{2} mv_1^2 - G \frac{mM_0}{r_e},$$

where v_1 is the speed needed for the spacecraft to enter the transfer orbit beginning at the earth. Solving for v_1, we find

$$v_1 = (0.84GM_0/r_e)^{1/2} = 0.91v_0 = 27.2 \text{ km/s}.$$

Therefore a rocket blast is needed to slow down the craft by 2.6 km/s, the difference between 29.8 and 27.2.

By Kepler's third law, the period of the transfer orbit can be found from the ratio

$$\frac{T^2}{T_E^2} = \frac{a^3}{a_E^3}.$$

This yields a travel time (one-half the period) $t = 0.4$ years $= 146$ days.

Using conservation of angular momentum, we can calculate v_2, the speed of the spacecraft when it reaches the orbit of Venus. Because

$$mv_1 r_e = mv_2 r_V,$$

where r_V is the radius of the orbit of Venus, we find $v_2 = 37.8$ km/s. The speed of Venus is 35.0 km/s, so the spacecraft needs to be slowed down by 2.8 km/s.

While the spacecraft spends 146 days traveling to Venus the planet moves through $(146/225)360° = 234°$. Therefore, as indicated in the diagram below, Venus should be

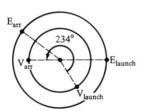

at point V_{launch}, which is 234° from the arrival point V_{arr}. During the trip, the earth moves through 144° and, as you can easily verify, is at the point E_{arr} on the diagram.

Example 2

Plan a transfer orbit to Mars, knowing that the radius of the planet's orbit (assumed to be circular) is $1.52r_e$ and that the orbital speed of Mars is 24.1 km/s.

The calculations for this voyage are identical to those in Example 1, with the exception that the spacecraft travels to a higher orbit. This means that at launch and arrival it will need to speed up.

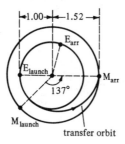

transfer orbit

As indicated in the diagram above, the length of the major axis is 2.52 AU $= 2.52r_e$. Using this to determine the energy of the transfer orbit and setting that energy equal to

$$E = \tfrac{1}{2} mv_1^2 - G \frac{M_0 m}{r_e},$$

we find that the necessary speed for the transfer orbit is

$$v_1 = 1.10v_0 = 32.8 \text{ km/s}.$$

Since the orbital speed of the earth is 29.8 km/s, the spacecraft needs to have its speed boosted by 3.0 km/s.

Conservation of angular momentum allows us to calculate the spacecraft's speed v_2 when it arrives at the orbit of Mars:

$$mv_1 r_e = mv_2 r_M,$$

which tells us that $v_2 = 0.72v_0 = 21.6$ km/s. Therefore, to attain the speed of Mars, the craft needs to be boosted by 2.5 km/s when it reaches this point.

The travel time is one-half of the period of the transfer orbit, and is calculated from Kepler's third law:

$$t = \tfrac{1}{2}(1.26)^{3/2} \text{ years} = 0.71 \text{ years} = 259 \text{ days}.$$

While the spacecraft is voyaging to Mars, that planet moves through $(259/687)360° = 136°$ along its orbit. Therefore the launch opportunity occurs when Mars is at the point M_{launch} in its orbit, as indicated in the above diagram. You can show that when the spacecraft arrives, the earth is at the point E_{arr}, at an angle of 255° from its position at launch.

Questions

6. For the transfer orbit of Example 1 calculate the energy that must be supplied to a 3000-kg spacecraft (a) to enter the transfer orbit, and (b) to slow down to match the speed of Venus.

7. Some science fiction stories refer to a mysterious sister planet of the earth which shares the same orbit as the earth but is always opposite the sun and hence remains unobserved. Approximating the earth's orbit as a circle, qualitatively explain how a spacecraft could be sent from the earth to Counter-Earth.

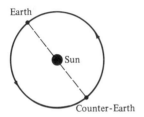

8. Using necessary data in Appendix D determine the following for a transfer orbit between the earth and Mercury:
(a) length of the major axis, (b) change in speed to enter orbit, (c) speed of spacecraft when it arrives at planet, (d) travel time, and (e) position of Mercury for launch opportunities.

9. Repeat Question 8 for a voyage from the earth to Jupiter.

10. The physicist Dr. Lee DuBridge posed the following question in an after-dinner speech to the American Physical Society on April 27, 1960: Suppose that two spacecraft are in the same circular orbit around the earth, but one is a few hundred yards behind the other. An astronaut in the rear craft wants to throw a ham sandwich to his partner in the other craft. How can he do it? Qualitatively describe the possible paths of transfer.

11. The orbit of Halley's comet can be approximated by a parabolic orbit having a closest approach to the sun of 0.4 AU. Assuming that the earth's orbit is a circle of radius 1 AU, find the angles θ shown in the figure.

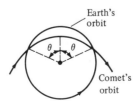

12. A satellite is in a circular orbit of radius r_0 about the earth. The velocity of the satellite is changed by firing of its rockets tangent to its orbit in such a way that the speed is increased to 1.1 times its original speed. Find (a) the eccentricity of the new orbit, (b) the apogee distance in terms of r_0.

13. A satellite is in a circular orbit of radius $2R$, where R is the radius of the earth. The spacecraft is to be boosted to a higher orbit having a radius $4R$. Using data in Appendix D for the mass and radius of the earth and taking the mass of the satellite to be 2500 kg, find the following for the transfer orbit:
(a) length of the major axis, (b) change in speed at the beginning and end of the orbit.

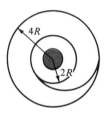

28.4 GRAVITY ASSIST

In 1977 *Voyager 2* was launched for a once-in-a-lifetime chance to tour the four outer gaseous giant planets – Jupiter, Saturn, Uranus, and Neptune – as illustrated in Fig. 28.5. Once every 175 years these planets line up so that one spacecraft can visit all of them on a "grand tour." Mission navigators used the gravitational fields of the planets themselves to provide extra boosts to *Voyager 2* in a technique known as *gravity assist*. Through a gravity assist from Jupiter, *Voyager 2* would visit Saturn, Uranus, and Neptune within 12 years. If the spacecraft had been launched directly to Saturn, the voyage would have taken more than 6 years. A trip to Uranus without gravity assist requires 16 years. And *Voyager 2* could never have reached Neptune at all.

Imagine a spacecraft traveling close to a large planet, for example, Jupiter. In this case, we actually need to take into account the gravitational force of both the sun and Jupiter. But if the satellite passes close enough to Jupiter, the attraction will temporarily become much stronger than that of the sun; therefore the force from the sun can be temporarily ignored. (This reduces a difficult three-body problem to a solvable two-body problem involving only the spacecraft and Jupiter.) Let's see what happens from the perspective of Jupiter.

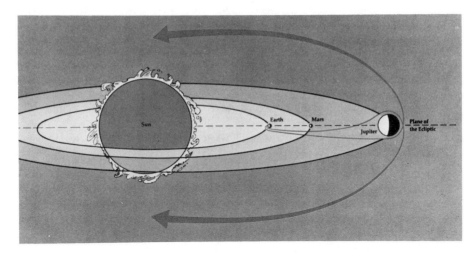

Figure 28.7 Cranking the orbit of a spacecraft through a gravity assist. (Courtesy JPL/NASA.)

28.5 A FINAL WORD

One of the spectacular revelations of the *Voyager 2* mission was the intricacies of the rings of Saturn, shown in the photograph of Fig. 28.8. There is an outer ring, called the F ring, which is a very thin, well-organized ring separate from the other rings. Even before *Voyager 2* flew past Saturn, calculations and theories described how such a well-organized, thin ring might have come about.

In order for the F ring to have become so narrow there must be two moons orbiting Saturn, one just inside the ring and one just outside it. How do the two moons shape the

Figure 28.8 *Voyager 2* photograph of the rings of Saturn. (Courtesy JPL/NASA.)

F ring? Suppose we examine the interaction between the F ring and one moon. If two bodies pass alongside each other in space, they interact gravitationally, and if one body is moving faster, it loses kinetic energy, whereas the slower one gains kinetic energy. We've just seen this in gravity assists. It is also true that the closer a moon is to a planet, the faster it moves. Therefore the inside moon travels faster than the outer moon. The ring material moves at an intermediate speed.

The material in the ring interacts with the fast, inner satellite and gains energy. The increase in energy pumps the material into a higher orbit. So through this interaction with the inner moon, material in the ring is pushed outward, away from Saturn, toward the center of the ring. On the other hand, material on the outer edge of the ring interacts with the outer moon, which is moving slower than it. Consequently, the particles lose energy to the slower satellite and move down to a lower orbit, toward the center of the ring.

The presence of the two moons on the outside and inside of the ring tends to compress the ring together into a narrower ring. The moons act like shepherds keeping their flock together; they keep the particles in a narrow ring gravitationally.

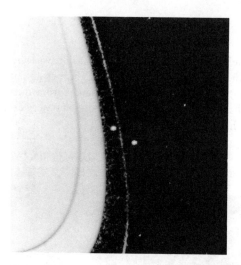

Figure 28.9 *Voyager 2* photograph of the F ring of Saturn and two shepherding moons. (Courtesy JPL/NASA.)

This theory was suggested by Dr. Peter Goldreich of Caltech before the *Voyager 2* fly-by of Saturn. He predicted that there would be two moons, which hadn't been observed yet, shepherding the F ring. And when *Voyager 2* flew past the rings, the two moons were found, just as predicted – a natural example of gravity assist.

CHAPTER 29

LOOSE ENDS AND BLACK HOLES

"The time has come," the walrus said,
"To talk of many things;
Of shoes – and ships – and sealing-wax –
Of cabbages – and kings –
And why the sea is boiling hot –
And whether pigs have wings."

Lewis Carroll, *Through the Looking Glass*

29.1 KEPLER'S THIRD LAW

In 1618 Kepler published the *Harmony of the Worlds*, the culmination of his lifelong obsession. In this book he attempted to reveal the ultimate secret of the universe in a synthesis of geometry, music, astronomy, and astrology – an ambitious undertaking which had not been attempted since Plato. The harmonies Kepler refers to are certain geometric proportions which he finds everywhere reflecting the universal order from which the planetary laws, the harmonies of music, the variations of the weather, and the fortunes of man are derived.

Hidden among the luxuriant fantasies of the *Harmony of the Worlds* is Kepler's third law of planetary motion, which states that the square of the period of revolution of a planet is proportional to the cube of the length of its semimajor axis. Mathematically we've written this law

$$T^2 = ka^3,$$

where k is a constant, the same for all planets. Kepler searched for such a law because, he thought, the universe would be hopelessly disharmonious without such a correlation. If the sun has the power to govern a planet's motion, then that motion must somehow depend on its distance from the sun. But how?

Veiled in the third law is the clue that led Newton to the universal law of gravity and the edifice of the solar system. No small achievement of Newton was that he spotted the three laws in Kepler's writings and plucked them from the numerous correct and incorrect laws. Kepler never realized their real importance.

Let's now derive Kepler's third law from Newtonian mechanics. To recapitulate the other laws briefly, the first law states that the planetary orbits are ellipses. In Chapter 26 we found that the semimajor axis a and eccentricity e of an orbit are related to the mass of the sun M_0, the mass of the planet M, and the planet's angular momentum L, as follows:

$$\frac{L^2}{DM} = a(1 - e^2), \tag{26.14}$$

where $D = GMM_0$.

The second law states that the radius vector of a planet sweeps out equal areas $A(t)$ in equal times:

$$\frac{dA}{dt} = \frac{L}{2M}. \tag{23.5}$$

To derive the third law, let's start with the second law and integrate it around one complete orbit of a planet. If T is the time to complete one orbit, then

$$A = \frac{L}{2M}T,$$

where A is the area enclosed by the elliptical orbit. Solving for T and squaring both sides, we have

$$T^2 = \left(\frac{2M}{L}\right)^2 A^2.$$

Substituting $A = \pi ab$ for the area of an ellipse, where $b = a(1 - e^2)^{1/2}$ is the semiminor axis, we get

$$T^2 = \left(\frac{2M}{L}\right)^2 \pi^2 a^2 b^2.$$

But from Eq. (26.14) we have

$$b^2 = a^2(1 - e^2) = \frac{aL^2}{DM},$$

and when this is used in the formula for T^2 we obtain

$$T^2 = \pi^2 a^3 \frac{4M^2}{L^2} \frac{L^2}{DM} = 4\pi^2 \frac{M}{D} a^3.$$

Because $D = GMM_0$, we have Kepler's third law:

$$T^2 = \frac{4\pi^2 a^3}{GM_0}. \qquad\qquad (29.1)$$

From our derivation we see that the constant of proportionality $k = 4\pi^2/(GM_0)$ depends only on the mass of the sun, and therefore is the same for all the planets. Since Kepler's third law is a consequence of the universal law of gravity and Newton's laws, it holds not only for planetary orbits, but also for the elliptical orbits of moons or satellites about planets. For the motion of satellites, the mass of the attracting planet replaces the M_0 in Eq. (29.1).

Questions (Use Appendix D for data on masses and distances.)

1. Determine the mass of the sun from the period and semimajor axis of the earth's orbit.

2. The period of the moon's orbit about the earth is 27.3 days and the semimajor axis has a length of 3.85×10^8 m. From this information calculate the mass of the earth.

3. What is the shortest possible period for an Earth satellite?

4. A communication satellite is placed in a circular "synchronous orbit" about the equator, where it remains overhead, having a period of one day. Determine the radius of this orbit.

5. The moon Ganymede of Jupiter orbits the planet with a period of 7.16 days and a semimajor axis of length 1.07×10^9 m. From this information calculate the mass of Jupiter.

29.2 THE EARTH–SUN AND EARTH–MOON SYSTEMS

In our discussion of Kepler's laws we've assumed that the sun is stationary and that the earth orbits it. This assumption is not correct. Just as the sun tugs on the earth with gravity which determines its motion and gives us the equations we've discussed, the earth too pulls on the sun. And the sun must move because of that force. It is this motion of the sun that we've ignored entirely.

The sun and earth both move and rotate about a fixed point between them, which is called the center of mass. In earlier chapters we encountered the center of mass as the point where we can treat all the mass of a body as being concentrated. An object suspended on a knife-edge at the center of mass will be balanced. Let's use this idea to determine the center of mass of the earth–sun system.

As illustrated in Fig. 29.1, we'll call r_0 the distance from the center of mass to the center of the sun; r_e will be the distance from the earth to the center of mass. The center of mass is the point at which we could theoretically balance a scale with the sun on one side and the earth on the other (if something else provides gravity to make the scale swing). The scale will balance if there is no torque about the point of suspension. Recalling that torque is force times moment arm, the scale will balance if the torque due to the sun is equal to that due to the earth,

$$M_0 g r_0 = M g r_e,$$

from which we get

$$r_0 = \frac{M}{M_0} r_e.$$

(29.2)

Figure 29.1 Locating the center of mass of the earth–sun system.

Inserting the masses of the earth (6×10^{24} kg) and sun (2×10^{30} kg) and the distance $r_0 + r_e$ from the sun to the earth (1.5×10^{11} m), we find that $r_0 = 4.5 \times 10^5$ m, which is about the distance from Los Angeles to San Francisco. But the radius of the sun is 7×10^8 m, so r_0 is about one-thousandth of that distance. Since the center of mass is so close to the center of the sun, we can safely ignore the error in assuming that the center of the earth–sun system is at the center of the sun.

On the other hand, the error in assuming that the earth is at rest in the earth–moon system may not be so small. Using Eq. (29.2), we can calculate the distance of the center of mass from the center of the earth. Inserting the moon's mass (7.3×10^{22} kg) and distance from the earth (3.8×10^8 m), we find that the distance from the center of the earth to the center of mass of the earth–moon system is 4.6×10^6 m. The radius of the earth is 6.4×10^6 m, so we see that the center of mass is inside the earth, about three-quarters of the way out from the earth's center. And both bodies, the moon and the earth, are rotating about that common center in their mutual motion.

For most purposes, like estimating the orbit or period of the moon, the error introduced by assuming that the center of mass of the earth–moon system is at the center of the earth is quite small, only one percent of the distance between the two bodies. But for certain purposes, the motion of the earth about the center of mass is much more significant. For example, it would be very important to you if you happened to be a mussel or a clam, or even if you just like eating mussels or clams, because it governs the nature of the tides.

Question

6. Using data in Appendix D, calculate the distance of the center of mass of the sun–Jupiter system from the center of the sun. How does this distance compare to that for the earth–sun system?

29.3 THE TIDES

The first person to present a modern theory of the tides was Galileo. He considered his theory, which is presented in his *Dialogues Concerning the Two Chief World Systems*, to be the crowning achievement of all his scientific theories. But it was completely wrong.

In order to understand why he thought his theory of tides was so important, we need to put it into historical perspective. At the time Galileo had become a staunch defender of the Copernican system in which the earth was moving around the static sun. At the same time the Church had gradually hardened its position to the point that the Copernican system was an unallowable heresy. So Galileo was locked in battle with the Church.

Galileo naively thought he could convince the Church with logical arguments, and so he said that the earth only seems to be standing still. The tides, he argued, are evidence that the earth is moving. The water is sloshing around as a result of the motion of the earth. Galileo based his theory on the motion of the earth around the sun and predicted that high tide would occur once a day everywhere on Earth at exactly high noon. Although his central idea – that tides arise from the motion of the earth – was correct, his prediction was a failure. There isn't one high tide per day, there are two, which don't necessarily occur at noon, but at slightly different times each day. This feature of the tides is probably the most puzzling to people when they first learn about tides, as it was to Galileo.

Basic to the cause of tides is the fact that the earth moves around the center of mass of the earth–moon system. Figure 29.2 shows the paths of the earth and the moon as they both rotate about the common center of mass. The cause of the earth's motion is the gravitational force of the moon on the earth, given in magnitude by

$$F = G\frac{MM_{\mathrm{m}}}{r_{\mathrm{m}}^2},$$

where M_{m} is the mass of the moon and r_{m} is the distance from the center of the earth to the center of the moon. The resulting acceleration of the earth is

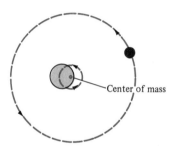

Figure 29.2 Motion of the earth–moon system about their center of mass.

$$a = G\frac{M_{\rm m}}{r_{\rm m}^2}.$$

There is only one point, however, where this is exactly the acceleration and that point is the center of the earth.

The force of gravity from the moon is stronger on the side of the earth nearer the moon. That greater force tends to cause that side of the earth to accelerate more than the center of the earth. On the far side of the earth, the pull of the moon is weaker than at the earth's center and the force is less than what is needed to keep that part of the earth moving with the same acceleration as the center of the earth.

If the earth were a rigid body, the difference in the gravitational force of the moon on opposite sides of it wouldn't matter. Forces inside the earth would be strong enough to keep the entire body moving with the same acceleration as the center. But the earth is not a rigid body, it's partly covered by a thin sheet of water. Consequently, the water is free to slosh around and respond to the unbalanced forces. The result is that, on the side near the moon where gravity is stronger, water is pulled toward the moon, as illustrated in Fig. 29.3. The water also bulges on the side away from the moon because there the force of the moon's gravity is too weak to keep the water in the circular orbit matching the motion of the earth.

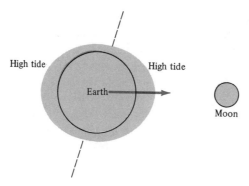

Figure 29.3 Tidal bulges produced by the gravitational force of the moon and the motion of the earth about the center of mass of the earth–moon system.

The oceans on the earth bulge because the water there can flow freely. Their flow, however, is impeded by the continents and other land masses. As a result, local factors determine the height of the tides; some regions experience very high tides whereas other places have much lower ones. In the Bay of Fundy in Canada, for example, the ocean floor is V-shaped, and the water is channeled into the bay, creating changes in water level between high and low tide by as much as 15 m. On the other hand, in the center of the ocean the tides are approximately 0.5 m in height.

In addition to the bulging effect described above, the earth is rotating on its axis, and the bulges are dragged eastward by friction from the land masses and ocean floor. Consequently, the position of the bulges is shifted eastward from the earth–moon line, as indicated in Fig. 29.4. The earth revolves under the bulges, which are stationary with respect to the moon, and each point on the earth passes under each bulge once a day. Therefore each point on the earth has two high tides a day and, correspondingly, two low tides.

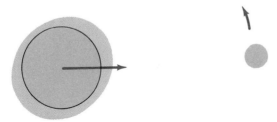

Figure 29.4 Tidal bulges are slightly displaced from the earth–moon line owing to the rotation of the earth.

The bulges are also slowly carried eastward by the moon's motion about the center of mass. The moon rotates about that point once every 27.3 days, and, as a result, there are more than 12 hours between successive high tides. Because the moon causes the bulges to advance ahead of the earth's rotation, it is actually about 12 hours and 25 minutes between high tides.

In addition, the sun also tugs at the earth and contributes to the tides, in the same way as the moon. Although the sun is much farther away from the earth than the moon is, it is also much larger. The net result of the effect of the sun is about one-half that of the moon.

Twice a month – once every two weeks – the sun is either in direct line with the earth and a "new" moon, or on the opposite side of the earth from a full moon. In those instances the effects of the sun and the moon combine to produce a tide one and a half times as big as it would be from the moon alone. At intermediate times (half moon) when the angular position of the sun and moon are separated by 90°, the sun tends to cancel the effect of the moon, and the tides are only one-half as large as they would be if the moon were acting alone.

Questions

7. In his *Dialogue*, Galileo cites the Aristotelian explanation of tides, which is as follows: The seas have various depths, and the deepest waters, being more abundant and therefore heavier, expel the waters of lesser depth. The waters of lesser depth, being raised up, then try to descend. The tides are derived from this continual interplay. How does this explanation describe or fail to describe observed tides?

8. Why do you suppose that lakes don't exhibit tides?

9. Do you think that there are tides in the atmosphere? If so, how are they different from tides in the ocean?

10. Explain why the tides are convincing evidence that the earth is in free fall.

11. As a result of the friction between the oceans and land masses, the rotation of the earth is gradually slowing down. Since angular momentum of the earth–moon system is conserved, what happens to compensate for the loss in angular momentum of the earth?

29.4 THE PRINCIPLE OF EQUIVALENCE

Let's return now to the starting point of this entire book and a subtle mystery – the law of falling bodies. We found that all bodies fall with the same acceleration, because when the force of gravity is set equal to mass times acceleration,

$$ma = -G\frac{mM_E}{R_E^2},$$

the mass of the falling object cancels out of the equation. But the mass plays an essentially different role on the two sides of the equation.

On the left-hand side of the equation, the mass stems from $F = ma$ and is a measure of the reluctance of a body to be accelerated. In this role it is called the *inertial* mass. The mass on the right-hand side of the equation comes from the universal law of gravity, $F = -GmM_E/R_E^2$, and measures the strength of the gravitational force a body exerts on all other bodies. Here the mass plays a role similar to charge in the case of electricity, and it is called the *gravitational* mass. It is not and never has been obvious why these two kinds of mass should be exactly the same for every body. This question bothered Newton and the line of eminent physicists who followed him. The very first question we raised still stands: Why do all bodies fall with the same acceleration? Now it can be rephrased: Why is inertial mass equal to gravitational mass?

There have been numerous experiments to determine to what extent inertial and gravitational mass are equal. Newton investigated the equivalence by studying the period of a pendulum with interchangeable masses. His experiment consisted of looking for a variation in the period of the pendulum using bobs of different composition. He found no such change, and from an estimate of the sensitivity of his method, he concluded that the inertial and gravitational mass cannot differ by more than one part in a thousand. Recent experiments, which are capable of detecting a variation of one part in 10^{11}, found no variation.

The answer to the problem of equivalence between inertial and gravitational mass forms the basis of Albert Einstein's general theory of relativity, his theory of gravity. Einstein did not believe in coincidence. Rather he thought that there must be some deep principle at work, a principle which would have the equivalence of gravitational and inertial mass as a simple, inevitable consequence. The principle he adopted is called the *principle of equivalence*.

The principle of equivalence states that there is no way locally to tell the difference between a gravitational field and an acceleration. To understand this principle, imagine a closed box that is a laboratory with a scientist inside who makes measurements but who can't see outside. In one situation the box is on Earth, as in Fig. 29.5a. In another situation, the box is somewhere in intergalactic space – very far from any body that could exert a gravitational force. However, this box is accelerated with a uniform acceleration in the upward direction equal in magnitude to g, as indicated in Fig. 29.5b.

Acceleration
$a = g$

Gravity g

(a)

(b)

Figure 29.5 The principle of equivalence states that motion inside a box resting on the earth (a) is indistinguishable from the motion inside a box which is accelerated upward with acceleration g (b).

The principle of equivalence says that there is no way the scientist inside the box can do any experiment that would determine what the situation is, whether the box is on Earth, or accelerating upward with acceleration g. (Obviously, the box must be too small to detect the fact that g decreases with height above the earth.)

Now take Newton's apple falling out of a tree. On Earth, gravity applies a force to the apple and it falls with acceleration g as in Fig. 29.5a. In intergalactic space the apple is released from the tree, but the tree is accelerated upward (as is the observer inside the box) with acceleration g. As soon as the apple is released it becomes inertial, which means it no longer is accelerated. The box, however, is accelerated upward, and so it seems that the apple falls with acceleration g. Thus any body must fall with exactly the acceleration g regardless of mass.

So the law of falling bodies and the apparent equivalence of inertial and gravitational mass are simple consequences of the principle of equivalence. This is fine for falling bodies, but what does the principle of equivalence imply for the propagation of light?

Imagine sending a light beam horizontally inside the box. The observer sees the light beam travel across the box. But while the beam is traveling across the box in space, the box is accelerated upward. As the beam crosses the box with speed c, the box moves up a distance $\frac{1}{2}gt^2 = \frac{1}{2}g(L/c)^2$, where L is the width of the box and t is the time it takes for light to travel across the box. As the light moves across, the box moves upward and the beam hits a lower point than it started from by just that distance, as depicted in Fig. 29.6. The principle of equivalence tells us that the observer can't do any experiment that will determine whether the box is accelerated or in a gravitational field. This means that the same results must be observed in a closed box on Earth: light curves downward as it crosses the box. In other words, a gravitational field bends light beams. This was the first prediction Einstein produced from his analysis. In 1919 an experiment was performed during a total solar eclipse to observe the bending of starlight. The success of that experiment made Einstein into a world-famous folk hero.

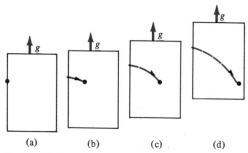

Figure 29.6 Successive stages in the path of a light beam traveling horizontally in a box which is accelerated upward.

29.5 EINSTEIN'S THEORY OF GRAVITY

If light itself travels in curved paths, then what do we mean by a straight line? Einstein said that it is meaningless to speak of straight lines – it is space itself which is curved. Not only is space curved, but the four-dimensional fabric of the universe – spacetime – is curved. This means that both rulers and clocks change their properties as they move through gravitational fields.

The idea of a straight line has meaning in the physics we know – in the law of inertia. According to the law of inertia, a body moving along at some speed continues to move *in a straight line* unless something interferes with it. If a straight line no longer has meaning in physics, then the law of inertia must be reformulated.

It was Einstein who made the reformulation. He realized that the idea of a straight line can be generalized for curved spacetime or curved space, or anything, by the idea of the optimal distance between two points. In a plane, the shortest distance between two points is along a straight line. On the surface of a sphere, the shortest distance between two points is not along a straight line but along a great circle, as illustrated in Fig. 29.7b. In any kind of curved geometry the shortest path between two points is called a *geodesic*; it is the most economical way to get from one point to another.

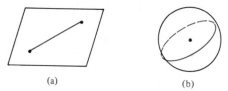

Figure 29.7 The shortest path between two points is (a) in a plane a straight line, and (b) on a sphere a great circle.

Einstein described the bending of starlight by the sun not as a gravitational force from the sun which makes the light alter its straight-line path, but rather by saying that light travels at a constant speed along a geodesic in the local spacetime curved by the sun. According to Einstein, the orbit of the earth does not have to be thought of, as it must in Newtonian physics, as a compromise between the earth's inertia, which wants to make it fly off in a straight line, and the force of the sun, which wants to keep it bound to the sun. Instead the earth can be described as moving inertially without any forces, along the geodesic in the local spacetime created by the presence of the sun. The earth follows its characteristic orbit because it is traveling along a geodesic. In the presence of an extremely massive object, such as the sun, the geometry of spacetime is locally disturbed, so geodesics which were once straight lines in flat spacetime become curved lines. What happens near a massive object, according to Einstein, may be interpreted not in terms of a gravitational field but in terms of curvature of spacetime.

In Einstein's theory of gravity forces may be done away with and replaced by the curvature of spacetime. There is a historical irony in this reformulation. Galileo thought that bodies, if unimpeded, would keep on moving not along a straight line, but in a perfect circle parallel to the surface of the earth. It was only later that Newton and Descartes discovered that inertia tends to move objects along straight lines. Einstein's reformulation is much closer to Galileo's picture. In Einstein's picture, the nearly circular orbit of the earth is itself inertial motion.

29.6   BLACK HOLES

How does mass cause local spacetime to curve? Einstein spent the most difficult seven years of his life, from 1909 to 1916, trying to answer this question. In the end he produced a consistent set of equations – Einstein's field equations. They are among the most mathematically difficult equations in all of physics. Nonetheless, there are intriguing predictions found in these equations.

The mass of an object causes spacetime in its vicinity to curve. An object in a curved spacetime has more energy than it would in flat space, and because it has more energy, it has more mass, since $E = mc^2$, another of Einstein's results. And because this object has more mass it causes more curvature in spacetime. There is a kind of unstable feedback effect built into Einstein's theory of gravity. Under ordinary circumstances the effect is extremely small – so small that only the most exquisitely delicate observations and experiments can detect the difference between Einstein's theory and Newton's theory. However, there are other circumstances in which the difference in theories is extremely large, and the feedback effect becomes important. When the feedback effect causes an increase in mass, which causes a corresponding increase in curvature, which creates more mass and more curvature, *ad infinitum*, the object collapses into what is known as a *black hole*.

It takes a great deal of mass in a very small space to create a black hole. In our present understanding of the universe, there are only two circumstances in which black holes might be created. One of them, the primeval black hole, arises from the extraordinary events that attended the birth of the universe. The other is when there is a very large mass, such as a star cluster or a galactic core or even a very large star, much larger than our sun, which runs out of nuclear fuel. A star is made up of gases that attract each other gravitationally, but are kept from collapsing by the fact that they're burning fuel inside, and radiation is pushing outward, preventing collapse. When the star runs out of nuclear fuel and goes dead, it will collapse gravitationally, and if it's big enough and does not blow off too much matter while collapsing, it will enter the state where it runs away with itself and falls into a black hole. It's not difficult to specify the conditions for having enough mass in a small enough space to have a black hole. From mechanics we know that in order for a body to escape from the earth, for example, it must have enough kinetic energy to overcome its gravitational potential energy

$$\frac{1}{2}mv^2 = G\frac{Mm}{R}\ .$$

There is an escape velocity for any body of any mass to escape from a given mass, like the earth. That's the solution of this equation. If we set that velocity equal to the speed of light, then $c^2 = 2GM/R$, which tells us the amount of mass within a given radius from which not even light has sufficient velocity to escape. Actually, when a body moves at nearly the speed of light, its kinetic energy is no longer equal to $\frac{1}{2}mv^2$. Nevertheless, the idea is right, and the result $R = 2GM/c^2$ is exactly correct. When there is that much mass in that small a distance, then a black hole is formed. The R that solves this equation is called the *event horizon*. Nothing can pass out of the event horizon – things can fall in, but nothing can ever get out. That's why it's called a black hole. Not even light can escape.

There are other consequences of the general theory of relativity, one of which is the current view of all cosmologists that the universe began with a cataclysmic event, known as the Big Bang. Before the Big Bang there was nothing, and then the universe exploded out of nothing, and it's still expanding. In a certain sense the Big Bang is related to the idea of a black hole. There is a kind of opposite solution of the same equation, which we might call a white hole. A black hole is something that everything falls into, a white hole is something that everything flies out of, a sort of black hole with a minus sign in front. The Big Bang resembles a white hole.

Once the Big Bang occurs, the universe is expanding and there are two possible things that can happen according to Einstein's equations. One is that it can go on expanding forever. The other is that there is so much mass in the universe that it is really gravitationally stable, and so it won't go on expanding forever. It will eventually lose all its kinetic energy against its own gravitational potential and start contracting. If it does that, then at the end of the universe there will be an event that is the reverse of the Big Bang, where everything comes back together again. In cosmology there's a name for that too; it's known as the Big Crunch. The Big Crunch is a little like the black hole, with the whole universe falling into nothing, but it differs technically from a black hole because the black hole is a kind of singularity in the curved space, where space itself is due to all the rest of the mass in the universe. The Big Crunch is the entire universe falling into nothing because there's nothing left to create curved spacetime.

29.7   A FINAL WORD

It is intriguing to try to imagine what it would be like inside a black hole. But physics says there is no way of finding out. There's no way of determining what it is like, because there's no way of communicating with the inside of a black hole. You could fall into it if you wanted to, but that would be it. You could never let anyone know what you saw once you got inside. So it sounds pretty hopeless for us to try to guess what it might be like. But there's one thing we might think about; that is, we might suppose that our universe is not going to expand forever – that instead it will fall back on itself. That would mean a closed universe. Technically, the question of whether the universe is closed or open depends on how much mass there is in it. If the mass of the universe is larger than some critical mass, it will eventually turn around and come back and collapse on itself. So the question of whether that can happen is an observational one; you have to weigh everything in the universe and find how much mass there is. Then we would know whether the universe is closed or open. Our astronomers and cosmologists find that an extremely interesting question and work very hard to estimate the total mass in the universe. The total mass we know of is smaller than the necessary critical mass, but not by very much. It's of the same order of magnitude as the necessary critical mass, and it's quite possible that, for various reasons we haven't discovered yet, a lot of the mass in the universe is hidden from us – we don't know about it. Many cosmologists, if you ask them to guess, would say that there probably is enough mass to make it a closed universe, which would mean that eventually it will fall back on itself and undergo the Big Crunch.

But even before that happens, while the universe is expanding, it is nevertheless true that nothing escapes from the universe. If light is sent outward it follows curved spacetime around and eventually comes back – because the universe has enough mass to make spacetime closed. That's exactly the reason that all mass itself is eventually going to collapse. So the universe is something from which nothing – not even light – can escape. But that's the description of the inside of a black hole. So it is possible that our universe is simply the inside of a black hole in somebody else's universe.

CHAPTER 30

THE HARMONY OF THE SPHERES: AN OVERVIEW OF THE MECHANICAL UNIVERSE

The heavenly motions are nothing but a continuous song for several voices (perceived by the intellect, not by the ear); a music which, through discordant tensions, through sincopes and cadenzas, as it were (as men employed them in imitation of those natural discords), progresses towards certain pre-designed, quasi six-voiced clausuras, and thereby sets landmarks in the immeasurable flow of time. It is, therefore, no longer surprising that man, in imitation of his creator, has at last discovered the art of figured song, which was unknown to the ancients. Man wanted to reproduce the continuity of cosmic time within a short hour, by an artful symphony for several voices, to obtain a sample test of the delight of the Divine Creator in His works, and to partake of his joy by making music in imitation of God.

Johannes Kepler, *Harmony of the World* (1618)

Nature and Nature's law lay hid in night.
God said: "Let Newton be"; and all was light.

Alexander Pope, "Epitaph Intended for Sir Isaac Newton"

30.1 WINDING UP THE MECHANICAL UNIVERSE

We've now arrived at the final chapter in our study of the mechanical universe. In our story we've introduced revolutionary ideas and heroes from Copernicus to Newton, and just as they did before us, we've linked the physics of the heavens to the physics of the earth.

Mankind has not always applied the same laws to the heavens and to the earth. When Copernicus yanked the earth from the center of the universe and sent it hurtling through space about the sun, new questions arose that his theory had to answer. With courageous imagination, Galileo destroyed old notions of how the world works and began to build new ones. And Newton gave the world new vision through his mechanics. Throughout our story, the focus has been the compelling question: How does the universe work?

Our task in this final chapter is to review the highlights and to present the underlying structure and connecting principles of *The Mechanical Universe*.

30.2 THE PHYSICS OF THE HEAVENS AND EARTH

Newton's laws are the key to the mechanical universe. They lead to a compelling and widely applicable description of how the world works. When combined with the universal law of gravity, they extend the physics of the earth to that of the heavens – celestial mechanics. Let's summarize exactly what you need to know to accomplish this.

Newton's first law states that every body continues in its state of rest or of uniform motion in a straight line unless an external force acts on it. The first law is a statement of a deep principle of nature – the law of inertia that Galileo had deduced from his fertile experiments with balls rolling on inclined planes. Newton, however, did more than describe motion; he explained the underlying causes of motion.

The science of dynamics is embodied in what is perhaps the most profound statement of classical physics – Newton's second law. Through the second law the causes of motion – forces – are related to changes in motion, the acceleration, of an object of mass m:

$$\mathbf{F} = \frac{d}{dt}(m\mathbf{v}).$$

(6.1)

Forces shape the motion of a system. In particular, if the mass of the system is constant, the law can be written as

$$\mathbf{F} = m\frac{d\mathbf{v}}{dt} = m\mathbf{a}.$$

(6.2)

Thus, knowing the forces that act on a system, you can determine the acceleration of the system.

In Chapter 6 we applied the second law to the motion of a projectile acted upon by the downward force of gravity. The result was the law of falling bodies and parabolic trajectories. Free fall, projectile motion, and many other problems are solved by applying $\mathbf{F} = m\mathbf{a}$. So the immeasurable success of this law is its widespread applicability: if you can isolate a system and identify all the forces acting on it, then you can determine its acceleration and hence its motion.

To apply Newton's second law to systems that have more than one component, such as masses connected by a string passing over a pulley, we invoked the third law: for every action there is an equal and opposite reaction. In Chapter 10 we developed a general procedure for applying the three laws to various systems: blocks sliding down inclined planes, masses and pulleys, and other mechanical arrangements. Through these examples we began to understand what the second law says and how to apply it in various situations.

Before Newton could develop a system of dynamics for objects on Earth, he needed kinematics, that is, a description of their motion. Galileo graciously gave that to the world. Similarly, before mankind could arrive at a dynamical description of the motion of heavenly bodies, a kinematical framework had to be set in place. Kepler provided that framework in his three laws of planetary motion. Let's recall those laws.

Kepler's first law states that the planets move in elliptical paths, one of the conic sections known to ancient Greek geometers. In Chapter 25 we found the mathematical description of ellipses in polar coordinates:

$$r = \frac{a(1 - e^2)}{1 + e \cos \theta}. \tag{25.3}$$

Kepler's second law of planetary motion states that the radius vector from the sun to a planet sweeps out equal areas in equal times. We now know that it's rooted in the deeper principle of conservation of angular momentum and can be expressed as

$$\frac{d\mathbf{A}}{dt} = \frac{1}{2} \mathbf{r} \times \mathbf{v} = \frac{\mathbf{L}}{2m} = \text{const.} \tag{23.5}$$

Kepler's third law states that the square of the period of a planet's orbit is proportional to the cube of its semimajor axis:

$$T^2 = \frac{4\pi^2 a^3}{GM_0}. \tag{29.1}$$

Kepler's laws describing the motions of the heavens and Galileo's earthly kinematics were explained by Newton's laws. When Newton's universal law of gravity,

$$\mathbf{F} = -G\frac{M_0 M}{r^2}\hat{\mathbf{r}}, \tag{8.1}$$

is combined with his second law of motion, all three of Kepler's laws and much more flow out. The solution of the Kepler problem is a crowning achievement of Western thought and the culmination of *The Mechanical Universe*.

The differential equation that results when the universal gravitational force is inserted into Newton's second law,

$$M\frac{d\mathbf{v}}{dt} = -G\frac{MM_0}{r^2} \hat{\mathbf{r}},$$

has for its solutions only the conic sections – the ellipse, parabola, or hyperbola:

$$r = \frac{ed}{1 + e \cos \theta}. \tag{25.7}$$

The orbits of planets are ellipses, but other heavenly bodies such as comets can travel along paths that are ellipses, hyperbolas, or even parabolas. The solution to the differential equation does not, by itself, reveal which type of conic the orbit would be.

Sorting out which orbits are ellipses and which are hyperbolas or parabolas is made much easier by an extremely important principle: the conservation of energy. For a heavenly body moving about the sun, its total energy is conserved. The sum of its kinetic energy

$$K = \frac{1}{2}Mv^2 \tag{13.7}$$

and potential energy

$$U = -G\frac{MM_0}{r}, \tag{14.5}$$

remains constant as the body moves from one point to another along its path:

$$E = K + U = \text{const.}$$

The precise path of a planet, asteroid, or comet is fixed by the laws of conservation of energy and angular momentum. The total energy E and angular momentum L that a heavenly body has determines the eccentricity e of its orbit to be

$$e = \sqrt{1 + \frac{2L^2E}{G^2M_0^2M^3}}. \tag{27.8}$$

Having thus solved the Kepler problem, we found that for negative energy, $E < 0$, the eccentricity is bounded by $0 \leq e < 1$ and the orbit is elliptical. In the case of zero energy, $E = 0$, the eccentricity equals one, $e = 1$, and the orbit is parabolic. And if a heavenly body has positive energy, $E > 0$, the eccentricity is greater than one, $e > 1$, which describes a hyperbolic orbit. All of this and much more is contained in the simple, concise laws of gravity and motion that form the basis of the mechanical universe. Thus did Isaac Newton transform the revolution of Copernicus, Galileo, and Kepler into a system of the heavens and earth. Indeed, as the opening quotation from Alexander Pope declares, after Newton ''all was light.''

30.3 THE LANGUAGE OF PHYSICS

The rise of modern science grew out of quantification, which tends to condense the ideas of physics in mathematical formulas. Thus, to better understand the vast phenomena of the universe and the laws that organize them, we had to develop a certain mathematical facility with differential and integral calculus, vector algebra, analytic geometry, and differential equations.

In the riddle of motion we found rhyme and reason in the language of mathematics in which, Galileo declared, the book of nature is written. We found that the derivative arises naturally when we try to describe how things change:

$$\frac{dy}{dx} = \lim_{\Delta x \to 0} \frac{\Delta y}{\Delta x} = \lim_{h \to 0} \frac{y(x + h) - y(x)}{h}. \tag{3.3}$$

From a theoretical concept to a practical tool, the derivative rose to determine the instantaneous speed and acceleration of a falling body once the distance fallen was known as a function of time:

$$s(t) = \frac{1}{2}gt^2. \tag{2.23}$$

$$v(t) = \frac{ds}{dt} = gt. \tag{2.22}$$

$$a(t) = \frac{dv}{dt} = g. \tag{2.21}$$

We further developed differentiation to calculate how one quantity changes in relation to another. A few simple rules – the sum rule, the product rule, the power rule, and the chain rule – became an essential part of our vocabulary:

$$\frac{d}{dx}[y(x) + z(x)] = \frac{dy}{dx} + \frac{dz}{dx}, \tag{3.4}$$

$$\frac{d}{dx}[y(x)z(x)] = y\frac{dz}{dx} + z\frac{dy}{dx}, \tag{3.5}$$

$$\frac{d}{dx}(x^n) = n\,x^{n-1}, \tag{3.6}$$

$$\frac{dz}{dx} = \frac{dz}{dy}\frac{dy}{dx}. \tag{3.7}$$

Nature, we believe, is best described by mathematical laws, and to understand and apply the laws of nature we need differential calculus.

Many of the mysteries in physics require determining some quantity from its rate of change. For example, how can the speed of an object be found if its acceleration is known at every instant? To answer this question, we had to understand how to reverse the process of differentiation. This is done by using another process, integration, which arises from the problem of quadrature, the calculation of areas of curved shapes. The area $A(t)$ under the graph of a function $f(x)$ from $x = a$ to $x = t$ is the integral

$$A(t) = \int_a^t f(x)\,dx.$$

The first fundamental theorem of calculus tells us that the derivative of the integral is the original function f:

$$\frac{d}{dt}\int_a^t f(x)\,dx = f(t),$$

so the integral $\int_a^t f(x)\,dx$ is also an antiderivative of $f(t)$. The second fundamental theorem gives another relation between differentiation and integration,

$$A(t) = \int_a^t \frac{dA}{d\tau}\,d\tau + A(a).$$

That is, a function is equal to the integral of its derivative, plus a constant. Thus, to recover velocity from acceleration, we have

$$v(t) = \int_0^t a(\tau)\, d\tau + v(0), \tag{7.10}$$

and to recover displacement from velocity,

$$s(t) = \int_0^t v(\tau)\, d\tau + s(0). \tag{7.11}$$

To solve the Kepler problem another mathematical tool is needed – analytic geometry. Many of its ideas were introduced in Chapter 5, in connection with components of vectors. In Chapter 25 we used analytic geometry to describe the conic sections and to show how they are predicted by Newton's laws.

To describe quantitatively how the world works, physicists and mathematicians need scalar quantities, such as distance and speed, as well as vector quantities, such as displacement and velocity, which have direction and magnitude. For example, forces are vectors, and therefore follow the rules of vector addition – an important property when we need to find the total force acting on a system in order to apply $\mathbf{F} = m\mathbf{a}$.

There are physical quantities that are best expressed as products of vectors. For example, the transfer of energy, the work W, is the line integral of the dot product of force and change in displacement,

$$W = \int_A^B \mathbf{F} \cdot d\mathbf{r}, \tag{14.1}$$

that provides a link between forces and potential energy (for a conservative force),

$$U_B - U_A = \int_A^B -\mathbf{F} \cdot d\mathbf{r}, \tag{14.4}$$

and between forces and kinetic energy,

$$W_{AB} = K_B - K_A. \tag{14.3}$$

These connections led to the law of conservation of energy,

$$E_A = K_A + U_A = E_B = K_B + U_B$$

The cross product of two vectors aided us in arriving at another conservation law – the conservation of angular momentum. The cross product summarized the geometric properties of torque and angular momentum. We found that for a central force, such as gravity, the torque, given by

$$\boldsymbol{\tau} = \mathbf{r} \times \mathbf{F}, \tag{23.6}$$

is zero. Consequently, the time derivative

$$\frac{d}{dt}(m\mathbf{r} \times \mathbf{v}) = \mathbf{0}, \tag{23.3}$$

and the quantity defined as angular momentum,

$$\mathbf{L} = m\mathbf{r} \times \mathbf{v}, \tag{23.4}$$

is conserved.

Whatever form the laws of physics take, we believe that they are the same everywhere in the universe. Vectors provide a mathematical device for expressing the law in a way that is the same for all coordinate systems. That's one reason why the vector law, $\mathbf{F} = m\mathbf{a}$ is so widely applicable: it has the same mathematical form everywhere. To apply it and work out the details of an object's motion, we choose a convenient coordinate system and apply the law to the components of forces and acceleration in that system.

But that's not the only reason Newton's laws are so powerful. Newton's second law is a particularly important example of a differential equation. Equations about derivatives describe more than algebraic equations do because their solutions are general; specific solutions to those equations are determined by appropriate initial conditions. For example, in Chapter 6 we found that $\mathbf{F} = m\mathbf{a}$ led to trajectories of all projectiles:

$$x(t) = x_0 + v_{x0}t, \tag{6.8}$$

$$y(t) = y_0 + v_{y0}t, \tag{6.9}$$

$$z(t) = z_0 + v_{z0}t - \tfrac{1}{2}gt^2. \tag{6.10}$$

The specific trajectory of a particular projectile is determined by the initial position (x_0, y_0, z_0) and the initial velocity (v_{x0}, v_{y0}, v_{z0}). Initial conditions are used to extract particular solutions from the general solution of a differential equation.

In the Kepler problem we used $\mathbf{F} = m\mathbf{a}$ to show that conic sections are the only possible orbits for heavenly bodies. The particular orbit of a specific object is given when its energy and angular momentum are specified.

Another important differential equation we studied was that of simple harmonic motion:

$$\frac{d^2}{dt^2} x(t) = -\frac{k}{m} x(t). \tag{20.1}$$

This differential equation describes systems in which motion repeats itself over time – a spring and mass, a marble in a bowl, a pendulum – systems in which a restoring force proportional to the distance from equilibrium is at play . Simple harmonic motion, nature's response to twanging any stable system, serves as a powerful model for many systems.

30.4 THE CONSERVATION LAWS OF THE MECHANICAL UNIVERSE

A third major theme of the course has been the development of the conservation laws of momentum, angular momentum, and energy. We found conservation laws to be useful bookkeeping devices for applying Newton's laws in complicated situations. They provide another tool for analyzing how the mechanical universe works. These laws tell us why a phenomenon occurs without giving us the details of the actual process. They tell us how things change by focusing our attention on physical quantities that don't change.

Each of the conservation laws has a direct link to dynamics. Newton's second law embodies the concept of conservation of momentum: when no external forces act on a system, the total momentum is constant:

$$m_1\mathbf{v}_1 + m_2\mathbf{v}_2 + m_3\mathbf{v}_3 + \cdots = \text{const.} \tag{19.2}$$

In Chapter 19 we found that this law gave us startling predictive power about what happens in collisions. No matter what the nature of internal forces that act between colliding objects, the total momentum of an isolated system is conserved. Armed with an understanding of conservation of momentum you can delve into the interactions of objects ranging from colliding billiard balls to fragmenting subatomic particles.

Similarly, in Chapter 23, we found that when no torque acts on a system, that is, when

$$\mathbf{r} \times \mathbf{F} = \mathbf{0},$$

then angular momentum is conserved:

$$\mathbf{L} = m\mathbf{r} \times \mathbf{v} = \text{const.} \tag{23.4}$$

Conservation of angular momentum is an important tool for applying Newton's laws in complex situations such as firestorms, hurricanes, spinning ice skaters, and the everyday bathtub vortex as well as giving us insight as to why galaxies are so often pancake-shaped. And Kepler's second law is rooted in conservation of angular momentum – a principle that was an invaluable idea in solving the Kepler problem.

Another principle we used to solve the Kepler problem is conservation of energy. The connection between energy and force was found in the idea of work. As already noted, work is the result of a force acting through a displacement,

$$W_{AB} = \int_A^B \mathbf{F} \cdot d\mathbf{r}. \tag{14.1}$$

It is the dynamical link between force and energy because work done on an object represents the transfer of energy either to or from the object by a force acting on it. In conservative systems, the work done on a system can appear as either kinetic energy, whence

$$W_{AB} = K_B - K_A, \tag{14.3}$$

or as potential energy,

$$U_B - U_A = \int_A^B - \mathbf{F} \cdot d\mathbf{r}. \tag{14.4}$$

From this we found that the total mechanical energy of an isolated, frictionless system is conserved:

$$U_A + K_A = U_B + K_B.$$

Although the idea of conservation of energy was implicit in Galileo's fertile experiments with inclined planes and intimately connected to Newton's dynamics, realization of the law did not come about until the nineteenth century with the development of the science of thermodynamics. Why? Conservation of energy looked like a law that didn't work. For example, if friction is present, the sum of the kinetic and potential energies of an object does not remain constant.

Not until heat was recognized as a form of energy did the law of conservation of energy emerge. The first law of thermodynamics relates the heat added Q, the work done by a system W, and the change in internal energy ΔU:

$$Q = W + \Delta U. \tag{16.3}$$

Like work, heat can be added to or taken from an object. Energy is always conserved, but it does change from one form to another.

Investigations in thermodynamics – a post-Newtonian development – led to a subtle, yet extremely important feature of nature, the entropy principle or second law of thermodynamics. Entropy measures the usefulness of heat energy either added or taken from an object:

$$\Delta S = \frac{Q}{T}. \tag{17.2}$$

The second law of thermodynamics states that in any process, entropy stays constant or increases. When any system reaches the maximum entropy it can have, it is in equilibrium, and no further work can be obtained from it. The entropy of the universe always increases, energy is degraded into more useless forms, and matter tends to less ordered states. The second law of thermodynamics is a principle so vast in range that the entire mechanical universe is subject to its reign.

30.5 A FINAL WORD

The Copernican revolution was 2000 years in the making. Classical mechanics was the result. Through the struggles and achievements of intellectual giants such as Galileo, Kepler, and Newton who followed the right clues, a deeper understanding of the universe emerged. The nature of this new science was determined by the questions that it asked, and these were the questions on which we have focused in *The Mechanical Universe*.

The great book of nature contains many more pages, full of new phenomena, new clues to the mysteries that still remain unsolved; and new questions arise as each old question is answered, so the book of nature may not even have a final chapter. Through this book we've developed a mathematical background and extensive principles such as the conservation laws that will serve as trusty guides as ever emerging questions together with new research take us beyond the mechanical universe.

APPENDIX

THE INTERNATIONAL SYSTEM OF UNITS

Basic Units	Definitions
Length	The meter (m) is currently defined as the distance that light travels in 1/299,792,458th of a second.
Time	The second (s) is the duration of 9,192,631,770 periods of the radiation emitted in a transition between two specified energy levels of the cesium-133 atom.

Mass	The kilogram (kg) is the mass of a particular cylinder of platinum–iridium alloy preserved in a vault at Sèvres, France.
Current	The ampere (A) is that current in two very long parallel wires 1 m apart that gives rise to a magnetic force per unit length of 2×10^{-7} N/m.
Temperature	The kelvin (K) is 1/273.16 of the thermodynamic temperature of the triple point of water.

Names and Symbols for the SI Units

Quantity	Name of unit	Symbol	Definition
length	meter	m	
time	second	s	
mass	kilogram	kg	
current	ampere	A	
temperature	kelvin	K	
force	newton	N	$1 \text{ N} = 1 \text{ kg m/s}^2$
work, energy	joule	J	$1 \text{ J} = 1 \text{ N m}$
power	watt	W	$1 \text{ W} = 1 \text{ J/s}$
frequency	hertz	Hz	$1 \text{ Hz} = \text{s}^{-1}$
electric charge	coulomb	C	$1 \text{ C} = 1 \text{ A s}$
electric potential	volt	V	$1 \text{ V} = 1 \text{ J/C}$
electric field	volt per meter	V/m	$1 \text{ V/m} = \text{V m}^{-1}$
electric resistance	ohm	Ω	$1 \text{ }\Omega = 1 \text{ V/A}$
capacitance	farad	F	$1 \text{ F} = 1 \text{ C/V}$
inductance	henry	H	$1 \text{ H} = 1 \text{ J/A}^2$
magnetic field strength	tesla	T	
magnetic flux	weber	Wb	$1 \text{ Wb} = 1 \text{ T m}^2$
entropy	joule per kelvin	J/K	$1 \text{ J/K} = 1 \text{ J K}^{-1}$
specific heat	joule per kg kelvin	J/kg K	
pressure	pascal	Pa	$1 \text{ Pa} = 1 \text{ N/m}^2$

APPENDIX B

CONVERSION FACTORS

Length

    1 in. = 2.54 cm

    1 ft   = 12 in. = 30.48 cm

    1 yd = 3 ft = 91.44 cm

$1 \text{ km} = 0.6215 \text{ mi}$

$1 \text{ mi} = 1.609 \text{ km}$

$1 \text{ Å} = 0.1 \text{ nm}$

Time

$1 \text{ min} = 60 \text{ s}$

$1 \text{ h} = 60 \text{ min}$

$1 \text{ d} = 24 \text{ h} = 1440 \text{ min}$

$1 \text{ yr} = 365.24 \text{ d} \approx \pi \times 10^7 \text{ s}$

Mass

$1 \text{ kg} = 1000 \text{ g}$

$1 \text{ slug} = 14.59 \text{ kg}$

$1 \text{ kg} = 6.852 \times 10^{-2} \text{ slugs}$

Area

$1 \text{ m}^2 = 10^4 \text{ cm}^2$

$1 \text{ in.}^2 = 6.4516 \text{ cm}^2$

$1 \text{ m}^2 = 10.76 \text{ ft}^2$

$1 \text{ acre} = 43{,}560 \text{ ft}^2$

Volume

$1 \text{ m}^3 = 10^6 \text{ cm}^3$

$1 \text{ L} = 1000 \text{ cm}^3 = 10^{-3} \text{ m}^3$

$1 \text{ gal} = 3.786 \text{ L}$

$1 \text{ gal} = 4 \text{ qt} = 8 \text{ pt} = 128 \text{ oz} = 231 \text{ in.}^3$

$1 \text{ ft}^3 = 1728 \text{ in.}^3 = 28.32 \text{ L}$

Force

$1 \text{ N} = 0.2248 \text{ lb} = 10^5 \text{ dyn}$

$1 \text{ lb} = 4.4482 \text{ N}$

Energy

$1 \text{ ft lb} = 1.356 \text{ J}$

$1 \text{ cal} = 4.1840 \text{ J}$

$1 \text{ Cal} = 1000 \text{ cal}$

$1 \text{ Btu} = 778 \text{ ft lb} = 252 \text{ cal}$

$$1 \text{ eV} = 1.602 \times 10^{-9} \text{ J}$$

$$1 \text{ erg} = 10^{-7} \text{ J}$$

$$1 \text{ J} = 1 \text{ W s}$$

Power

$$1 \text{ hp} = 550 \text{ ft lb/s} = 745.7 \text{ W}$$

$$1 \text{ W} = 1.341 \times 10^{-3} \text{ hp}$$

Pressure

$$1 \text{ atm} = 101.325 \text{ kPa}$$

$$1 \text{ atm} = 14.7 \text{ lb/in.}^2 = 1.01 \times 10^6 \text{ dyn/cm}^2 = 1.01 \times 10^6 \text{ erg/cm}^3$$

$$1 \text{ atm} = 760 \text{ mm Hg} = 38.8 \text{ ft H}_2\text{O}$$

$$1 \text{ lb/in.}^2 = 6.895 \text{ kPa}$$

$$1 \text{ torr} = 1 \text{ mm Hg} = 133.32 \text{ Pa}$$

$$1 \text{ bar} = 100 \text{ kPa}$$

Angles

$$\pi \text{ rad} = 180°$$

$$1 \text{ rad} = 57.30°$$

$$1° = 1.745 \times 10^{-2} \text{ rad}$$

FORMULAS FROM ALGEBRA, GEOMETRY, AND TRIGONOMETRY

Algebra

$$(a + b)^2 = a^2 + 2ab + b^2.$$

$$(a + b)^3 = a^3 + 3a^2b + 3ab^2 + b^3.$$

Quadratic formula: If $a \neq 0$ the roots of the quadratic equation $ax^2 + bx + c = 0$ are

$$x = \frac{-b \pm \sqrt{b^2 - 4ac}}{2a}.$$

Geometry

The slope of the straight line passing through points (x_1, y_1) and (x_2, y_2) with $x_1 \neq x_2$ is

$$\text{slope} = \frac{y_2 - y_1}{x_2 - x_1}.$$

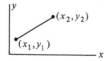

Equation of a straight line of slope m is

$$y = mx + b,$$

where b is the y intercept (the value of y when $x = 0$).

A circle of radius r has circumference $2\pi r$ and area πr^2.

A sphere of radius r has surface area $4\pi r^2$ and volume $\frac{4}{3}\pi r^3$.

Trigonometry

If r is the distance from the origin to a point (x, y) and θ is the polar coordinate angle, then

$$r = \sqrt{x^2 + y^2},$$

$$\cos \theta = x/r, \qquad \sin \theta = y/r, \qquad \tan \theta = y/x.$$

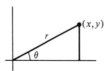

$$\csc \theta = 1/\sin \theta, \qquad \sec \theta = 1/\cos \theta, \qquad \cot \theta = 1/\tan \theta.$$

$$\sin^2 \theta + \cos^2 \theta = 1.$$

$$\sin(\theta \pm \alpha) = \sin \theta \cos \alpha \pm \cos \theta \sin \alpha,$$

$$\cos(\theta \pm \alpha) = \cos \theta \cos \alpha \mp \sin \theta \sin \alpha.$$

On a circular sector of radius r subtending an angle θ (measured in radians) the arc length of the circle is $s = r\theta$:

$$\theta (\text{in radians}) = \theta (\text{in degrees}) \frac{\pi}{180},$$

$$\theta (\text{in degrees}) = \theta (\text{in radians}) \frac{180}{\pi}.$$

APPENDIX

ASTRONOMICAL DATA

Earth

mass $\quad M_E = 5.975 \times 10^{24}$ kg

radius $\quad$ 6371 km

acceleration due to gravity $\quad$ 9.80665 m/s^2

Sun

mass 1.987×10^{30} kg
radius   696,500 km
Earth–Sun mean distance 1.496×10^{11} m $= 1$ AU.

Moon

mass 7.343×10^{22} kg
radius   1738 km
Earth–Moon mean distance   384,400 km

Planet	Semimajor axis (10^6 km)	Orbital period (d)	Mass/M_E	Eccentricity
Mercury	57.9	87.96	3.167	0.2056
Venus	108.2	224.68	4.870	0.0068
Earth	149.6	365.24	5.975	0.0167
Mars	227.9	686.95	0.639	0.0934
Jupiter	778.3	4,337	1900	0.0483
Saturn	1427.0	10,760	568.9	0.0560
Uranus	2871.0	30,700	86.9	0.0461
Neptune	4497.1	60,200	102.9	0.0100
Pluto	5983.5	90,780	5.37	0.2484

APPENDIX  E

PHYSICAL CONSTANTS

Gravitational constant	G	6.672×10^{-11} N m²/kg²
Speed of light	c	2.997925×10^8 m/s
Electron's charge	e	1.60219×10^{-19} C
Coulomb constant	K_e	8.98755×10^9 N m²/C²
Permittivity of free space	ε_0	8.85419×10^{-12} C²/N m²
Magnetic constant	K_m	10^{-7} N/A²

Permeability of free space	μ_0	$4\pi \times 10^{-7}$ N/A^2
Boltzmann's constant	k	1.3807×10^{-23} J/K
		8.617×10^{-5} eV/K
Avogadro's number	N_A	6.0220×10^{23} particles/mol
Gas constant	$R = N_A k$	8.314 J/mol K
		1.9872 cal/mol K
		8.206×10^{-2} L atm/mol K
Planck's constant	h	6.6262×10^{-34} J s
		4.1357×10^{-15} eV s
	$\hbar = h/2\pi$	1.05459×10^{-34} J s
		6.5822×10^{-16} eV s
Mass of the electron	m_e	9.1095×10^{-31} kg
		0.511 MeV/c^2
Mass of the proton	m_p	1.67265×10^{-27} kg
		938.28 MeV/c^2

SELECTED BIBLIOGRAPHY

Apostol, T. M., *Calculus*, Vol. 1, Second Edition (John Wiley and Sons, New York, 1967).

Aristotle, *Works*, ed. by W. D. Ross, Vol. II, *De Caelo*, trans. by J. L. Stocks (Clarendon Press, Oxford, 1930).

Atallah, S., "Some Observations on the Great Fire of London, 1666," *Nature*, pp. 105–6, 2 July 1966.

Bekenstein, J. D., "Black Hole Thermodynamics," *Physics Today*, Vol. 3, pp. 24–31 (1980).

Boorstein, Daniel J., *The Discoverers* (Random House, New York, 1983).

Boscovich, Roger Joseph, *A Theory of Natural Philosophy* (MIT Press, Cambridge, Massachusetts, 1966).

Boyer, Carl, *The History of the Calculus* (Dover Publications, New York, 1949).

Boyer, Carl, *A History of Mathematics* (John Wiley and Sons, New York, 1968).

Casper, B. M., and Noyer, R. J., *Revolutions in Physics* (W. W. Norton and Co., New York, 1972).

Cohen, I. Bernard, *The Newtonian Revolution* (Cambridge University Press, Cambridge, 1980).

Cornford, Francis Macdonald, *Plato's Cosmology* (Routledge and Kegan Paul, London, 1966).

Crowe, Michael J., *A History of Vector Analysis* (University of Notre Dame Press, Notre Dame, Indiana, 1967).

Dijksterhuis, E. J., *The Mechanization of the World Picture*, trans. by C. Kikshoorn (Oxford University Press, Oxford, 1961).

Drake, Stillman, *Galileo Studies* (University of Michigan Press, Ann Arbor, 1970).

Drake, Stillman, and MacLachlan, James, "Galileo's Discovery of the Parabolic Trajectory," *Scientific American*, Vol. 232, pp. 102–110.

Epstein, L. C., and Hewitt, L. C., *Thinking Physics Part 1 and 2* (Insight Press, San Francisco, 1979).

French, A. P., *Newtonian Mechanics* (W. W. Norton, New York, 1971).

Galilei, Galileo, *Dialogues Concerning Two New Sciences*, trans. by Henry Crew and Alfonso de Salvio (Macmillan, New York, 1914).

Galilei, Galileo, *Dialogue Concerning The Two Chief World Systems*, trans. by Stillman Drake (University of California Press, Berkeley, 1953).

Galilei, Galileo, *Two New Sciences*, trans. by Stillman Drake (University of Wisconsin Press, Madison, 1974).

Goldreich, Peter, "Tides and the Earth-Moon System," *Scientific American*, Vol. 226, pp. 42–52, April 1972.

Goldreich, Peter, "Toward a Theory of the Uranian Rings," Nature, Vol. 277, p. 97, 1979.

Haldane, Elizabeth S., *Descartes His Life and Times* (John Murray, London, 1905).

Heilbron, J. L., *Elements of Early Modern Physics* (University of California Press, Berkeley, 1982).

Jespersen, James, *From Sundials to Atomic Clocks* (National Bureau of Standards, Monograph 155).

Joule, James Prescott, *The Scientific Papers of James Prescott Joule* (Taylor and Francis, London, 1884).

Kepler's Somnium, trans. by Edward Rosen (University of Wisconsin Press, Madison, 1967).

Koestler, Arthur, *The Sleepwalkers* (Grosset and Dunlap, New York, 1963).

Lawrence, E. N., "Meteorology and the Great Fire of London, 1666," *Nature*, pp. 168–9, 14 January 1967.

McCloskey, Michael, "Intuitive Physics," *Scientific American*, Vol. 248, pp. 122–130.

Magie, William Francis, *A Source Book in Physics* (McGraw-Hill, New York, 1935).

Maxwell, James Clerk, *The Scientific Papers of James Clerk Maxwell*, ed. by W. D. Niven (Dover Publications, New York, 1966).

Millikan, R. A., *Electrons* (University of Chicago Press, Chicago, 1927).

Millikan, R. A., *Physical Review*, Vol. 46, p. 1023 (1929).

Millikan, R.A., Roller, D., and Watson, E. A., *Mechanics, Molecular Physics, Heat, and Sound* (Ginn and Company, Boston, 1937).

Newton, Isaac, *Mathematical Principles*, trans. by Florian Cajori (University of California Press, Berkeley, 1934).

Peters, Philip C., "Black Holes: New Horizons in Gravitational Theory," *American Scientist*, Vol. 62, pp. 575–583, September–October 1974.

Rosenthal, Arthur, "The History of Calculus," *American Mathematical Monthly*, Vol. 58, pp. 75–86 (1951).

Sambursky, Shmuel, *Physical Thought from the Presocratics to the Quantum Physicists* (Pica Press, New York, 1975).

Settle, Thomas B., "An Experiment in the History of Science," *Science*, Vol. 133, No. 3445, pp. 19–23, January 6, 1961.

Sibulkin, M., "A Note on the Bathtub Vortex and the Earth's Rotation," *American Scientist*, Vol. 71, p. 352, July–August 1983.

Tea, P. L., Jr., and Falk, H., "Pumping on a Swing," *American Journal of Physics*, Vol. 36, p. 1165 (1968).

Turner, D. M., *Makers of Science: Electricity and Magnetism* (Oxford University Press, London, 1927).

Walker, Jearl, *The Flying Circus of Physics* (John Wiley and Sons, New York, 1975).

Walker, J., "The Physics of the Follow, the Draw, and the Massé (in billards and pool)," The Amateur Scientist, *Scientific American*, Vol. 248, No. 7, p. 124, July 1983.

Westfall, R. S., *Forces in Newton's Physics* (Macdonald, London, 1971).

Westfall, Richard S., "Newton and the Fudge Factor," *Science*, Vol. 179, pp. 751–758, 23 February 1973.

Westfall, R. S., *Never at Rest: A Biography of Isaac Newton* (Cambridge University Press, Cambridge, 1980).

Whitt, Lee, "The Standup Conic," Texas A & M University, 1981.

INDEX

Absolute temperature, 294
Absolute zero, 294
Accelerated motion, 34
Acceleration, 34, 181, 190
  average, 9, 32
  centripetal, 186
  components, 181
  due to gravity, 34, 166
  instantaneous, 34, 181, 190
Action and reaction, 115
Addition, of vectors, 86
Adiabatic process, 309, 312
  for ideal gas, 313
Albert of Saxony, 15
Ampere, 564
Amplitude, 383
Angular frequency, 383
  natural, 385

Angular momentum, 441
  conservation of, 441
  and galactic formation, 450
  and torque, 445
Angular speed, 184
Angular velocity, 464, 502
Antiderivative, 130
Antidifferentiation, 130
Aphelion, 439, 485, 487
Apogee, 516
Appolonius, 177, 484
Archimedes, 134, 150
Area, 133
  of hyperbolic segment, 153
  of parabolic segment, 134
Aristotle, 1, 13, 70
Astronomical data, 571
Audibility, limits of, 429

Average acceleration, 32
Average speed, 25
Avogadro's number, 574

Barrow, Isaac, 40, 138
Black hole, 550
Body temperature, 300
Boiling, 300, 345, 349
Boiling point, 345
Boltzmann's constant, 296
Boscovich, Roger, 267, 284
Boyle, Robert, 293
Boyle's law, 293
Brahe, Tycho, 477
British units, 6, 117
Bubble chamber, 369

Cailletet, Louis Paul, 353
Calorie, 261
Cartesian coordinates, 96
Carnot, Nicolas Léonard Sadi, 263, 302
Carnot cycle, 319
Carnot engine, 319
Cathode-ray tube, 229
Cavendish, Henry, 196
Cavendish apparatus, 197
Celsius, Anders, 300
Celsius temperature, 288
Center of mass, 282, 542
Centripetal acceleration, 186
Centripetal force, 168
Chain rule, 54
Change of phase, 345
Characteristic time, 236
Charge, 194, 197
    of electron, measurement, 243
Charles, Jacques Alexandre César, 294
Charles, law of, 295
Circular motion, 176, 184
Clapeyron, Emile, 324
Clausius, Rudolph, 325, 327
Coefficient of friction, 210
    table, 211
Collisions, 369
    elastic, 371
    inelastic, 371
Components of vectors, 96
Conic sections, 483
Conservation
    of angular momentum, 441
    of energy, 254
    of momentum, 360
Contact force, 198
Conversion factors, 565

Coordinates
    polar, 184
    rectangular, 96
Copernicus, Nicolaus, 2, 69, 111, 191, 458, 497
Corpuscle, 231
Coulomb, Charles Augustin, 194, 210
Coulomb's law, 194, 218
Coupled oscillators, 415
Cranking, 536
Critical point, 346
Cross product, 106

Dalton, John, 286
Damped vibration, 404
Decibel, 428
Derham, William, 430
Derivative(s), 28
    and acceleration, 35, 120
    and slope, 44
    and speed, 28
    table of, 66
    and velocity, 120
Descartes, René, 96, 360
Determinant for cross product, 109
Dewar, James, 357
Diatomic gas, 298, 310
Differential equation, 120, 123, 237
Differentiation, 28, 40
    rules for, 57
Directrix, 484, 489
Displacement, 84
Dot product, 93
Dynamics, 112
Dyne, 117, 566

e (base of natural logarithm), 64
Earth's gravitational field, 222
Eccentricity of conic sections, 485, 489
    and energy, 512
Effective potential, 451, 517
Efficiency, 318
Einstein, Albert, 36, 68, 195, 547
Electric charge, 194, 197
Electric field, 221
Electromagnetic force, 195
Electromagnetic wave, 225
    speed, 225
Electron, 231
    charge of, 243
Electron-volt, 567
Ellipse, 483, 489
Energy, 246
    and eccentricity, 512
    conservation, 254

effective potential, 451, 517
gravitational potential, 275
internal, 298, 304
   of ideal gas, 298, 304
kinetic, 254
and mass, 549
potential, 246
in simple harmonic motion, 388
in thermodynamics, 297, 310
and work, 249
Energy conversion, 264
Engine
   Carnot, 319
   steam, 302
Entropy, 328, 330
   and disorder, 341
   and second law of thermodynamics, 338
Epicycle, 177
Equality, of vectors, 83
Equilibrium, 280
   neutral, 281
   phase, 344
   stable, 281
   unstable, 281
Equilibrium state, 347
Equivalence principle, 546
Erg, 248
Escape velocity, 277
Euler, Leonhard, 64, 114
Euler number e, 64
Exhaustion, method of, 150
Exponential function, 64
External force, 114

Fahrenheit, Daniel, 299
Fahrenheit temperature, 288, 299
Falling bodies, 14
   law of, 34
Faraday, Michael, 219, 286, 351
Farquharson, Bert, 409
Fermat, Pierre de, 40
Field lines, electric, 221
   magnetic, 220
Firestorms, 449
First law of motion, 113
First law of thermodynamics, 310
Focus, of a conic section, 485, 489
Foot-pound, 248
Force, 113
   centripetal, 168
   electromagnetic, 195
   gravitational, 195
   strong, 195
   weak, 195

Forced oscillation, 400
Free-body diagram, 203
Frequency, 384
   angular, 384
   natural, 385
   resonance, 406
   of wave, 419
Friction, 210
   coefficient of, 210
      table, 211
F ring of Saturn, 537
Fundamental theorems of calculus, 138, 139, 152

Galactic formation, 450
Galileo, 3, 15, 36, 72, 299, 393, 399, 504, 543
Galileo's law
   of inertia, 73
   of odd numbers, 18, 22
Gas constant, 574
Gas, kinetic theory, 292
Gassendi, Pierre, 74
Gay-Lussac, Joseph Louis, 295
Gay-Lussac, law of, 295
Gibbs, Josiah Willard, 82
Gilbert, William, 218
Grassmann, Hermann, 82
Gravitation, 160
Gravitational constant, 161
Gravity, acceleration due to, 166
Gravity assist, 534
Gyrocompass, 467
Gyroscope, 459
Gyroscopic precession, 459, 522

Halley, Edmund, 160, 474, 522
Halley's comet, 522
Hamilton, William Rowan, 82
Harmonic motion, 381
Harmonic oscillator, 386
Heat, 260, 298, 310
Heat engine, 312, 317
Heat of fusion, 345
   table, 345
Heat of vaporization, 346
   table, 345
Heat reservoir, 318, 319
Heat transfer, 318
Heaviside, Oliver, 82
Helium, 298
Helix, 190
Helmholtz, Herman von, 260
Hertz, 384
Hohman transfer orbit, 526
Hooke, Robert, 199, 389

Hooke's law, 199
Huygens, Christian, 389
Hyperbola, 152, 483, 489
Hyperbolic segment, 153

Ideal gas, 295
    adiabatic process, 312
    internal energy of, 298
    kinetic theory of, 292
Impetus, 77
Impulsive force, 377
Indefinite integral, 141
Inertia, 73, 113
Inertial reference frame, 76
Initial velocity, 125
Instantaneous acceleration, 33, 181, 190
Instantaneous speed, 26, 180
Integral, 138
    definite, 141
    indefinite, 141
    line, 269
Integrand, 138
Integration, 130
Internal energy, 298
    of ideal gas, 298
International System of Units, 7, 563
Irreversible process, 328
Isothermal process, 312, 315

Joule, James Prescott, 260, 265, 292, 309, 354
Joule, unit of energy, 248, 261
Joule's apparatus, 261
Joule–Thomson effect, 354
Jupiter, 449, 534

Kármán, Theodore von, 408
Kelvin, Lord (William Thompson), 294, 354
Kelvin temperature, 294, 296, 564
Kelvin temperature scale, 296
Kepler, Johannes, 150, 160, 434, 480, 504
Kepler's laws of planetary motion, 480
    first law, 481
    second law, 482
    third law, 482, 541
Kepler problem, 498
Keplerian system, 504
Kilogram, 564
Kinematics, 112
Kinetic energy, 254
Kinetic friction, 211

Latent heat, 346
Launch opportunity, 527
Launch window, 527

Leaning Tower of Pisa, 282
Leibniz, Gottfried Wilhelm, 36, 40, 47, 137, 150
Leibniz notation for derivatives, 45
Leonardo da Vinci, 17
Linde, Karl von, 356
Line integral, 269
Lines of force, electric, 221
    gravitational, 222
    magnetic, 220
Liquefaction, 351
Logarithms (base e), 142, 154
Longitudinal wave, 419
Loudness, 429

Magnet, 218
Magnetic field, 220
    of earth, 218
Magnetic field lines, 220
Magnetic force, 218
Magnetism, 218
Magnitude, of vector quantity, 83
Major axis of an ellipse, 485
Mars, 480
Mass, 546, 564
    center of, 282, 542
    gravitational, 546
    inertial, 546
Maxwell, James Clerk, 82, 194, 219, 224, 227
Mayer, Walther, 68
Mean distance from the sun, 572
Melting, 344
Meter, 563
Michell, John, 218
Michelson, A. A., 68
Millikan, Robert Andrews, 68, 239
Millikan oil-drop experiment, 239
Minor axis of an ellipse, 486
Momentum, 360
    angular, 441
    conservation of, 360
    linear, 444
Monatomic gas, 298
Motion, circular, 176, 184
    constant-acceleration, 33
    in a plane, 178
    of planet, 481
    of projectile, 119
    simple harmonic, 386
    uniformly accelerated, 33

Natural frequency, 385
Newton, Isaac, 4, 36, 40, 47, 112, 150, 159, 172,
    414, 431
Newton, unit of force, 564

Newtonian synthesis, 4
Newton's law of gravitation, 161
Newton's laws of motion, 112
    first law, 113
    second law, 113
    third law, 115
Noise levels, 429
Normal force, 200

Oersted, Hans Christian, 218
Orbit, of planet or satellite, 504
Oresme, Nicole, 16, 33
Oscillation, 381
Oscillatory motion, 381

Parabola, 134, 483, 489
Parabolic segment, 134
Parabolic trajectory, 119
Parallelogram law, 87
Parking orbit, 527
Pendulum, 393
    simple, 394
Perigee, 516
Perihelion, 439, 485, 487
Period, 384
    of orbital motion, 482, 528, 540
    of wave, 384, 419
Periodic motion, 381
Phase, 344
Phase change, 344
Phase diagram, 345
Phase equilibrium, 344
Pictet, Raoul Pierre, 353
Plane motion, 178
Planetary motion, 480
Plato, 1, 175, 183
Platonic circles, 176
Polar coordinates, 184
Position vector, 178, 189
Potential energy, 246, 249, 270
Pound, 566
Power, 567
Precession, 459
    of Earth, 474
    of equinoxes, 458
    gyroscopic, 459
Prefixes, in SI units, 8
    for units, 8
Pressure, 290, 567
Principle of equivalence, 546
Products of vectors
    cross product, 106
    dot, or scalar, product, 93
Projectile motion, 119

Proton, 197
Ptolemy, 177, 191
Pumping in space navigation, 536
Pythagoras, 5, 20
Pythagorean theorem, 101, 105

Quadrature, 133
    of a hyperbolic segment, 153
    of a parabolic segment, 135
Quantification, 10
Quasistatic process, 304
Quaternions, 82, 93

Radial acceleration, 186
Radian, 59
Ramsay, William, 357
Range, of projectile, 125
Refrigeration, 343
Reservoir, heat, 317
Resonance, 400, 405
Resultant, 87
Reversible cycle, 331
Right-hand rule, 96
    for cross product, 107
    for magnetic field, 221
Right-handed coordinate system, 96
Roemer, Olaus, 299
Rotation, 184
Rumford, Count (Benjamin Thompson), 309

Satellite, 519
Scalar product, 93
Scalar quantity, 83
Second law of thermodynamics, 318
Sections of a cone, 483
Simple harmonic motion, 386
Simple pendulum, 393
Sinusoidal motion, 383
Sinusoidal wave, 384
Sliding friction, 211
Slope
    of a curve, 44
    of a line, 42
Slug, 117
Sonic boom, 429
Sound intensity level, 428
    table of, 429
Sound waves, 426
Speed, 9
    angular, 184
    average, 9, 25
    instantaneous, 26, 180
    of longitudinal wave, 422

Speed (*cont.*)
  of sound wave, 427
  of water wave, 423
Stability, 278
  and potential energy, 281
Standards, 6
Static friction, 211
Steam engine, 302
Stiffness, 199
Stokes, George Gabriel, 233
Stokes's law, 233
Strong force, 195
Sublimation, 345
Subtraction of vectors, 89
Superposition principle, 162
Surface, *PVT,* 347
Système International, 7
Systems of units, 6, 118

Tacoma Narrows Bridge, 409
Tangent line, 44
Temperature, 288
Temperature scales, 296
Tension, 200
Terminal velocity, 234
Thermal efficiency, 318
  Carnot cycle, 320
Thermal equilibrium, 337
Thermodynamics
  first law, 310
  second law, 318
Thermometer, 299
Thilorier, 351
Thompson, Benjamin (Count Rumford), 309
Thomson, Joseph John, 230
Thomson, William (Lord Kelvin), 294, 354
Threshold
  of audibility, 429
  of pain, 428
Tides, 543
Torque, 444
  and angular momentum, 445
Transfer orbit, 526
Transverse waves, 419
Trigonometric functions, 59
Triple point, 346
  table, 346
Tychonic theory, 479

Uniform circular motion, 176, 184
Uniformly accelerated motion, 33
Unit conversion factors, 7
Unit vectors, 97, 499
Units
  conversion of, 7, 565
  International System of, 7, 563
  systems of, 5
Universal gas constant, 296, 574
Universal gravitational constant, 196, 573

Van der Waals, Johannes Diderik, 297, 354, 357
Van Marum, Martin, 351
Vector product, 106
Vector quantity, 83
Vector sum, 87
Velocity, angular, 464, 502
  average, 179
  in polar coordinates, 500
  instantaneous, 179, 190
  terminal, 234
Viscosity, 202, 233
  table, 234
Volt, 564
Vortices, 448

Water waves, 418
Watt, James, 302
Watt, 567
Wave
  frequency, 419
  longitudinal, 419
  sinusoidal, 419
  speed, 419
  transverse, 419
  water, 418
Wavelength, 419
Weak force, 195
Weighing the earth, 196
Weight, 117
  and mass, 117
Work, 247, 249, 268
  done by a varying force, 249, 268
  and energy, 249, 270
  in isothermal expansion, 312
  in volume change, 305

Zero vector, 85